OFFSHORE POWER

OFFSHORE POWER

BUILDING RENEWABLE ENERGY PROJECTS IN U.S. WATERS

Markian M.W. Melnyk
Robert M. Andersen

PennWell®

Disclaimer: The recommendations, advice, descriptions, and the methods in this book are presented solely for educational purposes. The author and publisher assume no liability whatsoever for any loss or damage that results from the use of any of the material in this book. Use of the material in this book is solely at the risk of the user.

PennWell Corporation
1421 South Sheridan Road
Tulsa, Oklahoma 74112-6600 USA

800.752.9764
+1.918.831.9421
sales@pennwell.com
www.pennwellbooks.com
www.pennwell.com

Marketing: Jane Green
National Account Executive: Barbara McGee

Director: Mary McGee
Managing Editor: Stephen Hill
Production Manager: Sheila Brock
Production Editor: Tony Quinn
Book Designer: Susan E. Ormston
Cover Designer: Jesse Bennett

Library of Congress Cataloging-in-Publication Data
Melnyk, Markian M. W.
Offshore power : building renewable energy projects in U.S. waters / Markian M.W. Melnyk and Robert M. Andersen.
p. cm.
Includes bibliographical references.
ISBN 978-1-59370-211-3
1. Electric power-plants, Offshore--United States. 2. Ocean energy resources. I. Andersen, Robert M. II. Title.
TK1193.U5M45 2009
333.790973--dc22 2009024232

Printed in the United States of America

1 2 3 4 5 13 12 11 10 09

To Alejandra, Katerina, and Camilo,

who provided me with companionship and distraction

during the writing of this book,

and to Joseline, patient beyond measure.

Markian M.W. Melnyk

To my family and future generations

and beneficiaries of clean,

renewable power.

Robert M. Andersen

Contents

Preface

Why write a book about developing offshore renewable energy projects in the United States when not one commercial project has yet been built? It should be obvious to anyone who is paying attention the United States cannot continue to make and use energy with the same practices of the last 50 years. We cannot continue to use the same technologies, the same financing methods, or the same modes of public consultation and dispute resolution, while expecting a more secure, stable, and economically and environmentally sound energy system to emerge. There is no better time than now to examine the emerging offshore renewable energy industry and to call attention to the practical hurdles of building offshore.

We hope to engage the readers, whether project developers, lenders, regulators, or environmentalists, in thinking about how offshore renewable energy projects can be built under our present system, and how they can be built better. The two cannot be separated. We need to start building now, with the technology we have, while supporting technological advancement to make harvesting renewable energy more efficient, economical, and safer for the environment. Yet we also must evolve our social mechanisms, how society carves up rights to the submerged public commons and decides whether or not to proceed with projects. Rapid deployment of offshore renewable energy, on the scale needed to appreciably reduce greenhouse gas emissions and provide for sustainable economic growth, is not possible if as a society we require projects to run through a multiyear regulatory gauntlet. We must make decisions about how the offshore environment is used in a more collaborative and less confrontational fashion. Rapid growth also will not occur if the price offered for offshore renewable energy does not signal developers that there is money to be made in taking on the challenges of building offshore. The innovations in policy and process that we need are just as significant a contributor to making clean energy a viable option for the United States as are innovations in technology.

The United States is at a critical moment in history where it must choose to act to reduce energy dependence and greenhouse gas emissions, improve national security, and transform our economic system into one that creates jobs and economic growth while also achieving environmental sustainability. These choices will affect our competitiveness as a nation and could set the groundwork for sustainable economic growth in many countries around the world. It is our view that developing offshore renewable energy in the United States is an essential part of starting on the right path toward a more sustainable and stable world economy.

Acknowledgments

This book would not have been done without the help of Debra Sargent, who spent many hours creating figures, soliciting permissions, and generally keeping me on track, and Doug Ross, an associate at the firm of Dewey & LeBoeuf, whose genuine interest in this field is matched by sharp attention to detail. Dewey & LeBoeuf, a remarkable firm that is committed to developing excellence among its lawyers and staff, also gets my sincere thanks for tolerating my coffee-slurping presence for so long.

Markian M.W. Melnyk

Acronyms and Abbreviations

AEP	American Electric Power
CAA	Clean Air Act
CAISO	California Independent System Operator
Call	Call for Information and Nominations
CBP	U.S. Customs and Border Protection
CCS	carbon capture and storage/sequestration
CE	categorical exclusion
CEQ	White House Council on Environmental Quality
CHP	combined heat and power
CO_2	carbon dioxide
COP	construction and operations plan
Corps	U.S. Army Corps of Engineers
COWRIE	Collaborative Offshore Windfarm Research into the Environment
CTDEP	Connecticut Department of Environmental Protection
CVA	certified verification agent
CWA	Clean Water Act
CZMA	Coastal Zone Management Act
DLWC	deep lake water cooling
DNV	Det Norske Veritas
DOD	U.S. Department of Defense
DOE	U.S. Department of Energy
DOT	U.S. Department of Transportation
DSCR	debt service coverage ratio
EA	environmental assessment
EEZ	exclusive economic zone
EFH	essential fish habitat
EIA	U.S. Energy Information Administration
EIS	environmental impact statement
EPA	U.S. Environmental Protection Agency
EPAct	Energy Policy Act of 2005
EPRI	Electric Power Research Institute
ESA	Endangered Species Act of 1973
FAA	Federal Aviation Administration

FERC	Federal Energy Regulatory Commission
FERN	Fishermen's Energy of New Jersey, LLC
FIFRA	Federal Insecticide, Fungicide, and Rodenticide Act
FOIA	Freedom of Information Act
FONSI	finding of no significant impact
FPA	Federal Power Act
FTR	financial transmission right
FWS	U.S. Fish and Wildlife Service
GAP	general activities plan
GIS	geographical information system
GL	Germanischer Lloyd
GOW	Galveston Offshore Wind
GW	gigawatt (equivalent to 1,000 MW)
HAWT	horizontal axis wind turbine
HG	mercury
HVAC	high-voltage alternating current
HVDC	high-voltage direct current
IEC	International Electrotechnical Commission
ILP	integrated licensing process
IMO	International Maritime Organization
IPCC	Intergovernmental Panel on Climate Change
IRS	U.S. Internal Revenue Service
ISO	independent system operator
ITC	investment tax credit
KKR	Kohlberg Kravis Roberts & Co.
kW	kilowatt (equivalent to 1,000 watts)
kW/m	kilowatts per meter (of wave front)
kWh	kilowatt hour (equivalent to 1,000 watt-hours)
LIPA	Long Island Power Authority
LNG	liquefied natural gas
LOA	letter of authorization
LOLP	loss of load probability
MARPOL 73/78	International Convention for the Prevention of Pollution from Ships, 1973, as modified by the Protocol of 1978
MBTA	Migratory Bird Treaty Act

MEC	Marine Energy Challenge
MLLW	mean lower low water
MMbd	million barrels per day
MMPA	Marine Mammal Protection Act
MMS	U.S. Minerals Management Service
MOU	memorandum of understanding
MPRSA	Marine Protection, Research and Sanctuaries Act of 1972
MTC	Massachusetts Technology Collaborative
MW	megawatt (equivalent to 1,000 kilowatts)
MWh	megawatt-hour (equivalent to 1,000 kilowatt-hours)
NDBC	National Data Buoy Center
NEPA	National Environmental Policy Act of 1969
NHPA	National Historic Preservation Act
NIMBY	not in my backyard
nm	nautical mile (equivalent to 1.15 statute miles or 1.85 km)
NMFS	National Marine Fisheries Service
NMSA	National Marine Sanctuaries Act
NOAA	National Oceanic and Atmospheric Administration
NOI	notice of intent
NO_x	nitrogen oxides
NPDES	National Pollutant Discharge Elimination System
NREL	National Renewable Energy Laboratory
NWF	National Wildlife Federation
NWP	numerical weather prediction
NWS	National Weather Service
NYISO	New York Independent System Operator
NYSERDA	New York State Energy Research and Development Authority
O&M	operations and maintenance
OATT	open access transmission tariff
OCS	outer continental shelf
OCS Lands Act	Outer Continental Shelf Lands Act
OMB	Office of Management and Budget
OTEC	ocean thermal energy conversion
OWC	oscillating water column

PILOT	payment in lieu of taxes
PM	particulate matter
PPA	power purchase agreement
PTC	production tax credits
REC	renewable energy credit (or certificate)
RFP	request for proposals
RGGI	Regional Greenhouse Gas Initiative
ROC	renewable obligation certificate
ROW	right-of-way
RPS	renewable portfolio standard
RTO	regional transmission organization
RUE	right-of-use and easement
SAP	site assessment plan
SDC	Sustainable Development Commission (UK)
SLA	Submerged Lands Act of 1953
SMUD	Sacramento Municipal Utility District
SOS	standard offer service
SO_x	sulfur oxides
SPE	single-purpose entity
STB	Surface Transportation Board
SWPPP	storm water pollution prevention plan
TADS	thermal animal detection system
TBT	tri-butyl-tin oxide
TMDL	total maximum daily load
TPG	Texas Pacific Group
TWh	terawatt-hours (equivalent to 1 billion kilowatt-hours)
UCAP	unforced capacity
USGS	U.S. Geological Survey
VAWT	vertical axis wind turbine
VLA	vertical load anchor
WAsP	Wind Atlas Analysis and Application Program
WEC	wave energy converter
WEDI	wave energy development index
WEST	Wind Energy Systems Technology, LLC
WTG	wind turbine generator

1 Stepping off Dry Land: The Case for Offshore Renewable Energy

I'm gonna ask again, do you want to put your country first? Then let's reduce our dependency on foreign sources of oil and promote oil and gas production at home. Let me make it very clear. Drill, baby, drill! And drill now!

—former Maryland Lieutenant Governor Michael Steele

I've been an oilman my whole life, but this is one emergency we can't drill our way out of.

—T. Boone Pickens

Oil dependence has imposed serious costs on the U.S. economy. It threatens U.S. security and has depleted our wealth. According to the first rule of holes, when a person finds himself in one, he should stop digging. Drilling anywhere cannot make the United States less oil dependent or protect the American consumer from dramatic, budget-sapping swings in gasoline and home heating oil prices. Drilling will not lessen the airline industry's vulnerability to fuel cost increases or insulate the automobile industry from crippling shifts in consumer demand when the price of gasoline doubles. Drilling does not lessen global warming. It is time for something different.

Oil Volatility, Gas Volatility, and Economic Vulnerability

The U.S. economy runs on oil to fuel transport and to make chemicals used in so many ways. It also runs on electricity produced primarily with coal but, increasingly over the last two decades, with large quantities of natural gas, too. Recent electricity price shocks in many parts of the country have been caused by rising natural gas prices. Unwise energy decisions could cause the United States to compound its economically crippling oil addiction with a growing natural gas dependency. This is a particular danger because the obvious, immediate path to reducing utility carbon dioxide emissions involves switching from coal-based electricity to less-polluting power from natural gas.

This book explores new energy alternatives based on abundant domestic renewable resources found right off the U.S. coasts. If developed, these alternative energy resources could improve security, stanch the wealth transfer to oil-rich nations, create domestic jobs, and contribute meaningfully to solving the climate challenge.

Could renewable sources some day satisfy the energy needs of the United States, and perhaps the world? In Europe, a fortunate intersection of geography, economics, technology, and government policies has driven a resurgence of wind power, leading to wind farms being built in the North Sea and North Atlantic. Engineers in Great Britain and Portugal are testing wave and water current power technologies that may soon be practical for widespread deployment. The United States likewise has tremendous, largely untapped, renewable energy resources, from the windy plains of West Texas, and the steam and hot rock found under California's rugged valleys, to the litter produced in poultry farms on Maryland's eastern shore. To power the world with renewable energy, each niche of possibility should be explored. One of the most important niches lies in the expansive, and hugely energetic, reaches of the marine environment.

The submerged lands of the United States and the untapped energy in the wind, waves, and currents offshore are immense. In fact, the challenge is not finding the energy, it is building machines that can survive the pounding of a winter storm yet still operate cost effectively. The challenge also entails building projects responsibly, acknowledging the perspectives of coastal residents, respecting the livelihood of fishers, and minimizing the impact on the marine environment. Crafting projects that make money for developers and investors in an environment of uncertain energy prices and shifting clean energy incentives is a high-stakes puzzle. The social case for renewable energy is strong, but the financial case for these projects must become clearer if there is to be hope of attracting the very large capital investments needed to move with alacrity toward a renewable energy future.

This practical guide to developing marine renewable energy projects in the United States does not dwell on policy issues, yet attention to policy is needed to unlock the future of the renewable energy industry. Society must determine how much "free" pollution traditional energy sources should emit and whether the United States can remain strong if it continues to cede enormous amounts of national wealth for oil and natural gas purchased overseas.

Unlike traditional energy sources, renewable energy is not extracted from concentrated stores within the earth. Renewable energy is manufactured energy collected with machines that are built and maintained to capture a portion of the solar energy that rains down from above and is transformed into the wind and waves. Policies that promote renewable energy reflect support for employment and investment in building and maintaining collector facilities that, over time, will lead to a deeper and more sustainable economy, not one based on depleting resources.

Much has been made of the accelerating growth of China and India. These Asian giants' economic engines have lifted millions from poverty. Yet the environmental implications of a growing middle class in two countries with a combined population of approximately 2.5 billion are sobering. In *Hot, Flat, and Crowded*, Thomas Friedman ruminates about the 1 billion new people who will join the planet in just the next 12 years. If each new person is given a 60-watt light bulb, and they use it just four hours each day, that would require 10,000 MW of new generating capacity, or about 20 new 500-MW coal-fired power plants "just so the next billion people can turn a light on!"[1]

There is little sense in begrudging the residents of the developing world the benefits of modern life. That aspiration cannot be stopped. Widespread middle-class consumption modeled on American or Western European society, however, is fraught with environmental risk. The United States can help bridge the growing environmental conflict between the developed and developing worlds by finding the smartest ways to satisfy everyone's needs. Policies that advance renewable energy in the United States accelerate technology development, drive down its cost, and make clean technology more available for deployment in developing countries. Some developing countries have installed cellular telephone networks before ever having built extensive landline telephone networks. If the United States and other nations help provide the needed technology, developing countries can similarly skip dirty energy technologies and leapfrog directly to clean energy. Such a strategy would advance the U.S. national interest in a secure world while it builds an export market for U.S. renewable energy technology. Sustainable development is the best chance that the United States has to promote the conditions under which democracy, prosperity, and security flourish everywhere.

Energy Insecurity

The United States' singular dependence on one fuel—oil—is a clear threat to its national security. The United States imports 58% of the oil that drives our national economy, fuels transportation, provides the key feedstock for chemical production, and makes modern, highly productive agriculture possible. Imagine for a moment that the United States imported more than one-half of its food from abroad—from various politically unstable countries. One would expect a focused, nationwide effort to secure the U.S. food supply. Unfortunately, the absence of a concerted effort to achieve energy security has cost the United States dearly and puts us at risk. Prolonged military engagements and an imported oil habit of 13.5 million barrels per day (MMbd), costing about $540 billion each year when oil was at $110 per barrel, are but two of the obvious consequences of the U.S. oil "addiction."

Columbia's efforts to secure just one pipeline demonstrate how difficult it can be to attain energy security. Since 2002, the United States has provided approximately $99 million in military aid, such as U.S. Special Forces training and helicopters, to the Columbian government to defend the Caño Limón oil pipeline from guerilla attack.[2] In 2001 alone, the pipeline was the subject of 170 attacks designed to slow oil production and pressure the Columbian government.

The pipeline is partly owned by Occidental Petroleum, a U.S. corporation. More than 55% of the oil transported on the pipeline is exported to the United States. Amazon Watch, a nonprofit advocate for the rights of indigenous people, estimated that the U.S. aid package subsidized Occidental's security expenses at a cost of $3.70 per barrel. Yet the effort is probably fruitless. As one Columbian general has commented, "We would need to have a soldier for every three feet of pipeline to keep it from being attacked."[3] Even if protecting the pipeline itself were possible, it would be insufficient because guerillas continue to adapt their tactics. Attacks may be targeted at other structures, equipment, and even software essential to operate the pipeline system. The U.S. General Accountability Office reported that "in recent years, the insurgents have attacked the electrical grid system that provides energy to the Caño-Limón oilfield. Without electricity, the oil wells cannot operate."[4]

In the Middle East, U.S. troops fight to protect full access to oil supplies, and in the Pacific, the U.S. Navy patrols tanker routes and key shipping choke points, such as in the Strait of Malacca, to protect the petroleum that fuels the U.S. economy. By some pre-Iraq war estimates, the U.S. military spends up to $60 billion per year securing stability in the Middle East, yet the results of these expenditures have been mixed.[5] Increasingly, oil price stability is pursued at the expense of other national security interests such as supporting democracy and fighting terrorists. Despite U.S. efforts abroad, the fundamental problem of energy dependence remains unresolved. U.S. Department of Energy (DOE) projections indicate that U.S. oil demand will increase steadily through 2030, and world oil demand will grow even faster (fig. 1–1). To meet these demands over the decades ahead, the Middle East's share of world oil production must increase substantially, because that is where 56% of the world's proved reserves lie.[6]

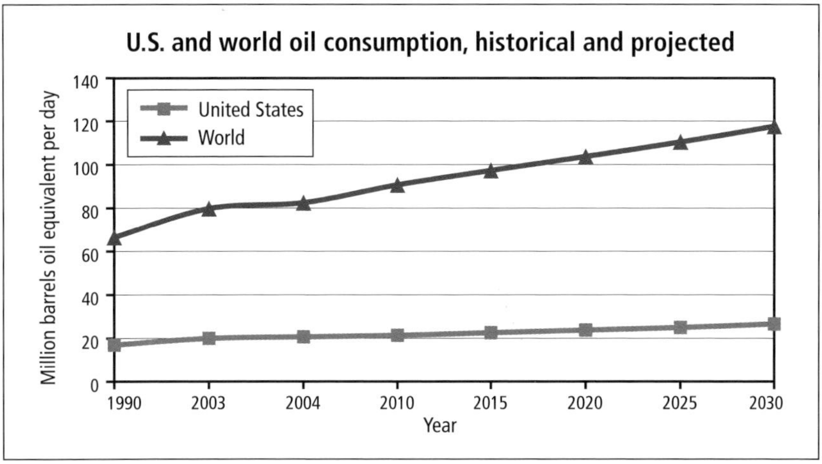

Fig. 1–1. U.S. and world oil consumption growth

Energy vulnerability is a concern shared by Frederick Smith, chairman, president, and CEO of FedEx Corp., and retired General P.X. Kelley, commandant of the Marine Corps and member of the Joint Chiefs of Staff under President Ronald Reagan. As cochairs of the Energy Security Leadership Council, they have led an effort to forge a path toward greater energy security. What they learned in the process is alarming. The world's oil production capacity is running full tilt. Very little spare capacity exists, fewer than 2 MMbd in the context of 82 MMbd consumption, to compensate for supply disruptions. The world primarily relies on the politically unstable OPEC countries and Russia to supply the majority of the additional quantity demanded. Oil production in the developed countries is falling.[7]

Given the tight global production outlook, it is not difficult to imagine a scenario in which the energy supply infrastructure is disrupted, leading to a dramatic spike in the price of oil. "It was surprising to learn the extent to which seemingly small disruptions in world oil supplies could inflict serious economic damage and alter the global security environment," commented Robert Gates, U.S. secretary of the Department of Defense and former director of the Central Intelligence Agency, after participating in an oil shock scenario exercise.[8] An obvious target is the Saudi oil shipment terminal at Ras Tanura, the world's largest oil port, handling approximately 4.5 MMbd. Striking a blow to the heavily guarded terminal would not be easy, but saboteurs have demonstrated a capacity for creativity and adaptability. In February 2006, another major Saudi oil facility in Abqaiq, which processes two-thirds of the country's output, was the target of an unsuccessful suicide bomb attack. In Iraq, terrorist sabotage of difficult-to-guard pipelines has been successful in limiting Iraq's oil production to a fraction of its prewar level.

Terrorists have targeted oil and gas production, processing, storage, and transportation facilities around the world. Al Qaeda operatives have been captured with plans and diagrams for U.S. energy facilities.[9] Further attacks are not a matter of if, but simply when and where, according to most homeland security reports. Many terrorists trained in the Iraq war will eventually return to their home countries. They may continue with sabotage and other attacks in an attempt to topple regimes throughout the Middle East and inflict pain on oil-dependent economies such as that of the United States.

Iran, the second largest oil-producing country in OPEC after Saudi Arabia, is also feared to be developing nuclear weapons technology and is known as a sponsor of terrorist groups in the Middle East. The United States' ability to discourage Iran's development of nuclear weapons is limited by the knowledge that sanctions against Iran could result in a reduction in Iranian oil exports, currently at about 2.5 MMbd. In 2002, Iran's supreme leader Ayatollah Ali Khamenei warned, "If the West did not receive oil, their factories would grind to a halt."[10] In an environment of political instability, growing long-term oil demand, and limited swing capacity among the oil producing nations, the United States must plan ahead for supply disruptions and take corrective action to lessen its exposure to oil shocks.

The oil shock exercise conducted by Robert Gates and high-ranking former U.S. officials found that an event causing a mere 4% shortfall in oil production resulted in a 177% increase in the price of oil (from $58 to $161 per barrel at the time of the exercise).[11] Consumer demand for oil is inelastic in the short-term because there are few immediate substitutes for oil. When demand is inelastic, a small drop in supply causes large price increases. Experts also found the key energy infrastructure, both in the United States and abroad, too dispersed to be effectively protected and that the only real defense is to reduce oil demand and promote fuel diversity by developing alternatives to oil.[12] Short-term options, such as releasing supplies from the Strategic Petroleum Reserve, are of limited use.

The economic implications of oil dependence are serious, as Americans living through the current depression can attest. A 2006 forecast by Robert Wescott predicted that doubling oil prices (from $60 to $120 per barrel) would increase the median U.S. household's annual energy bill to approximately $6,000, or about 15% of the median family's total annual income.[13] Disposable income falls, leading to cutbacks in spending for other goods and a 2.3% reduction in national GDP. Industries such as automobile manufacture, chemicals, travel, and transportation would face sharply reduced demand and begin to cut back and lay off workers, further multiplying the initial oil price shock. Stock market values could fall and interest rates could rise as fears of energy cost–driven inflation take hold.[14]

The United States currently is struggling to right the effects of dramatic increases in oil prices that helped to expose an underlying weak economic foundation. As this book is being written, oil prices are

Strategic Petroleum Reserve

The Strategic Petroleum Reserve (SPR) is located on the Gulf Coast in four underground salt caverns. These large geologic formations have a capacity of 727 million barrels and are currently storing approximately 703 million barrels of oil. At the U.S. daily consumption of approximately 20 million barrels per day (MMbd), the volume in storage represents a little more than a one-month supply. Oil imports of 12 MMbd, however, are unlikely to be completely interrupted. Since the United States imports oil from a variety of countries, including countries outside the Middle East such as Mexico, Venezuela, and Canada, and also relies on domestic production, in practice the SPR provides protection for a longer period. The SPR can be drawn down at a rate of 4.4 MMbd for 90 days, after which the rate would slow. Other commercial stocks of oil in the United States provide additional emergency capacity of perhaps an additional 90 days.

Source: National Commission on Energy Policy and Securing America's Energy Future. 2005. Oil ShockWave: Oil Crisis Executive Simulation. June 23. p. 8; Energy Security Leadership Council. 2006. Recommendations to the Nation on Reducing U.S. Oil Dependence. p. 55.

low, but a doubling again of oil prices is an entirely realistic scenario. Riding the oil price roller coaster is leaving the U.S. economy queasy and unsteady. Investments and job creation are held hostage to energy price uncertainty. Higher oil prices act like a tax on all goods and services that use oil and its derivatives as fuel or feedstock. As with other taxes, patterns of behavior and investment will change over time to accommodate the more expensive product. When oil prices decline, behavior changes again, whipsawing the investments in energy efficiency and alternative energy technologies made when oil was expensive. An important difference between increased oil prices and taxes, however, is that tax dollars largely remain in the United States, where they support domestic consumption and investment, while much of the money the United States spends on oil leaves the U.S. economy and contributes to a growing international trade deficit.

The U.S. oil addiction, though self-inflicted, is aided and abetted by OPEC. In contrast to terrorists, who aim to severely disrupt the U.S. economy, OPEC favors moderate but consistent oil price increases. Saudi Arabia, the most influential member of OPEC, has become adept at managing production in order to maintain the price of oil at profit-maximizing levels, yet not at prices so high that efforts to develop and implement alternatives to oil might succeed. As Philip Auerswald, director of the Center for Science and Technology Policy at George Mason University, has noted, "From the producers' standpoint, a particularly strong motivation exists to dissuade research investments by customers with an advanced ability to develop alternatives—the United States, for example."[15] Auerswald's view is echoed in the words of Saudi Crown Prince Abdullah's foreign policy advisor, Adel al-Jubeir: "We've got almost 30% of the world's oil. For us, the objective is to assure that oil remains an economically competitive source of energy. Oil prices that are too high reduce demand growth for oil and encourage the development of alternative energy sources."[16] Oil producers in the United States and elsewhere also benefit from this behavior, so the U.S. failure to develop and implement alternative energy sources should not be wholly attributed to OPEC.

Given the long-term trend of accelerating oil demand and tight production capacity, Saudi Arabia may not be able to play the role of swing supplier and shock absorber indefinitely. Sustained U.S. policies to develop alternative technologies and improve the resilience of its economy are needed now to promote long-term energy security. Increasing domestic renewable energy helps to lessen oil vulnerability. Even though, for the time being, developing domestic renewable energy supplies is more expensive than traditional fossil fuel alternatives, renewable energy investments are reasonable and necessary to ensure domestic economic security and provide a hedge against future price volatility.

In the long run, promoting enhanced efficiency and expanded alternative energy supply options increases domestic investment opportunities, spurs domestic employment, and improves U.S. competitiveness. In 2005, expenditures on imported oil equaled one-third of the U.S. current account deficit of $800 billion.[17] Shifting even a fraction of this expenditure to U.S. investments could have a dramatic, positive effect on the U.S. economy.

We are already seeing the benefits of domestically-produced renewable energy in the farm sector. Spurred by ethanol demand, corn prices have risen significantly. Developing markets for crops that could be converted into ethanol and biodiesel have provided alternative income sources for farmers. Perhaps these changes will lessen the need for the billions spent annually on U.S. farm subsidies. Former Iowa Governor Tom Vilsack (now secretary of the Department of Agriculture) might be expected to defend a subsidy system that brings money to his state. Nevertheless, he said emerging markets for crops, such as corn-based ethanol, make it more likely there will be sustained demand and good prices "without the necessity of subsidizing crops to the extent we have."[18] Over time, corn-based ethanol is likely to give way to cellulosic ethanol made with a wide variety of inedible plant crops and residues, but farmers will still remain at the base of this production system.

End use efficiency—improvements in the way energy is used in industrial, commercial and residential settings—involves many technologies and benefits virtually all segments of the economy. According to the Energy Security Leadership Council, "Measures to reduce domestic oil demand offer the greatest potential to improve America's oil security. In a global market, the benefits of each barrel of increased domestic production are shared by all consumers in all nations through a lower equilibrium price, whereas the benefits of each barrel of reduced consumption accrue fully to those who have lessened their oil use."[19] The adage "Give a man a fish, and you have fed him for today. Teach a man to fish, and you have fed him for a lifetime" also is true for energy. Efforts to increase production have only a temporary return, while investments in efficiency pay long-term dividends.

The Connection between Oil and Power Generation

Transportation is responsible for 67% of U.S. oil consumption. Residential oil use, mainly for heating, takes up another 4%. Power generation is only 3% of total U.S. oil consumption. How would electricity from renewable sources reduce oil dependence?

In the long term, plug-in hybrid electric vehicles or hydrogen fuel cell vehicles could substitute electricity for oil used in transportation. The Argonne National Laboratory estimates that a 60% penetration of plug-in hybrid electric vehicles could reduce gasoline consumption by 43% and that most of the increased electrical load could be provided by increased wind power.[20] Wind power is often generated in the evening and at night when electric vehicles could be plugged in for charging. Continued work on high-capacity batteries is needed to provide the desired driving range for plug-in vehicles. The transformation to plug-ins also will be gradual due to slow turnover in the U.S. automobile fleet. The median age of cars on U.S. roads today is nine years.

Hydrogen produced with renewable electricity is another long-term alternative. Building a hydrogen fueling infrastructure and perfecting onboard hydrogen storage technology are among the challenges facing the adoption of hydrogen-fueled vehicles.

Today, natural gas is the bridge between the economics of oil and electricity. T. Boone Pickens' plan for improving America's energy security calls for an expansion of the electric transmission infrastructure to support the growth of wind power. Pickens' vision is for wind energy to displace a large portion of the natural gas used in electric generation. Natural gas could then be redirected to the transportation sector to substitute for petroleum-based fuels. The plan is simple—substitute wind-generated electricity for gas-generated electricity where possible and then take the gas and substitute it for oil used in transportation. No technological breakthroughs are required, just the political will to induce the required investments.

Substituting one energy resource for the other is critical to avoiding gas supply constraints. Joseph Kelliher, the former chairman of the Federal Energy Regulatory Commission (FERC), the nation's electricity and gas regulator, has thought extensively on the role of natural gas in the economy. Kelliher states,

> *My concern is we can't use gas for everything. And right now, because of uncertainty in climate change policy, we, through decision or indecision, we have committed, subconsciously perhaps, to using natural gas to*

supply most of our additional electricity supply for the next 10 years or longer. And I think if we use gas for everything, there already is very strong upward pressure on natural gas prices and we use it for transportation as well as agriculture, industry, electricity, we're going to assure high gas prices continue for a long time.[21]

Natural gas is today a major fuel for electric generation, just as oil plays a critical role in the transportation, chemical, and agricultural industries. Natural gas has been the fuel of choice for new electric generating plants in the United States for several reasons, including low capital costs and relatively clean emissions. Between 2000 and 2004, the 20% increase in U.S. electric generating capacity was almost all from gas-fired plants.[22] In many electricity markets, the cost of gas-fired generation sets the market clearing price—the price paid to other generators selling into the market. Increasing natural gas prices generally result in higher electricity prices.

Since natural gas is more difficult to transport than oil, natural gas prices tend to be set in regional markets rather than in the world market. But this is beginning to change. Similar to its usage of oil, the United States has only 3% of world gas reserves and it uses about 20% of world supply. The Middle East and Russia control three-quarters of world gas reserves. To close the supply-demand gap, a market in liquefied natural gas (LNG) has emerged and is becoming increasingly important. Long-distance gas transportation, via liquefaction, transoceanic LNG tankers, and subsequent regasification at the port of destination, is turning natural gas into a world-traded commodity. According to Daniel Yergin, "The result of all this could be a much larger, flexible, global business, with tankers responding even on the high seas to changing price signals among markets."[23]

Natural gas as a transportation fuel is better than oil, but it is far from perfect. Gas reserves worldwide have held relatively stable since 2004, implying that producers are able to replace consumed gas with new supplies. The ratio of reserves to annual production also is relatively high (100 years and 78 years in the Middle East and Russia, respectively). Unconventional gas resources from shale deposits and coal bed methane are making up for the decline in conventional gas supplies in the United States. Despite the current good news, the United States should be wary of developing the same vulnerability to interruptible foreign gas supplies that it currently suffers with oil. Global natural gas demand will probably rise significantly in the decades ahead. The United States is not alone in viewing increased reliance on natural gas as a way to reduce greenhouse gas emissions.

Natural Gas Facts

- U.S. consumption of natural gas is projected to increase at 0.1% per year on average through 2030.
- Developing nation consumption is rising (5.5% and 4.6% on average per year in China and India, respectively).
- A large portion of North America's remaining technically recoverable natural gas resource base consists of unconventional sources, which include tight sands, shale, and coal bed methane. These costly sources of supply must make up for declines in conventional production.
- U.S. net imports are increasing.
- U.S. net imports from Canada and Mexico are decreasing, while imports of LNG from the Middle East are projected to increase substantially.

Source: Energy Information Administration. 2008. International Energy Outlook 2008. Ch. 3.

It is clear that increased reliance on natural gas as a transportation fuel will require the alleviation of the pressure on gas supplies exerted by using natural gas for electricity generation. Renewable energy is one way to reduce pressure on gas supplies. Long-term energy security comes from a sustained commitment to diversify away from both oil and natural gas. Securing a modern standard of living with cleaner energy requires multiple energy sources that are better insulated from supply disruption and price volatility than oil and natural gas.

Trending Toward Renewables

The sun's power transformed into wind, waves, and ocean currents, and the moon's gravitational force expressed in the tides, is vast. The National Renewable Energy Laboratory (NREL) has estimated a 90.1 gigawatt (GW) wind energy potential in the New England, Mid-Atlantic, and Great Lakes regions for projects built in water depths of 30 meters (m) or less (figs. 1–2a and 1–2b). One gigawatt is 1,000 MW, or the approximate output of a large nuclear power plant. Compare that to New England's total installed generating capacity of 33 GW, and the potential impact of offshore renewable energy becomes clearer.

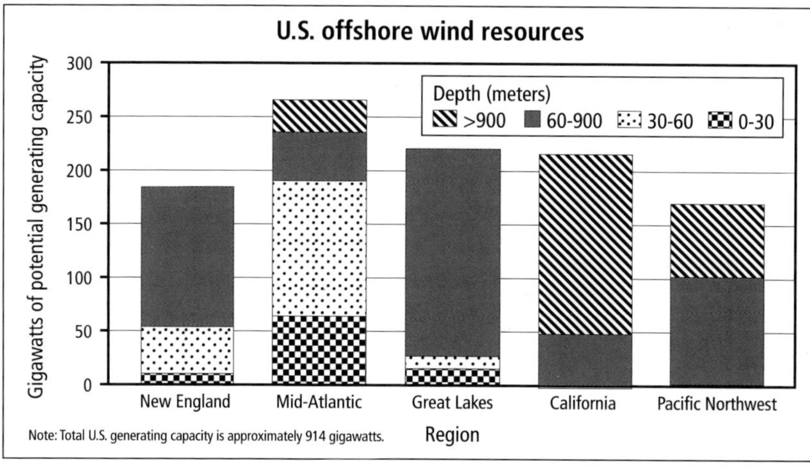

Fig. 1–2a. U.S. offshore wind resources

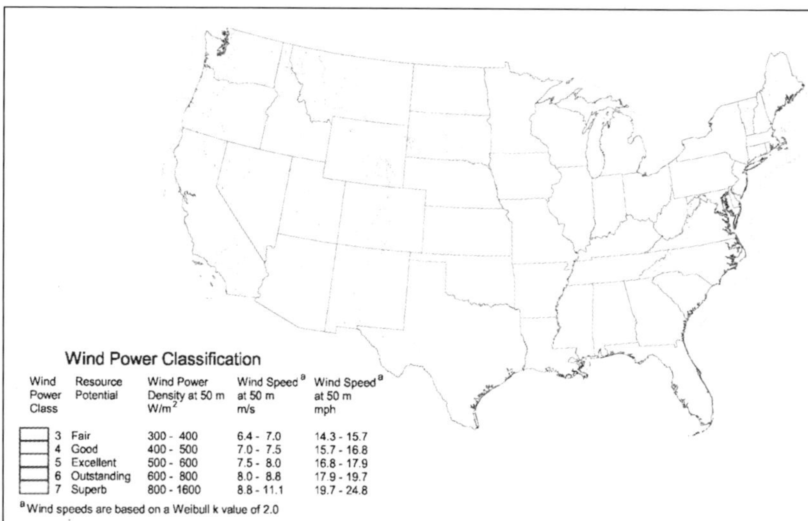

Fig. 1–2b. Terrestrial and offshore map of U.S. wind resources
Source: National Renewable Energy Laboratory

Numbers can deceive; so what is really behind NREL's estimate? By no means does the estimate contemplate a seascape covered with wind turbines. Note also that wave, tidal, and ocean current energy opportunities were not included in the estimate. All areas within 5 nautical miles (nm) of the shore were excluded, as were sensitive habitats, areas that should be off-limits due to avian and marine mammal use, shipping routes, and certain viewshed areas.[24] Based on conservative assumptions, the potential offshore wind resource in shallow water is extremely large. If technology evolves to permit the installation of floating wind platforms or other kinds of cost-effective bottom-mounted turbines in deeper waters, the total estimated offshore wind capacity could be 10 times larger, comparable to the total installed U.S. generating capacity.

Being land constrained, Europe is leading the way on offshore wind. Europe's installed offshore wind capacity is over 1,000 MW, and that capacity continues to grow at a healthy pace. In the United Kingdom alone, a boom in offshore wind projects could provide enough capacity to power one in six households there.[25] Aggressive development of Europe's offshore wind resource could supply roughly one-third of Europe's present day electricity load by 2020, according to the consulting group Garrad Hassan.[26]

Waves, ocean currents, and tides are other sources of ocean renewable energy that are further behind the development curve. Moving water also is approximately 835 times denser than air, making for a more concentrated form of energy. By government estimates, harnessing only one-thousandth of the energy in the Gulf Stream could supply as much as 35% of Florida's energy consumption.[27]

Tidal energy is similarly promising. Verdant Power, LLC is developing one of the more interesting projects, testing a field of submerged free-flow turbines in the East River near Roosevelt Island in New York City. Locally generated power is preferable to power imported from outside the city along congested transmission routes. This renewable power would be reliable (as predictable as the tides), clean, and located in the heart of New York City.[28] This trifecta of benefits should allow Verdant Power to sell the energy from its project at a premium price.

Wave energy varies across the ocean, but is strongest in the areas with the strongest wind currents, typically between 30° and 60° latitude.[29] Oregon and Hawaii, for example, have excellent wave energy potential. The total annual average wave energy off the U.S. coastline is estimated at 2,100 terawatt-hours (TWh). To put that in perspective, each year approximately 11,200 TWh of primary energy are used in the United States to meet electrical demand.[30]

A predictable energy system

Renewable energy generation is often accused of being intermittent, not predictable, and therefore impractical. Traditional energy advocates argue that a lull in the wind will leave consumers without power, insinuating that when we flip a switch, the lights may not go on. Studies refute the common claim that renewable energy intermittency requires utilities to keep 1 MW of traditional generating capacity on standby for each megawatt of renewable capacity built. Renewables could make up 10% or more of the installed generating capacity with little or no additional capacity required for standby service. Consumers can have both clean and reliable energy.

Intermittent supply is an operational challenge for utilities but is far from an insurmountable obstacle. In truth, the contrast between renewable and traditional fossil-fueled generating plants is not so stark. Traditional energy is volatile in terms of cost and vulnerable to fuel supply shortages. Utilities with coal-fired plants can face tight fuel supplies due to rail capacity shortages and coal freezing in rail cars in the winter, and limited gas transport capacity can keep gas-fired plants off-line. Traditional generators are not 100% operationally reliable either. They can go out of service unpredictably, and the electrical system must be built with reserves to respond at times of operating failures. Cost and operating unpredictability both pose challenges for utilities and customers. Renewable energy systems generally do not require fuel, so they provide an automatic hedge against the risk of fuel price volatility. Adding renewable generation to utility supply portfolios can help moderate rates.

Puget Sound Energy in Washington State is acquiring new wind and hydroelectric generation, refurbishing coal and hydroelectric generation units, and conducting energy efficiency and peak-demand reduction programs. These efforts aim to reduce exposure to volatile fossil fuel prices. "Over the long term, we're trying to shield our customers, as much as we can, from the financial stings an unstable energy market can inflict. The way to do that, we believe, is to plan ahead—to invest now and secure the energy resources and infrastructure needed to serve our region long into the future."[31]

A renewable energy project's capital costs, insurance, site rent, and royalties will be known or reasonably predictable. The European experience offers a relatively good idea of the cost of construction, maintenance and repair of offshore wind energy facilities. As we acquire field experience with other marine renewable technologies, our ability to predict long-term costs for the newest marine technologies also will improve.

Fossil fuel prices will likely continue to be volatile, to increase, and to drive the cost of electricity higher. Without debating whether the world has reached *peak oil*—the point at which newly discovered reserves do not replace depleted reserves—it seems clear that new reserves of oil and gas will become more difficult and expensive to find and develop. New reserves will increasingly be located in remote or politically unstable areas, deep waters, or in shale or oil sands that require extensive processing. Existing wells that are declining in their productivity will require investments in enhanced recovery technology to extract the remaining hydrocarbons. Deepwater production facilities, sometimes accessing oil 20,000 feet (ft) below the surface, can cost $1 billion to construct, compared to $100 million in 100 ft of water, or $10 million for a land-based well.[32] Newly discovered natural gas resources will increasingly require expensive liquefaction and transport facilities to move gas from remote fields to consumers.

Increasing fossil fuel prices, driven by higher exploration and production costs and greater demand, will reduce the price gap between renewable energy sources and traditional depletable energy sources. At the same time, economies of scale and learning efficiencies will drive renewable energy costs down. California, the first state to develop wind power on a large commercially viable scale, began in the 1980s with turbines averaging 100 kW in size. Today it is not uncommon to see turbines ranging from 1.0 MW to 2.5 MW used for onshore wind farms. Turbines in the range of 3.0 MW to 3.6 MW are common in offshore applications. A 5-MW wind turbine has recently been erected by REpower in the Scottish North Sea in 44 m of water as part of a pilot project. As turbine size increases, installed cost per megawatt of capacity and cost per kilowatt-hour of energy generated should decrease.

There is a trade-off in renewable energy economics. It takes a large up-front capital investment to obtain the advantage of low renewable energy operating costs. This is particularly true in the case of marine renewable energy technologies, which must be able to withstand the energetic and corrosive marine environment and are more expensive to build. Traditional low capital cost fossil-fueled generation is, however, beginning to lose its advantage as the added cost of newer pollution control technologies increases capital costs and reduces operating efficiency. A state-of-the-art integrated gasification combined-cycle coal plant that employs carbon sequestration technology has very high capital costs. "Clean" coal plants also have higher operating costs than traditional coal-fired plants due to the losses associated with coal gasification, carbon separation, and carbon sequestration.

Since coal-based electricity is produced domestically, it is credited with helping foster energy security. Coal, and to a lesser extent other fossil fuels, however, are linked to global warming, regional air and water pollution, the depletion of increasingly scarce surface and ground water, thermal pollution, and associated environmental and human health effects, from asthma to mercury poisoning. The cost of litigation, increased regulation, and technological fixes to address the externalities of fossil fuel plants will gradually push utilities to reevaluate their investment decisions. As utilities are forced to bear the externalities that society now shoulders, the cost of nonrenewable power will rise, allowing renewable energy to compete on a more level field. According to Lester Brown, "The central challenge, the key to building the new economy, is getting the market to tell the ecological truth. The dysfunctional global economy of today has been shaped by distorted market prices that do not incorporate environmental costs. Many of our environmental travails are the result of severe market distortion."[33]

Corporate peer pressure

The oil shock and the mortgage shock would hit later; in 2005, it was a weather shock that hit the world's insurance companies. Driven by Hurricane Katrina, insured weather-related losses worldwide were approximately $80 billion, compared to the approximately $20 billion of insured property losses associated with the September 11, 2001 terrorist attacks.[34] Compared to the recent financial carnage on Wall Street, insurance industry losses in 2005 seem quaint, but the magnitude of the losses was a stunning blow to the confidence of an industry dependent on correctly modeling the future.

Global warming appears to be contributing to severe weather events, and little doubt remains that man's actions are at least partly the cause. In almost 1,000 peer-reviewed scientific articles mentioning "climate change" published between 1993 and 2003, not one disputed that human activity influenced this phenomenon.[35] According to A.M. Best Co., a leading insurance ratings provider, storms even more costly than Katrina are possible, perhaps inevitable, and associated losses could bankrupt numerous insurers. Hoping to avoid this fate, proactive insurers are seeking to understand the impact of climate change and to mitigate its effects. Insurers also are beginning to pressure their clients to change their behavior and climate risk exposure. Helping develop improved building standards that will lead to reduced storm damage losses, and linking auto insurance premiums to miles driven, which has been shown to reduce driving and emissions by 10% to 15%, are just two examples. Insurers also are getting the attention of corporate America by increasing premiums and becoming active shareholders in the companies in which the insurers have invested reserves.

One business sector with an unhealthy climate exposure is the U.S. electric utility industry. The utilities listen to their largest shareholders, and when a leading investment bank opines that carbon emissions could create liability similar to asbestos, shareholders get nervous.[37] Some shareholders have asked management to disclose their company's climate change risk and to explain management's strategy for mitigating the risk. In response, certain utilities have begun to voluntarily reduce carbon emissions and have sought government controls over carbon dioxide emissions in an effort to bring a measure of certainty and predictability to the utilities' environmental exposure.

That exposure could be substantial. Southern Company, a large southeastern electric utility, has estimated its costs from moderate greenhouse gas regulations at approximately $780 million annually.[38] The giant Midwest utility American Electric Power (AEP) recently agreed to pay a record $4.6 billion, including the cost of installing pollution control equipment, to settle claims that its coal-fired power plants contributed to acid rain and pollution across the Northeast and Mid-Atlantic regions in violation of the Clean Air Act. Now that the Supreme Court has ruled that carbon dioxide is a pollutant that should be regulated under the Clean Air Act, the business case for investments in carbon-emitting technologies must reflect additional potential liability.

In October 2007, a $3.6 billion proposal to build twin 700-MW coal-fired generators was stopped by the Kansas Department of Health and Environment. According to Roderick Bremby, the agency's secretary, "It would be irresponsible to ignore emerging information about the contribution of carbon dioxide and other greenhouse gasses to climate change and the potential harm to our environment if we do nothing."[39] This was the first time in U.S. history that a plant was stopped based on carbon dioxide emissions. This radical step did not happen in a stronghold of environmental activism such as California or Massachusetts. It happened in Kansas, where farmers are welcoming wind power as a real alternative to coal-fired electricity. "The debates over acid rain to a certain extent have been resolved," said New York Governor Elliot Spitzer who, as attorney general of New York, participated in the AEP lawsuit. "Carbon is the real issue now. Do we have to move away from all fossil fuels? These are tough issues, from a policy standpoint and an economic standpoint."[40]

What does the future hold for AEP, the country's number one industrial emitter of carbon dioxide? The combustion of fossil fuels, principally coal, at power plants is responsible for a significant percentage of total U.S. emissions. In 2006, the electric utility industry was responsible for 40% of carbon dioxide, 70% of sulfur dioxide, 20% of nitrous oxides, and 68% of mercury air emissions in the United States.[41] The electric utility industry is the largest emitter of carbon dioxide in the United States; more than any other industrial source or the transportation sector (figs. 1–3a and 1–3b). The 100 largest electric power producers in the United States accounted for roughly 86% of the utility industry's emissions.[42] The concentration of emissions responsibility within the utility industry strongly suggests that these companies will be the focus of future regulations and that they also will bear the lion's share of control costs.

Utilities with a high-carbon fuel mix (i.e., those whose sources of supply are predominantly coal-fired power plants) may suffer earnings decreases of 24%–83%, while companies that are prepared for regulation with less polluting fuel mixes (e.g., renewables, nuclear, and gas) could be winners with 4%–139% earnings increases.[43] There are powerful industry incentives, therefore, to act now to control future climate change costs.

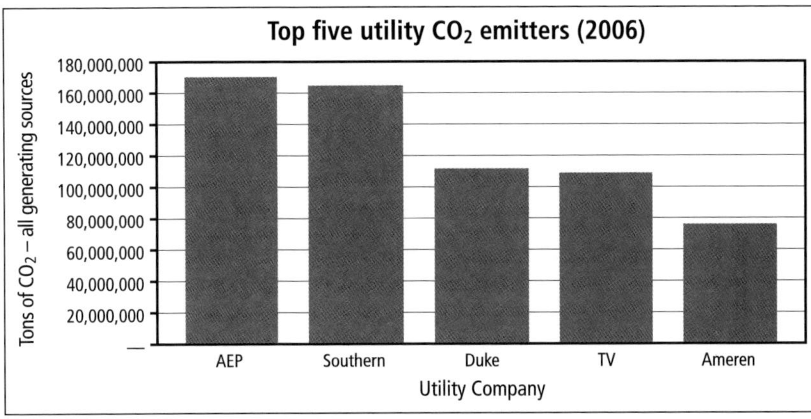

Fig. 1–3a. Top five utility CO_2 emitters

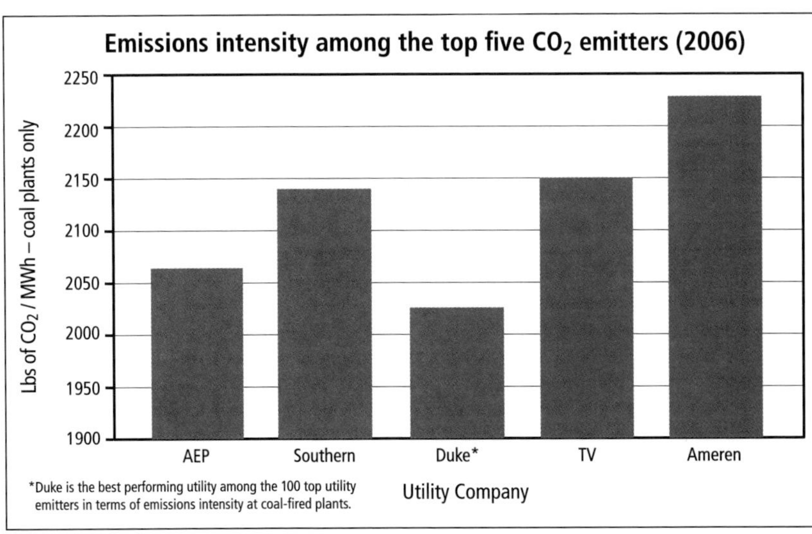

Fig. 1–3b. CO_2 emissions intensity

Source: Van Allen, et al. 2008. Benchmarking Air Emissions of the 100 Largest Electric Power Producers in the United States. Tables 1 and 2.

Where some utilities face losses and seek to resist shareholder and government pressure to limit CO_2 emissions, other companies see an opportunity to profit from the large capital expenditures that will be spurred by climate regulations. One example comes from General Electric, which started Ecomagination, a program to voluntarily reduce GE's greenhouse gas emissions by 1% by 2012. Given GE's growth rate, by 2012 its emissions without this commitment would have increased by over 40%.[44] While easily dismissed as mere public relations, Ecomagination expresses GE's vision that future competitive advantage will come from the production of environmentally significant technologies. GE would prefer to shape that future than be shaped by it. As Jeff Immelt, GE's chairman and CEO put it, "We are living in a carbon constrained world where the amount of CO_2 must be reduced. And there are two ways to address that reality: head on and driven by innovation or by getting pulled into it through regulation . . . I believe that we must have a proactive business policy or we will get a reactive government policy."[45] GE's offshore wind turbines and more efficient locomotives and airplane engines are part of the strategy. The company proposes to more than double its investment in cleaner technologies to $1.5 billion per year by 2010, compared to $700 million in 2004.

Leadership at companies like GE is driving other companies to adopt more environmentally sustainable practices. GE has stated that it will enter into partnerships with its customers to address their environmental challenges, and it will exert pressure on suppliers to make sure that they adhere to strong environmental standards. Gaz de France provides another environmental leadership example. The company encourages its suppliers to implement, at least in part, the Global Compact, a U.N. initiative to bring companies and U.N. agencies together to support universal principles, including environmental stewardship. Principle Nine of the Global Compact calls on businesses to encourage the development and diffusion of environmentally friendly technologies. Competitive pressure, regulatory pressure, customer pressure, investor pressure, market pressure, and environmental pressure are driving the adoption of new, cleaner energy solutions.

Momentum grows for carbon reduction

Ratified by 166 nations, the Kyoto Protocol requires mandatory greenhouse gas emissions reductions by the developed countries. The treaty is a political and social force that encourages conservation, efficiency, and an increased reliance on renewable electric generation. Under the treaty, participating developed countries must reduce their greenhouse gas emissions compared to 1990 levels by 5.2% on average calculated over the five-year period 2008–2012. Participants are attempting to reach this goal through physical changes to industry such as increased efficiency, the purchase of carbon credits on financial exchanges, or with carbon credits earned in carbon offset projects in developing nations. In practice, participating countries have established or joined country-specific or regional cap-and-trade systems to manage emissions allowances. A country's emissions target is assigned to specific industries and companies through emissions allowances that may be used by the company or traded. To the extent a company's emissions exceed its allowances, the company must purchase additional carbon credits to cover the deficit. Companies that implement efficient technologies that reduce emissions may profit by selling their excess emissions allowances or by increasing their ability to make products within their existing emissions cap.

Kyoto Protocol compliance will be expensive. In 2006, the European Union's carbon emissions trading market was valued at approximately $30 billion.[46] Individual credits fluctuated in trading from less than 10 euros to more than 30 euros per ton of carbon dioxide during 2005 to 2006.[47] Since a coal-fired electric plant emits about a ton of carbon dioxide for each megawatt-hour (MWh) generated, carbon emissions restrictions have the potential to significantly impact energy prices.

Mandatory limits on U.S. carbon emissions seem likely under the administration of President Obama. The United States is the only developed country that has not ratified the Kyoto Protocol and, as the leading emitter of greenhouse gases, the failure of the United States to abide by mandatory limits has become a sore point in international relations.

Meanwhile, state governments are occupying the void left by federal inaction. The most notable of these is the Regional Greenhouse Gas Initiative (RGGI). The objective of RGGI is to reduce carbon dioxide emissions by 10%, relative to 2006 levels, by 2019. RGGI's main component is a plan to cap carbon dioxide emissions from power plants in the participating states and to issue allowances to individual companies that would in the aggregate equal the cap. The companies could then trade the allowances on a market-based emissions trading system, turning them in to the government at year-end as evidence of compliance with their individual limits. Over time the cap would be reduced. States currently participating in RGGI include Maine, New Hampshire, Vermont, Massachusetts, Rhode Island, Connecticut, New York, New Jersey, Delaware, and Maryland. Although the initiative is expected to add \$3–\$21 to the average household's annual cost of electricity, it also is expected that the substantial investments in clean and efficient technologies required to reduce carbon emissions would provide an economic boost to the region.

The California Public Utility Commission requires utilities to include a *greenhouse gas adder* of \$8 per ton of carbon dioxide when they evaluate competing energy supply options. The adder is intended to reflect the cost of carbon emissions regulation on the price consumers pay for power.[48] California's Global Warming Solutions Act requires the California Air Resources Board to implement restrictions on greenhouse gas emissions that would, by 2020, reduce emissions to 1990 levels. The act also is notable for establishing a California carbon credit market that may be integrated with programs implemented in other western states, as well as the RGGI credit market in New England and the Mid-Atlantic states, to establish a more liquid U.S. market. Fabian Nuñez, a coauthor of the law, stated, "For us, this is not just about California. This is about making a push from the bottom up to get the Congress to take action."[49]

Reducing other pollutants and improving public health

Despite the increased attention to global warming on the world stage, pollution from fossil fuel combustion, including particulates, nitrogen oxides, sulfur dioxide, and mercury, also remains a concern. Pollution has a direct effect on public health and national productivity. Pollution from fossil fuel combustion has been linked to mental development problems in children and increased incidences of asthma and heart disease.[50] These effects translate into higher educational costs, lower achievement, missed work days, increased hospitalization expenses, and premature death.

Residents of one unfortunate town in rural Ohio complained of raspy throats, burning eyes, and mouth blisters brought on by a blue haze that hung over their homes. American Electric Power, the local utility whose plant was thought responsible, purchased the entire town to dispense with the problem. The town's unfortunate (or fortunate) residents received checks in an amount several times the value of their homes to move away.[51]

A blue cloud is visible and has immediate effects. Invisible, long-term effects from widespread mercury pollution are another side of the bargain Americans have unwittingly struck in exchange for "cheap" power. Methylmercury is the main form of organic mercury that accumulates in the environment, particularly in the flesh of fish as it biomagnifies up the aquatic food chain. The EPA has labeled methylmercury a developmental toxicant and found that it causes human germ cell mutations.[52] Mercury poisoning weakens the immune system, alters genetic and enzyme systems, and damages the nervous system, including coordination and the senses of touch, taste, and sight.[53] Solid waste and fossil fuel combustion contribute 87% of U.S. mercury emissions.[54] Each year U.S. coal-fired electric plants emit 48 tons of mercury.

As of December 2003, 45 states had issued fish advisories for mercury covering more than 13,000,000 lake acres and more than 750,000 river miles.[55] Eliminating mercury and other pollutants from the emissions of conventional power plants requires the installation of expensive pollution control technology, the use of cleaner fuels with lower concentrations of mercury and sulfur, or new processes like coal gasification that remove pollutants from fuels before combustion. In each case, the consumers' cost of conventional energy sources increases. In 2005, EPA issued its Clean Air Mercury Rule that required coal-fired power plants to reduce mercury emissions by 70% from current levels by 2018. The rule was overturned in 2008 by a federal appeals court in response to a petition by various states claiming that EPA had acted improperly under the Clean Air Act in promulgating the rules.[56] EPA must now issue new and possibly more stringent mercury rules. Pending EPA's rules, several states have imposed their own stringent mercury emissions restrictions.

If state pollution regulations are the stick that provides utilities with an incentive to clean up their emissions, state renewable portfolio standards (RPS) are a carrot offered to generators of clean power. These programs require utilities and power marketers to use renewable energy for a portion of the electricity they supply to consumers. Because the demand for renewable energy is mandated and the supply is limited, the purchasers typically pay the renewable energy generator extra for the clean power. Twenty-nine states and the District of Columbia have an RPS or a nonbinding renewable energy goal (fig. 1–4). There is as yet no federal RPS.

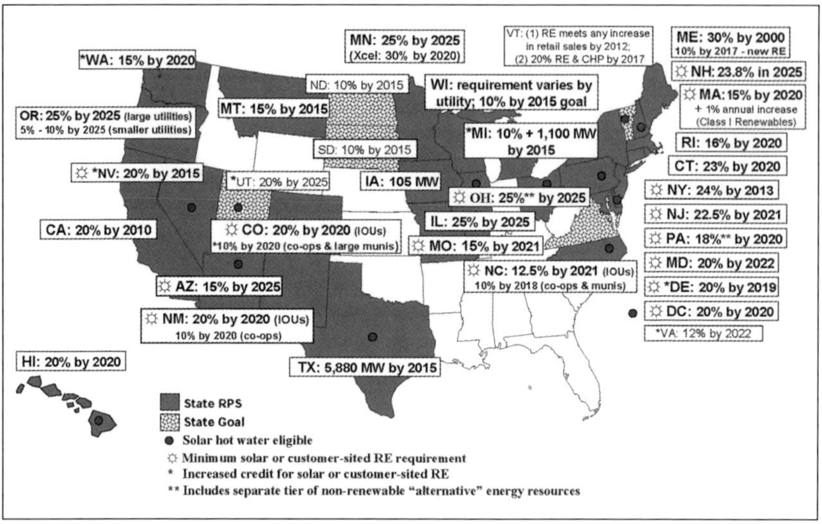

Fig. 1–4. States with renewable portfolio standards or goals
Source: Interstate Renewable Energy Council. www.dsireusa.org

Coal, Nuclear, and Renewable Energy Alternatives

Where then should the United States look for abundant, clean energy? Some believe that "clean" coal is the answer. It is often said that there is more energy available in U.S. coal deposits than in nearly all the oil in the world. There may be a 250-year coal supply. Yet even if significant air pollution and carbon emissions from coal combustion could be avoided through clean coal technology, and that is a big "if," coal mining and waste disposal present their own difficult problems. Mountaintop removal coal mining practiced in Appalachia, for example, has covered at least 700 miles of streams with rock rubble, polluted drinking water, promoted flooding, and scarred the landscape.[57] Surprisingly, as mountaintop removal has expanded, employment in the Appalachian coal industry has declined, because this method relies more on explosives and gargantuan machines to reach coal seams than it relies on miners.

As emissions regulations attempt to limit the damage from coal combustion, researchers have sought ways to capture coal's impurities, including sulfur and heavy metals, before they reach the smokestack. Coal gasification and carbon sequestration technology is one potential solution under development. Plants are now being built that will demonstrate these technologies. BP and Edison Mission Group have a joint project to build a $1 billion hydrogen-fueled power plant at an existing oil refinery in Carson, California. The plant will gasify petroleum coke refinery residue into hydrogen and carbon dioxide. Hydrogen burned in turbines will power electric generators, while the carbon dioxide will be transported to nearby mature oil wells, where it will be pumped underground to increase the production from those wells. While this project would use coke, the gasification technology is adaptable to coal. Although promising, coal gasification and carbon sequestration on a large scale are untested. Geologic formations that would keep large amounts of gas trapped for a long time must be identified and evaluated. Capturing the gas, compressing it, and pumping it into storage also reduce the overall efficiency of converting coal to electricity.

Nuclear-fueled generating plants supply approximately 20% of the electricity consumed in the United States. Nuclear fission does not produce harmful air pollutants. The nuclear energy option, many argue, promotes energy security because enriched uranium fuel can be produced from plentiful domestic sources. As concern about the climate grows, the solid opposition to nuclear energy among environmentalists is beginning to thaw. Patrick Moore, one of the cofounders of the staunchly antinuclear Greenpeace, surprised many when he publicly announced his support for nuclear power as the only large-scale, cost-effective source of power that can reduce U.S. carbon dioxide emissions.[58]

Nuclear plant construction, however, is an expensive and lengthy process. It has been so long since a new nuclear plant was built in the United States that the nuclear supply chain has withered. The U.S. would need to rely on Japan and France for major components for new construction, and industry experts expect a new plant to cost several billion dollars and to take the better part of a decade to complete. A single investment of this size represents a significant risk even for the largest U.S. utilities. Despite recent government incentives aimed at reducing the risks of construction delays, a history of nuclear plant licensing litigation and construction cost overruns has made the electric utility industry hesitant to build new nuclear units unless they can pass the costs through to ratepayers. Moreover, the problem of safe and secure permanent disposal of nuclear waste has not yet been solved.

America currently relies on nuclear, coal, natural gas, oil, hydro, and wind-powered plants for its energy. The future energy mix is not an either/or set of choices. Even if the United States is the "Saudi Arabia of coal," there is no excuse to ignore other viable, and perhaps better, solutions. Given the utility industry's present commitment to large-scale fossil and nuclear technologies, there probably will be numerous, cleaner coal plants and new nuclear plants built in the future. This is the technology that utility managements, bankers, and investors understand, and the technology around which government programs and policies have already developed. The next generation of coal and nuclear plants, however, will be more costly than the last. Paying for them will require higher electricity prices and government subsidies such as loan guarantees. The already narrowing price gap between fossil-derived or "brown" electricity and renewable-derived or "green" electricity will close further.

National economic resiliency comes from cultivating a diverse mix of energy sources. Relying more on renewable energy makes the economy less susceptible to disruption due to volatile oil, natural gas, and coal prices. Offshore renewable energy has the potential to become a significant contributor to supply diversity.

The advantages of offshore siting close to population centers

Fortunately, a utility seeking to manage its environmental exposure has a variety of renewable energy options, many of which are ideally suited to specific geographic niches. Maine's large forests and timber and paper industries provide the right ingredients for biomass-generated electricity. Biomass-fueled combustion is reliable and provides a good baseload power source. The major challenge of biomass projects has been to procure an adequate supply of biomass fuel at a reasonable cost. Poultry litter, wood from forest thinning, and *bagasse* (the plant fibers left after sugar cane is processed) are examples of biomass fuels. These fuels are not as energy dense as fossil fuels and generally must be brought to the biomass-fueled electric generating plant by the truckload. Thus the cost of biomass may be correlated with the cost of diesel fuel, making biomass energy subject to volatility in oil prices. To manage fuel transport costs, these plants are best in small, distributed applications.

The western, midwestern, and northwestern states enjoy the greatest abundance of land-based wind resources. Texas and California lead the nation in installed wind generating capacity, and other states are rapidly expanding capacity. Iowa, Minnesota, and Oklahoma also are large wind power states.[59] The Southwest benefits from consistently high levels of *insolation* (exposure to the sun) and therefore is a good location for solar energy.

Offshore renewable energy has a developing niche in the Northeast and Mid-Atlantic states. Wind energy projects have been proposed for various sites from Maine to Georgia and off the Gulf Coast. Tidal current projects are being planned for New York's East River and for San Francisco Bay. Wave energy technology is being tested in Washington,

Oregon, and Hawaii. Finally, developers are exploring ocean current energy opportunities in the Gulf Stream off Florida's Atlantic Coast.

Offshore wind resources are attractive for several reasons, especially on the East Coast, where terrestrial wind sites are scarce. An ideal wind-generating site is large and sparsely populated, has strong, relatively persistent winds, and is located close to transmission lines. Acceptably strong winds in the eastern states tend to be found along the ridges of the Appalachian Mountains and beside the seashore. Aesthetic concerns often block projects in these locations.

The struggle to develop an abandoned radar base in a sparsely populated corner of Vermont illustrates the difficulties of developing terrestrial East Coast locations. Although the site seemed ideal for a small (6-MW) wind farm, the five-year effort to site this project ended in failure. "It doesn't bode well for Vermont or the region's energy future, if we can't site four turbines on a previously developed site deep in the Northeast Kingdom," commented Christopher Killian of the Conservation Law Foundation.[60]

Offshore sites have the potential to address many of the limitations of terrestrial renewable energy facilities. The power available in wind increases with the cube of wind speed. Winds are stronger, more persistent, and less turbulent offshore than on land, for greater energy output per turbine.[61] Offshore wind also has a clear advantage by virtue of turbine size. Practical obstacles to moving the gargantuan components of modern wind turbines over roads limit their size for terrestrial use. Turbine blades may be 40 m or longer, and together with the nacelle that houses the gearbox, generator, and other electrical components, the complete unit may weigh more than 250 tons. Ships, barges, and marine cranes can move very large components, allowing taller offshore turbines to capture stronger winds aloft with larger diameter rotors. GE's 3.6-MW offshore model has rotor blades that sweep an area as large as a football field, and an overall height comparable to a 40-story building; still larger models are under development.

Acquiring site leases and permits for a terrestrial project is time-consuming. Building access roads, particularly in forested areas, may have significant environmental impacts. Turbines may reduce a landowner's ability to use land for farming, housing, or other purposes. Nearby residents also may object on noise and aesthetic grounds. The developer of an offshore wind project has the advantage of being able to obtain access to a very large tract under a leasing process that is centralized under the U.S. Minerals Management Service (MMS). Though the MMS process is unlikely to be shorter or less expensive, it may prove more straightforward than cobbling together terrestrial development rights through lease negotiations with multiple land owners and braving permitting hearings in multiple jurisdictions.

Shell Wind Energy is planning a $2 billion, 1,000-MW wind farm in the North Atlantic near the U.K. coast. This is a giant project even by offshore standards, and it would dwarf the average terrestrial wind farm. Large offshore projects may benefit from economies of scale as fixed costs such as project development and transmission infrastructure are spread over a large installed capacity. Rarely could wind farms of such size be built on land in the eastern United States.

Offshore projects also may benefit from the production of renewable energy close to the consumer. The population of the United States is concentrated along its coasts (including the shores of the Great Lakes), near offshore renewable energy resources. Compared to wind farms in Wyoming, West Texas, and Iowa, offshore wind power can be generated close to more densely populated areas. In the Northeast, for example, wind resources offshore are exceptionally strong, and the outer continental shelf has a gradual slope. The comparatively shallow waters of the eastern continental shelf permit the construction of wind turbines on towers embedded in the seabed. This area holds numerous large sites where the wind is strong and relatively persistent, the water is shallow, and the load is proximate. Still more sites will become available as engineers develop economical ways to locate facilities in deeper waters.

Offshore renewable projects with direct access to East Coast load centers would alleviate chronic congestion in the transmission system and may reduce the need for additional transmission line construction. Many of the energy markets along the East Coast have a shortage of transmission capacity and locally-sited generating capacity. The ability to sell renewable energy into high-priced, transmission-constrained coastal markets is driving offshore development. A wind farm sited off Long Island's southern shore, for example, could moderate that region's high energy prices by relieving Long Island's isolated position "behind" transmission-constrained New York City. Similar opportunities abound up and down the U.S. coasts.

Public attitudes

Conventional wisdom holds that aesthetic concerns are the main objection to offshore wind energy projects. A proposed project in Massachusetts' Nantucket Sound caused a public and national controversy. Massachusetts' former governor, Mitt Romney, Senator Ted Kennedy, and the Alliance to Protect Nantucket Sound, a well-funded and vocal opposition group, are among the opponents that have succeeded in delaying the project for over seven years.

A New Jersey blue ribbon panel, however, concluded that the state faces a growing energy crisis and that it must lead with bold action in developing renewable energy resources. The panel recommended a carefully monitored offshore wind farm test project to gather environmental and economic data, including insights into the effect of offshore wind energy facilities on the state's $32 billion tourism industry. Subsequently, New Jersey's energy master plan called for the construction of 3,000 MW of offshore wind energy capacity by 2020, finding this to be one of the few viable avenues to greenhouse gas reductions and reduced reliance on volatile fossil-fueled generation. The state has awarded up to $4 million in assistance to one developer that is proposing a 346-MW project 16 to 20 miles off the southern New Jersey coast.

In Texas, two wind energy developers have had little difficulty entering into offshore site leases with the Texas General Land Office. Considering the Gulf of Mexico's history of oil and gas development, and that energy lease proceeds are dedicated to school funding, it is not surprising that Texans welcome offshore renewable energy. Texas could earn $34 million to $100 million in royalty payments for its 30-year offshore wind leases. Other states around the country similarly are moving forward with offshore renewable energy, and nowhere has the opposition approached the level encountered in Nantucket Sound.

Opportunities for the Intrepid

Iceland sits right on the Mid-Atlantic Ridge, a slowly-spreading fissure in the earth's crust where molten rock bubbles forth. The country benefits from rich geothermal energy resources close to the surface. Today, more than 70% of Iceland's energy is provided by hydropower and geothermal energy. Iceland plans to eventually replace imported fossil fuels used for transportation with hydrogen made with electricity generated from these renewable resources.[62] A can-do spirit, combined with renewable energy technology appropriate for Iceland's specific renewable energy resources, have placed Iceland well along on the path to a sustainable and affordable energy future.

In the late 1930s and early 1940s, the same spirit of determination infused American oilmen who ventured out of the marshes and bayous of southern Louisiana and into open water. A geologist of that time observed, "It may be tentatively assumed that the Gulf of Mexico is a potential source of salt-dome oil... Whether or not it will ever be economically feasible to explore these waters for the domes that must exist is a question for the future to answer."[63] Intrepid scientists, engineers, and investors produced in excess of 10 billion gallons of oil from platforms in Gulf water as deep as 7,500 ft in the years that followed.

Offshore renewable energy is today at a similar junction of technology, economics, and possibilities. A new generation of audacious energy prospectors is again looking offshore to explore for, plan, and build a renewable energy future for the United States.

A Roadmap of This Book

Developing sustainable, clean, renewable energy projects in U.S. waters is the fascinating challenge addressed in the pages ahead. The discussion begins with a brief description of the marine environment to orient readers to offshore conditions and energy resources. A look at the marvelous machines created to harvest marine renewable energy follows. This chapter describes the basic engineering problem—building energy harvesters capable of surviving the harsh ocean environment while producing competitively priced power. The next chapter guides a prospective developer through the considerations inherent in choosing an offshore project site. Several chapters that follow provide an overview of legal and regulatory issues, from obtaining rights to a project site to ensuring the project's environmental soundness and collecting all the permits needed to start construction. Since offshore projects will be located on public lands and may conflict with other uses of the marine environment, working with stakeholders to build understanding and support for a proposed project is key. A chapter on the public consultation process covers the issue. With the exception of rooftop solar panels, no renewable energy project works unless it is connected to the electric transmission grid. Transmission issues as they may affect offshore renewable energy projects are discussed in a separate chapter. Practical issues in financing and insuring offshore renewable energy projects are covered next. Identifying and allocating project risks among investors, project owners, contractors, the government, and others is critical to assembling a project that will attract financing. Lastly, the economic drivers of offshore renewable energy projects are discussed. This chapter identifies barriers that must be overcome if marine renewable energy is to play a substantial role in the energy future of the United States.

2 An Unfamiliar Environment: Offshore Conditions and Energy Resources

We usually find oil in new places with new ideas. When we go to a new area we can find oil with an old idea. Sometimes also we find oil in an old place with a new idea, but we seldom find much oil in an old place with an old idea.

—*Parke Dickey*

In an age of space exploration and advanced technology, it may be surprising that most of the oceans are still a mysterious place. They remain largely inaccessible and unexplored because 60% of the planet is covered by water that is at least 1 mile deep.[1] Depths in the Pacific, the world's deepest ocean, are as great as 7 miles (11.3 km) and average 2.3 miles (3.7 km). The peaks of the midocean ridges are on average 1.5 miles (3.7 km) below the surface. Where the midocean ridges join the ocean floor, it is 3.5 miles (5.6 km) deep on average. Even the shallower depths of the oceans present obstacles and challenges. Building rigid, bottom-mounted structures for renewable energy machines in most of the ocean is impractical, if not impossible. It is more likely that floating machines tethered to the bottom with cable or chain would be built. The cost of anchors and line will nevertheless be significant, placing a premium on access to sites with good energy characteristics, in modest depths, and close to shore.

To build and operate marine renewable energy technologies, and to appreciate the risks that accompany an investment in this industry, requires an understanding of the ocean environment. Seabed geology, marine ecology, bird migration patterns, wind, wave, and current behavior, and many other factors will influence design and siting decisions. Understanding these factors permits more efficient energy capture and allows the projects to be built for survival (without overbuilding) and to avoid environmental harm. Fortunately, European offshore renewable energy project experience and learning from other marine endeavors provide a solid foundation for progress.

Offshore Oil and Gas Industry Experience

At 5:00 a.m. on August 28, 2005, the National Data Buoy Center's (NDBC) station number 42003, located in the central Gulf of Mexico, reported a maximum significant wave height of 34.7 ft (10.6 m), maximum 10-minute wind speed of 64 mph (28.6 m/s), and maximum 5-second gusts of 87 mph (38.9 m/s). Then the buoy, 33 ft (10 m) in height, capsized and went silent. In the 30-year history of the NDBC, never before had such a buoy capsized in the Gulf of Mexico.[2] The next morning, nearshore buoy station number 42007 broke from its mooring and went adrift under similar conditions. Several hours later, anemometers at Grand Isle, Louisiana reported 5-second gusts of approximately 115 mph (51.17 m/s) just before they failed. Scientists would later learn from information gathered by other battered equipment in the Gulf that the storm produced maximum sustained winds of 175 mph (78.2 m/s), gusts up to 215 mph (96.1 m/s), and monster waves as high as 105 ft (32.1 m).

Before landfall, Hurricane Katrina destroyed 46 oil and gas platforms and four drilling rigs. Four weeks later, Hurricane Rita would barrel through the Gulf, destroying a further 63 platforms and one drilling rig. Interior Secretary Gale Norton commented, "We have no official estimate of the dollar value of the damage and the amount that it will cost to repair those facilities, but it will clearly be in the billions of dollars."[3]

Some oil industry observers have labeled the Katrina experience a success, noting that the newer platforms built to modern standards escaped catastrophic structural damage. If that is the case, does the history of offshore oil and gas exploration and production offer important information about the future of offshore renewable energy technology? Today, more than 4,000 offshore oil and gas platforms in U.S. waters produce energy in what can be an intensely hostile environment. The success in penetrating the depths of the Gulf and thousands of feet of rock below the seabed is an achievement of science and technology comparable to the exploration of space. Some of this technology and experience must be transferable to marine renewable energy.

The story behind today's gargantuan deepwater oil rigs begins with the oil prospectors, engineers, scientists, and tradespeople who adapted a combination of land- and water-based technology. They applied the technology first to the lakes, marshes, and bayous of southern Louisiana and then ventured to shallow sites in open water. Challenged with erecting drilling derricks on soft marsh soils, the industry's offshore pioneers adapted flat-bottomed barges to the task by linking the barges together and partially sinking them over the drilling site to form a stable drilling pad. Navy divers returning from World War II also contributed their skills and daring to advancing construction techniques and technology. Divers experimented with different diving gas mixtures to lengthen time spent underwater, employed decompression chambers and tables to make diving safer, and improved underwater welding equipment. Developments in underwater communications and photography promoted better coordination between surface crews and the divers below. Lengthening the time spent underwater and improving diver productivity were essential to reducing the costs of offshore construction, because the divers' work often dictated the speed at which the entire offshore construction project could proceed.

Offshore oil production began in earnest in 1937 near the small town of Creole, Louisiana. There, about 1 mile offshore, prospectors built a wooden platform 180 ft × 300 ft (54.9 m × 91.4 m). Owing to the gentle slope of the shelf, the water was only 15 ft (4.6 m) deep at the platform. Brown & Root, the engineering firm that designed the platform, decided that it should be 15 ft (4.6 m) above water level. Since they had little information about wave heights and storm severity, Brown & Root built the platform on wooden pilings, figuring that if it was struck by large waves, the platform could separate from the pilings and float away without damaging the piling foundation.

Over the succeeding years, consultants studied weather data and developed theoretical models of wave formation. Each successive generation of offshore oil and gas platforms incorporated lessons learned, and the most vivid lessons came from powerful hurricanes. Experts had opined in 1947 that in water depths of 100 ft (30.5 m), "waves will probably seldom, if ever, exceed 20 feet in height."[4] As hurricanes hit the Gulf in the ensuing years, engineers and designers gradually collected more information about platform performance in storms. In 1949, for example, a hurricane near Freeport, Texas generated waves as high as 40 ft (12.2 m) and caused more than $200,000 in damage to one platform with a deck that was 26 ft (7.9 m) high.

Gulf hurricane activity tends to follow cycles of relatively calm years followed by active years. Several major hurricanes in the 1960s, Hilda (1964), Betsy (1965), and Camille (1969), seriously challenged the oil and gas industry to think differently about offshore platform design. Hilda and Betsy, labeled 100-year storms, and Camille, listed as a 400-year storm, exposed the flaws in the engineering assumptions of the time that platforms should be built to withstand 25- or 50-year storms. Indeed, with so many powerful storms in such a short period, it was clear that the concept of a 25-, 50-, or 100-year storm itself needed rethinking.

Hilda damaged 18 platforms, causing an estimated $100 million loss. Camille, a category 5 hurricane, stunned the industry by producing waves as high as 75 ft (22.9 m) and winds up to 200 mph (89.4 m/s). Camille also exposed weaknesses in platform design other than deck height. A Shell Oil platform in 300 ft (91.4 m) of water collapsed in part due to churning seas that caused the liquefaction of unconsolidated sediments at the platform's foundation.

Offshore Challenges Drive Technological Progress

The intense storms of the 1960s gave insurers a troubling new perspective on the vulnerability of offshore structures and the difficulty of forecasting insurance policy underwriting exposure with precision. Realizing the need for a more sophisticated approach to design and hoping to avoid an insurance crisis, the oil industry organized the American Petroleum Institute's Offshore Committee. The committee was tasked with developing uniform design standards to limit future hurricane damage.[5] The rigorous design standards that emerged were based on the belief that it was cheaper to build a wide safety margin into offshore facilities than to roll the dice on the possibility of significant platform damage, lost production, and loss of life. The best practices and design standards developed for offshore oil and gas projects are today the foundation for offshore renewable energy project design.

Many of the early offshore renewable energy projects built in the United States will likely be sited in the shallow coastal waters of the Gulf of Mexico and the Atlantic Ocean. Historically, the strongest hurricanes have formed over the warm tropical waters of the Gulf and the waters off the southeastern U.S. coast. When a hurricane reaches the cooler waters of the northeast, it usually begins to dissipate in strength, although very strong storms are still possible in the northern latitudes (fig. 2–1). Marine renewable energy structures must be designed to withstand hurricane-force wind and waves in order to avoid repeating the trial-and-error history of offshore oil and gas extraction in the Gulf.

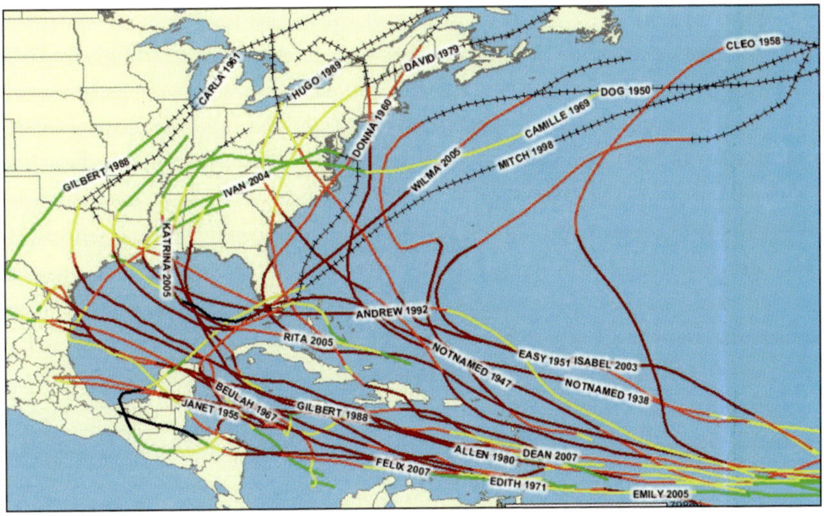

Fig. 2–1. Atlantic basin category 5 hurricanes, 1937–2007
Source: National Oceanic and Atmospheric Administration

The offshore oil and gas experience has other similarities to current efforts to exploit offshore renewable energy. Oil industry prospectors started with technology and methods developed for terrestrial applications and adapted them for the marine environment. Marine renewable energy pioneers are doing much the same. Many technologies used in terrestrial wind turbines have direct application to marine-based wind energy. Run-of-the-river hydroelectric technology also is a precursor to harvesting strong ocean and tidal currents. The first offshore wind projects have used turbines mounted on steel towers set on monopile foundations embedded in the seabed in relatively shallow water. Monopile foundation technology is limited to water depths of approximately 100 ft (30 m) or less. This adaptation of terrestrial tower technology is merely the beginning, and offshore wind turbine foundation technology must advance to enable deepwater wind resource development. Offshore wind project developers are now studying jack-up barge platforms, tripod towers, and floating tower designs, much as the offshore oil and gas industry moved from platforms built on wooden pilings toward floating platforms for deepwater production.

Time savings improve profitability. When oil production is brought online sooner, oil revenues are accrued earlier, and capital costs for construction work in progress are minimized. Ferrying workers and materials to remote marine sites is cumbersome and time-consuming. Work offshore is frequently interrupted by bad weather. In rough seas, workers cannot safely transition from boats to platforms, increasing hazards and work delays.

The oil and gas industry addressed these challenges with platform landing pads and helicopters to speed transit. More fundamentally, the industry increasingly built platforms on shore in dry docks, finding that this lowered costs and improved construction quality. The finished platform was then towed to the production site and moored as a floating platform, partially sunk, or attached to piles embedded in the seabed.

The industry also developed jack-up vessels. These ungainly flat-bottomed ships have retractable legs that can telescope down to the seabed and lift the vessel out of the water to make a stable platform in relatively shallow water. A much better understanding of wave forces also led to design advances. Compliant towers, for example, can withstand significant lateral forces and were developed for oil production in depths of 1,500 to 3,000 ft (457 to 914 m). Some oil and gas platforms in the northern Atlantic have serrated outer edges designed to withstand icebergs. Clearly, technology has adapted to provide solutions for various niche environments. A similar process of technological evolution can be expected for offshore renewable energy machines.

Offshore wind power projects constructed in Europe today use foundations fixed to the seabed. The turbines are generally erected at the offshore project site using specialized construction vessels. Engineers also are testing a variety of floating wind turbine platforms that can be built wholly or partially on shore and towed by conventional vessels to their moorings. Other designs incorporate both wave energy and wind energy captured in the same structure to minimize construction costs and maximize the use of expensive electric transmission infrastructure.

The learning that took place over decades in the Gulf of Mexico as hurricanes exposed the weaknesses of successive designs highlights the importance of reliable environmental data. Platform deck height is crucial for storm survival, but initial wave height estimates and wave models were poor. Large losses resulted when storm waves crashing into the deck of a platform imparted wave energy into the structure with explosive force. Offshore wind turbine designers benefit from these lessons. They make the cross section of the foundation that occupies the wave zone slender to minimize the area exposed to the most energetic wave forces. An improved understanding of sea bottom geology and the behavior of the seabed under storm conditions also has helped in the design of strong foundations and moorings. Scour protection used to maintain the integrity of the seabed around wind turbine foundations is an example of the application of this knowledge. Mooring designs that effectively and inexpensively secure wave energy machines is another area where the renewable energy industry can learn from and adapt offshore oil and gas technology.

Offshore construction design standards developed with industry participation have contributed to the industry's growth by reducing project risks and creating a common language used by project participants. Capital-intensive projects often rely on external debt financing and generally will not be built if they cannot be insured. Insurers and lenders will demand evidence that a project is designed and constructed to withstand reasonably predictable risks. It took decades for the offshore oil and gas industry to develop effective offshore design standards. The marine renewable energy facility standards developed by risk management consultants Det Norske Veritas and Germanischer Lloyd have been written in much less time. Offshore oil and gas experience clearly has paid dividends.

Legal Boundaries in the Marine Environment

The ocean and seabed are divided for legal purposes into several zones. The land meets the water in the *tidal zone*, which is the area between the high water mark and the low water mark that is affected by the movement of the tides. The tidelands, together with inland waters such as rivers and lakes, and bays that open to the coast, are considered state territory. The mark made by the tide at its low point is generally referred to as the *baseline*. (As a technical matter, official U.S. nautical charts show the mean low water mark as the baseline.) Other ocean boundaries are measured by reference to the baseline. Seaward of the baseline extending 3 nautical miles (nm) (5.6 km) lies the marginal sea or the *3-mile limit*. (One nautical mile is equal to approximately 1.15 statute miles, or 1.85 km.) The territory of most U.S. states extends seaward to the 3-mile limit. The exceptions are Texas, the Gulf coast of Florida, and the Commonwealth of Puerto Rico, which extend to a 9-nm limit (16.7 km).

The territorial waters of the states are part of a larger area known as the *territorial sea*, which occupies a band extending from the baseline seaward 12 nm (22.2 km). This area is part of the sovereign territory of a coastal nation under the United Nations Convention on the Law of the Sea.[6] Beyond the territorial sea, from 12 to 24 nm (22.2 to 44.4 km), is the contiguous zone in which a nation may exercise control necessary to enforce its laws concerning issues such as immigration, customs, and pollution. Specific statutory language, however, may limit the extent to which U.S. law applies in this area. The Federal Water Pollution Control Act and the Comprehensive Environmental Response, Compensation, and Liability Act, for example, both limit the contiguous zone applicable to those statutes to 12 nm. Other terms such as *navigable waters*, *ocean waters*, *territory and waters*, and *waters subject to the jurisdiction of the United States* may be used in U.S. law. When not defined, these terms may become a source of ambiguity.

The *exclusive economic zone* (EEZ), which extends from 12 to 200 nm (22.2 to 370 km), overlaps with the contiguous zone. The United States exerts control over ocean resources such as fisheries and minerals found in the EEZ. Federal regulation over ocean renewable energy applies up to the most seaward extent of this zone. Beyond the EEZ lies the open sea, the use of which is subject to international law, including international treaties (fig. 2–2).

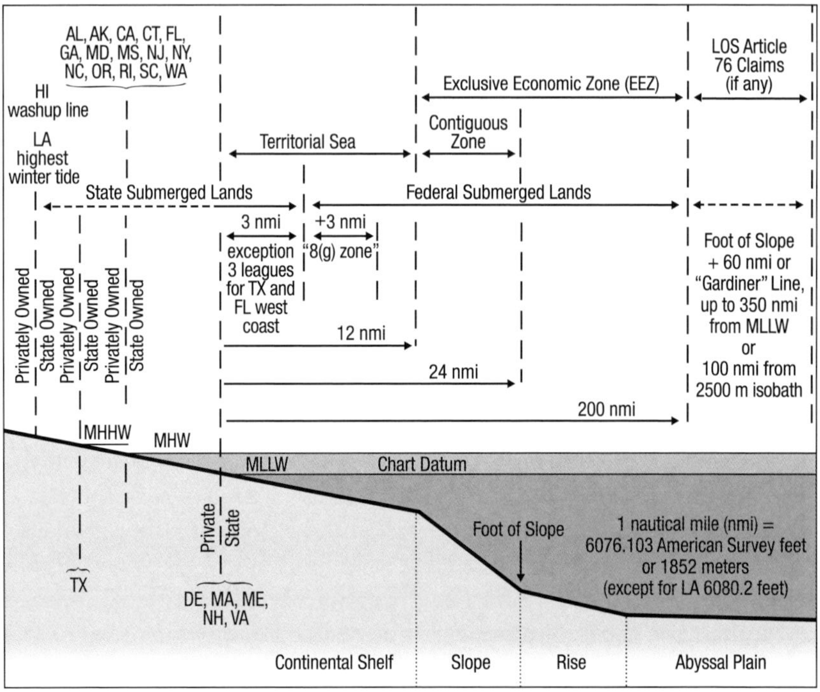

Fig. 2–2. U.S. lines of authority over ocean waters

Seabed Geology and Its Implications for Offshore Renewable Energy Projects

Seabed geology is an important factor in determining where marine renewable energy projects can be effectively located. Underwater features may focus currents or tides and provide superior conditions for free-flow current devices, and shoals may provide shallow areas ideal for wind energy foundations. As on land, where mountain ridges funnel winds and make certain areas ideal wind farm sites, undersea geologic features can have the same effect on currents. Off the Georgia coast a *scarp* (or cliff) 300 ft (91.4 m) high associated with the Blake Plateau intercepts the Gulf Stream and changes its path. This feature creates areas with high currents that may be exploited for energy production.

Over time measured on a geologic scale, the movement of the earth's plates also has produced areas of the seabed better suited to some renewable energy technologies than others. The earth's crust is composed of numerous slowly moving plates that float on the mantle, a thick layer of molten rock. New rock emerges from the mantle at rifts on the ocean floor, creating a range of undersea volcanic mountains that dwarfs the largest mountains on land. Convection currents within the mantle continue to spread the ocean plates apart at the ridges. The continents float on top of the oceanic plates.

Continental crust is thicker and less dense than oceanic crust. When continental and oceanic plates collide, the heavier and thinner oceanic plate tends to get pushed underneath the continent. As new crust forms at the expanding oceanic ridges, old oceanic plates are consumed at deep ocean trenches found at the plate boundaries opposite the midocean ridges. In these *subduction zones*, the edge of an ocean plate is driven below the continental plate and is eventually reabsorbed into the mantle as it continues downward.

Geologists classify continental margins as *passive* or *active*. The U.S. Atlantic continental shelf is a passive margin where the continental and oceanic crusts are joined together far from the volcanic spreading forces of the Mid-Atlantic Ridge. The Atlantic margin is stable, geologically old, and notable for its broad and shallow slope and thick sediment deposits. The East Coast and the Gulf of Mexico are where bottom-mounted wind power has been proposed in the United States. These areas are characterized by a gently sloping submerged continental shelf. The section closest to shore has a slope of about 0.5°. The shelf then gives way at the shelf break to the steep and narrow continental slope (with an approximate 3.0° gradient), which yields to the less sloped and wider continental rise. The shelf break is located at a fairly uniform depth of approximately 430 ft (131 m). The continental rise, which is formed by sediment that rolls off of the continental slope, ends at the relatively flat, deep seabed known as the *abyssal plain*.

Not surprisingly given its history of earthquakes, the U.S. Pacific Coast is an active continental margin. In this zone the continental crust ends abruptly at an ocean trench where the geologically new oceanic crust is subducted under the continent. The Mariana Trench, located in the western Pacific north of Indonesia, is the deepest known ocean trench at 35,798 ft (10.9 km) below sea level. Not all trenches appear deep, however. The Cascadia subduction zone, located off the coast of Vancouver, British Columbia and extending southward to California, is geologically a trench, but it has been filled by sediments flowing in from the rivers of this region.

Passive continental margins will typically have a large shelf. The New Jersey shelf, for example, extends approximately 92 miles (148 km) from the shoreline; further north off of Maine's coast, it is approximately 5 times that width (fig. 2–3). Active margins, such as the Pacific Coast, are usually narrower. Indeed, the Pacific Coast generally lacks a continental rise because the continental slope leads directly into a trench where the Pacific oceanic plate is subducted under the continental crust.

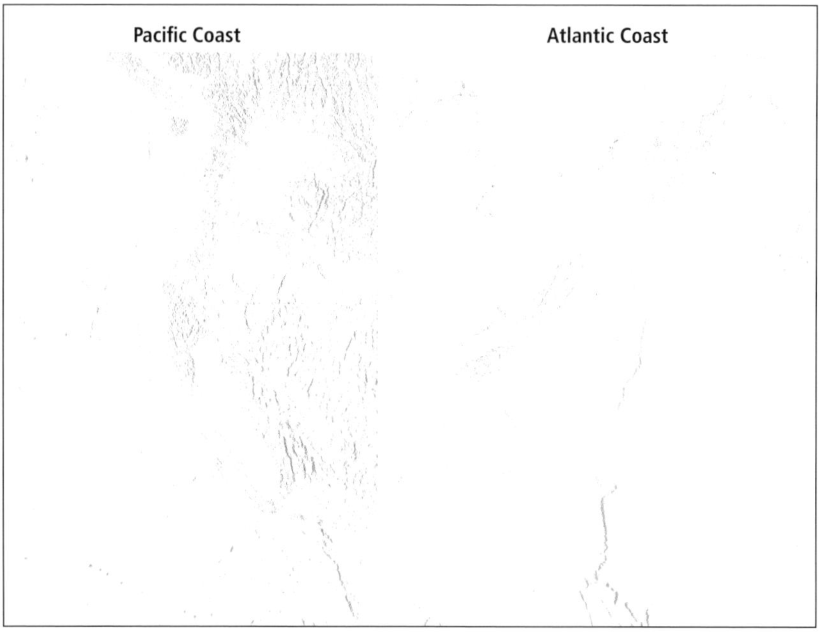

Fig. 2–3. Pacific and Atlantic continental margins compared
Source: U.S. Geological Survey Coastal and Marine Geology InfoBank

The implication for ocean renewable energy is plain. The foundations currently used for offshore wind turbines are restricted to shallow waters. Bottom-mounted ocean renewable technology that requires a shallow site can be located in fewer locations off of the Pacific Coast than on the Atlantic or Gulf coasts. Conversely, wave energy converters generally require water of significant depth where waves are at their full strength. The powerful waves in the deep waters found off the Pacific Coast provide ideal sites for floating wave energy generators; the relatively shallow Atlantic Coast waters have weaker waves.

Foundation and mooring options also will vary with seabed geology. Monopile foundations are more difficult and expensive to install into a rock seabed than into consolidated sediments. A rocky seabed also provides habitat for many ocean species and may give rise to more environmental obstacles than a sandy bottom site. Marine canyons may generate local currents that must be factored into foundation design.

Underwater marine landslides may occur on slopes and in canyons and present hazards to renewable energy projects. Landslides, known as *turbidity currents*, can scour sea slopes and cause far-reaching and violent changes to the seabed. These underwater avalanches of mud and rock can sever electric transmission cables and upset moorings. In 1929, an earthquake triggered a turbidity current on the U.S. Atlantic continental shelf that severed two transoceanic communications cables. Knowing the distance between the cables and the exact time each was severed, marine scientist Maurice Ewing was able to calculate that the turbidity current traveled at up to 50 mph (80 kph). Even more impressive was that the turbidity current traveled a distance of approximately 400 miles (640 km). Efforts such as Ewing's to solve the problems associated with laying subsea communication cables during the last century have contributed greatly to modern ocean science.

The Oceans As Energy Collectors and Conveyors

Earth's broad expanses of ocean and land collect the sun's light and heat. Solar energy produces the wind, waves, and ocean currents. Knowledge of these natural phenomena allows an understanding of the opportunities and risks inherent in designing machines for the marine environment.

Winds behave differently over land and water and at the margins between the sea and land. These differences play a role in siting decisions and, due to vibration and fatigue caused by wind turbulence, may even affect the expected life of turbines installed offshore. Wave behavior has implications for the siting and design of wave energy converters (WECs) and offshore wind turbine foundations. Designers might, for example, want to locate a WEC in the most energetic part of the water column and, to avoid damage, be able to submerge the device during storms. In the case of wind turbines, designers must understand the forces that waves will apply to a foundation. Considerations include whether the junction of the foundation and seabed will be weakened by storm waves or strong currents and whether a wind turbine tower can resist destructive harmonic resonance that may be caused by wind gusts and wave impacts acting in concert on the structure. Ocean and tidal currents arise from different forces and vary in strength throughout the marine environment. Knowing where currents are strongest and how they vary over time is crucial to designing the machines that will harvest this energy.

The wind and ocean currents are earth's system for heat redistribution. As the sun heats the earth, air warmed at the surface collects moisture, becomes less dense, and rises as dry, cool, and dense air flows into the void. In the Northern Hemisphere, warm air from the southern latitudes rises high into the atmosphere. This warm air travels north where it cools, drops its moisture as precipitation, and sinks to return south at low altitude to gather more heat and moisture to repeat the cycle.

Ocean currents are similar in origin. In the Northern Hemisphere, tropical seas are heated, while northern waters cool, increase in density, and sink. This process powers a conveyor of warm water flowing northward close to the surface and cold water returning southward at depth. Earth's rotation deflects both atmospheric and oceanic currents to the right relative to their direction of motion in the Northern Hemisphere, and to the left in the Southern Hemisphere. This is known as the *Coriolis effect* and is responsible for the generally clockwise motion of the air and water in the Northern Hemisphere and counterclockwise motion in the Southern Hemisphere. In the Northern Hemisphere, the Coriolis effect causes the northeast trade winds that blow westward between 30°N latitude and the equator, and the westerlies that blow eastward between 30°N and 60°N latitudes. By convention, winds are named according to the direction from which they blow.

Other effects occur due to the difference in capacities for heat absorption between land and water. It takes 1 calorie to raise 1 gram of water 1°C, compared to only 0.24 calories required to raise the temperature of 1 gram of sandy soil by 1°C. Consequently, land heats more rapidly during the day than water at the same latitude, while at night land cools quickly, and water tends to retain heat. One result of this is that small-scale effects of differential heating and cooling of land and water can be observed along the coasts. During the daylight hours land is heated, and the air density and pressure over land generally decreases. By afternoon the air mass over land is rising, and cool ocean breezes blow onto the land. At night the land cools, increasing the local air density and pressure, while the heat retained by the water warms the offshore air, creating low pressure at sea. At night and into the morning, breezes blow from land out to sea.

Wave behavior

Potential wave energy is on the order of 8,000 to 80,000 terawatt-hours (TWh), in comparison to the world's annual electricity production of approximately 18,000 TWh. (One terawatt-hour is equal to 1 billion kilowatt-hours.) Estimates of the annual economically exploitable resource vary. One estimate finds only 140 to 750 TWh can be captured annually using current technology.[7] Another study of the worldwide wave energy resource, which considered practical and economic factors that would make development commercially attractive, found that wave power could provide between 2,000 and 4,000 TWh/year.[8] This is on the scale of the combined output of the world's large hydroelectric power plants.

Waves are most often the result of wind friction along the surface of water, but they also are created by tides and earthquakes. As the wind blows across the water, air molecules from the wind interact with water molecules, stretching the water's surface and creating small ripples. These ripples roughen the water surface and provide greater traction for the wind, contributing to the growth of larger waves.

The interaction of wind, water's surface tension, and gravity create the familiar wave shape. The highest part of a wave is its crest, the depression between two crests is the *trough*, and the distance between any two successive crests or troughs is the *wavelength*. A wave's h*eight* is the vertical distance between crest and trough (fig. 2–4). The distance between crest or trough and the level of calm, undisturbed water (i.e., one-half of a wave's height) is a wave's *amplitude*. If someone in a boat starts timing at the crest of one wave and stops the stopwatch when he is on the crest of the next wave, the elapsed time is the wave's *period*. A wave's *frequency* is the inverse of its period and refers to the number of waves that pass a point over a given time.

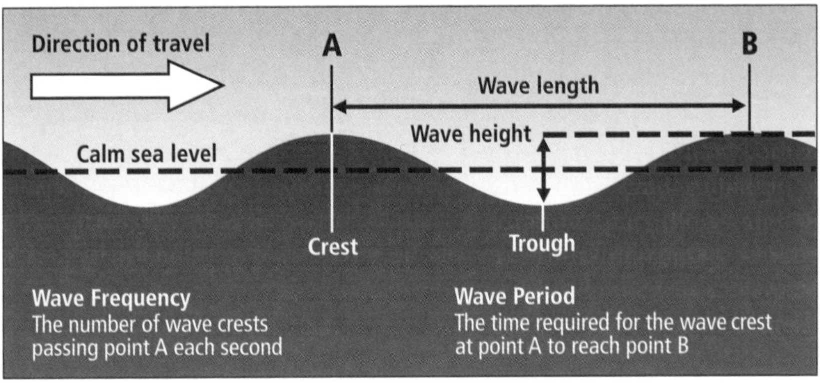

Fig. 2–4. Anatomy of a wave
Source: National Weather Service

Waves will continue to increase in height, length, and speed as long as the energy input continues. Wave energy, therefore, depends on wind speed, wind duration, and fetch, which is the distance over which wind blows. In a large storm system acting over an open sea, fetch and duration are generally not the limiting factors, and wave height tends to be determined by wind speed alone.[9]

A familiar experience to anyone who has played in the surf is being caught off guard by an unexpectedly large wave. The wave environment at any point in time is usually a mixture of waves of various sizes and strengths. The general sea state is expressed by the *significant wave height*. To calculate this measure, the height of a series of waves is measured, and they are ordered by height. The average height of the highest one-third of the waves gives the significant wave height.

Wave energy density is expressed in terms of kilowatts per meter of wave front (kW/m). The North Atlantic, Pacific, and Southern oceans have strong wave energy environments, in the range of 40 to 100 kW/m, while locations closer to the equator are substantially less energetic (in the range of 10 to 20 kW/m). The northeastern coast has the greatest wave energy potential on America's East Coast (20 to 50 kW/m), but energy levels are generally superior on the northwestern coast (40 to 60 kW/m). In comparison, average solar and wind power densities are an order of magnitude smaller, ranging in the hundreds of watts per square meter of panel area or rotor swept area.

Unlike ocean currents in which water flows from one place to another, waves are a flow of energy, not water (personal experience watching waves crash onto a beach notwithstanding). A swimmer bobbing up and down as waves pass experiences the circular motion of a particle of water caught in a wave. Water particles rise and move forward with the passing wave crest, and they fall and move backward as the trough passes. The orbital motion of each water particle decreases with distance from the surface. At a depth equal to one-half the wavelength, the motion of the water, and the wave energy, is almost zero. This explains how waves with very long wavelengths associated with a major storm can liquefy seabed sediments and undermine the foundations of an oil platform sitting in 300-ft-deep water.

Waves can generally be divided into deep and shallow water waves. A wave that does not reach bottom because the water depth is greater than one-half of the wave's wavelength is a *deepwater wave*. Such a wave will travel at a speed equal to its wavelength divided by its period. As a wave enters shallow water and begins to touch bottom, it slows. The wavelength shortens and wave height and steepness increase as the wave's energy is compressed in a smaller volume of water, but the wave period does not change. When a wave's steepness exceeds 1:7 (a ratio of height to length), the wave breaks and dissipates its energy. The loss of wave energy in shallow water is one reason why most wave energy converter designs are intended to work with deepwater waves.

In the trade wind latitudes, steadily blowing winds form regular waves. In contrast, irregular storm winds generate waves with varied speeds, wavelengths, periods, heights, and direction of travel. This jumble of confused waves is called a sea. The waves in a sea sort themselves into a *swell*, which is the type of regular harmonic wave oscillation that is most useful to wave energy production. The sorting occurs as waves meet. If the crests and troughs of two waves are in phase, then they will increase the amplitude of the combined wave. Alternatively, waves that are out of phase produce destructive interference that dissipates their energy. Over time this sorting process produces groups of waves with similar characteristics called *wave trains*. The uniform waves in a wave train (i.e., the swell) can move great distances through the ocean, retaining much of their energy.

A wave train travels at one-half of the speed of the individual waves forming its group. Thus information about storms and waves at sea, even at great distances from a wave energy farm, can help operators forecast the wave conditions that will be present at a wave farm hours and days later. Operators may use this information to predict and schedule energy output and maintenance activities.

Sea states are generally described by significant wave height, as noted, but also by the dominant or peak wave period. This measure identifies the wave train in a given sea that carries the most energy. Wave energy converter design often emphasizes making the machine tunable for the dominant wave period and adjustable over minutes as different wave trains arrive. Because successive wave trains vary in their energy, wave energy converters also may use devices such as flywheels or hydraulic accumulators to temporarily store power, perhaps only for several minutes. Smoothing the power flows to the electrical components allows for more efficient equipment sizing and stable output.[10]

The maximum wave height at any particular location matters when designing for survivability and is usually about twice the significant wave height at any point in time. A marine weather forecast that warns of "10-ft seas" is stating the average height of the top one-third of the wave spectrum. Yet within that average sea state, it would be possible to see waves twice as high. And rules of thumb are just that. Intersecting waves that are in phase can combine their force, resulting at times in a series of very high waves known as *episodic* (or *rogue*) waves. These extreme waves may reach heights of 65 to 100 ft (20 to 30 m) or more.

The North Atlantic, the North Sea, and the Southern Ocean around Cape Horn are known to develop episodic waves, but they also can occur elsewhere. In 1933 a U.S. Navy tanker ship in the Pacific encountered a wave measured at 112 ft (34.1 m) high with period of almost 15 seconds and a speed of 61 mph (27 mps). Given the possibility of waves of this size, it is not surprising that offshore wind turbines in unprotected locations must be built with robust foundations and sufficient clearance between the level of the undisturbed sea and the turbine blades.

Comparing average to extreme storm waves to rank sites

Prominent U.S. wave energy researcher George Hagerman has studied the resource potential in southern New England, from Portland, Maine to New York City, in an attempt to identify the most promising stretch of coastline for wave energy projects.[11] Hagerman considered the annual significant wave height, the annual average wave power density, and the 20-year maximum significant wave height. This latter figure represented the extreme storm hazard for design purposes. Potential sites were ranked according to the variability of the wave environment, proceeding from the basic assumption that the best locations would have (a) strong steady waves that allow the equipment to operate close to design capacity, yet (b) minimal exposure to extreme storm waves that would require more robust and expensive structures and moorings. Hagerman developed the wave energy development index (WEDI), an indicator of promising wave energy sites calculated by dividing annual average wave power density by the wave power density associated with the maximum storm wave environment.

Wave energy levels vary considerably over the course of a year and from year to year. In New England, winter storms tend to generate monthly average wave energy levels that are 3 to 5 times greater than during the summer. Wave energy levels also can vary as much as 50% from year to year. Hagerman's study showed that in exposed offshore locations, annual average significant wave heights of approximately 6.6 ft (2 m) could be expected, while during a storm, significant wave heights of 39 to 49 ft (12 to 15 m) were possible. Sheltered and nearshore locations had reduced average and storm wave heights, as would be expected.

The WEDI comparison showed that the coastline from Boston north to Portland was exposed to more extreme storm energy than the coast south of Nantucket Island. A wave energy converter located north of Boston must be designed and built with a foundation or mooring capable of resisting storm forces packing 100 times the energy of normal operating conditions. In contrast, a converter located south of Nantucket

Island would need to be built to withstand storms with 40 to 50 times the annual average wave energy level. This study demonstrates the value of thoroughly evaluating potential sites and matching the WEC design to the anticipated wave energy environment.

Predicting Wave Heights with Limited Direct Measurements

Collecting reliable wave information can be a challenge. The National Oceanic and Atmospheric Administration has operated weather data buoys for over two decades, but there are only seven buoys in the waters off of southern New England, and stations go out of service from time to time. To fill in information gaps, Hagerman's study used a mathematical model called a *wave hindcast* that related atmospheric data, such as the location of high- and low-pressure systems, to wind and wave creation. Hindcasts have been used by the U.S. Army Corps of Engineers to predict wave environments in support of harbor and other construction projects. The Corps' database, known as the Wave Information Study, is available on the Internet (http://chl.erdc.usace.army.mil/chl.aspx?p=s&a=DATA;1).

Tsunamis: A threat to ocean renewable energy projects?

On December 24, 2006, a magnitude 9.0 earthquake in the Indian Ocean raised a 750-mile (1,200 km) stretch of seafloor by as much as 26 ft (8 m), displacing hundreds of cubic kilometers of water. The resulting tsunami affected the entire Indian Ocean basin and was measured as far away as Canada's Pacific coast. Close to the epicenter, the Sumatran coast was stricken by waves over 33 ft (10 m) tall, and more than 225,000 people perished in only moments.

Tsunamis, which are often mistakenly called tidal waves, are formed by an underwater earthquake or landslide, not the gravitational attraction that produces the tides. Rapid movement at a seabed fault or landslide site energizes an entire column of water. In this respect, a tsunami wave is fundamentally different from wind-generated ocean waves that affect only the surface of the water to a depth of one-half their wavelength.

Tsunami waves are notable for their extremely long wavelength, perhaps 54 to 108 miles (100 to 200 km) long, and periods of 10 to 20 minutes. The wave travels from the point of the initial disturbance at about 400 mph (180 mps). In the open ocean a tsunami wave is generally not recognizable because its height may be only 3 to 6 ft (1 to 2 m). Like other waves, when a tsunami approaches shore and begins to slow, its wavelength shortens, and its height increases dramatically.

Tsunamis are most prevalent in the Pacific Ocean, and waves as high as 100 ft (30 m)—the height of a nine-story building—are not uncommon.[12] In 1971, Japan's Ryukyu Islands experienced one of the largest tsunami waves on record. The wave measured 279 ft (85 m) above sea level. Given that substantial wave heights do not build until a tsunami approaches shore, it is unlikely that deepwater-based marine renewable energy systems would be destroyed by a tsunami. Shallow water or shore-based marine renewable energy systems located along the geologically active Pacific Rim, however, would be at risk of tsunami damage. Tsunamis are unlikely to affect installations along the geologically inactive Atlantic coast.

Ocean Currents

Global ocean currents are produced by the differential heating of the tropical oceans and cooling that occurs closer to the poles. Ocean currents also are driven by the wind. In the northern Atlantic, the northeast trades blow westward between 30°N and the equator, and the westerlies blow eastward from about 30°N to 60°N. The Coriolis effect deflects the moving water to the right of the wind direction, resulting in a clockwise flow, or *gyre*, in the northern Atlantic basin. The flow on the western side of a basin is more intense than on the basin's eastern side. This phenomenon gives rise to the Atlantic's powerful Gulf Stream. The same physics operates in the northern Pacific basin, resulting in the Kuroshio Current, which flows northward along Japan's coast. On the eastern side of the Pacific basin the current flowing across from Japan splits. One branch, called the Alaska Current, heads north. The second branch, the California Current, moves south from Washington to Oregon and California.

The Florida Current makes a clockwise turn through the Gulf of Mexico and exits between Cuba and the southern tip of Florida. There it converges with the westward-flowing North Equatorial Current to form the Gulf Stream. The Gulf Stream moves an enormous volume of water, about 1,000 times the flow of the Mississippi River, along a path about 31 to 47 miles (50 to 75 km) wide and 1.2 to 1.9 miles (2 to 3 km) deep.[13] In the Strait of Florida, the Florida Current may reach speeds in excess of 3.4 mph (1.5 mps). In some areas, the Gulf Stream also may reach these speeds. The proximity of the Florida Current and the southern segment of the Gulf Stream to land make these currents an attractive resource for ocean energy production. The Gulf Stream widens and veers to the east at the latitude of North Carolina, and for this reason, the northern extent of the Gulf Stream would probably not be a practical renewable energy resource (fig. 2–5).

Given the role of ocean currents in redistributing nutrients throughout the sea and heat around the planet, there is some concern that extracting energy from these currents may have negative environmental consequences. Some modeling suggests that turbines may extract as much as 10,000 MW from the Gulf Stream, an amount within the current's normal variability, without causing harm.[14]

Fig. 2–5. Benjamin Franklin's map of the Gulf Stream

Heating and evaporation of surface water and variations in precipitation also change the salinity of seawater. A higher salt content corresponds to greater density. The movement of salt water of different densities contributes to currents in a process known as *thermohaline circulation.* The confined waters of the Mediterranean Sea and the Red Sea are highly saline. Where those bodies flow into the Atlantic and the Indian Oceans, respectively, the highly saline water sinks to middepths at about 3,000 ft (1,000 m). The thermohaline circulation within the ocean is typically very slow, on the order of 0.3 to 5 ft

(0.1 to 1.5 m) per day. In comparison to ocean surface currents, which are much faster and measured in miles per hour or meters per second, thermohaline circulation seems to have no potential at this time for ocean energy production.

Since currents are produced by large global-scale forces, such as the trade winds and the earth's rotation, it might be expected that ocean currents would be a steady and reliable source of renewable energy. In reality, like a winding river, an ocean current develops meanders, eddies, and other variations that may last for days or weeks. In addition, climate oscillations such as El Niño may affect wind and precipitation patterns and influence currents over the long term.[15]

According to Mills and Lecomte, the "Great Atlantic currents known as the 'North Atlantic Oscillation' (part of which is the Gulf Stream) may be slowing, an eventuality previously believed to be more than a millennium away."[16] Sufficient information about the variability of a current at the site of a potential energy installation should be obtained in the project planning stage.

Tides

Tides are the result of a complex interaction between the gravitational forces of the moon and sun, centrifugal forces created by earth's rotation, and local conditions such as a funnel-shaped bay or a narrow passage between islands. The centrifugal force of the rotating earth causes a bulge of water about the equator and away from the poles. As the moon orbits the earth, the gravitational force of the moon attracts the water that is closest to the moon. The sun's gravitational force also influences the tides to varying degrees as the earth rotates on its axis and during the course of the earth's yearly orbit about the sun. Since the sun is distant, the gravitational force that it exerts is less than one-half of the tidal force created by the moon. When both the moon and the sun are aligned on the same side of the earth, maximum tides occur. These are called *spring tides*. When the sun, moon, and earth form a right angle, a *neap tide*, the lowest tide, occurs.

In some areas the tides follow a diurnal pattern, with one high tide and one low tide each day. In other areas the tides follow a semidiurnal pattern in which the high water/low water sequence is repeated twice a day. In this case both high tides and low tides reach about the same point each day. There also are locations where the twice-daily high and low tides reach different levels. This is known as a *mixed tide* pattern. In Boston, Massachusetts a semidiurnal tide makes two daily swings of approximately 10 ft between the high water and low water marks.

In contrast, farther to the south, the diurnal tides in Galveston, Texas might exhibit a difference of little more than 1 ft between the high and low water marks.

During the course of the earth's yearly orbit around the sun, the sun moves from its position above latitude 23.5°N (the Tropic of Cancer) to 23.5°S (the Tropic of Capricorn). The moon moves through these latitudes on an 18.6-year period. These oscillations cause corresponding changes in the tides. For this reason, tidal power station planners should consider tide data covering the moon's full 18.6-year period. The National Oceanic and Atmospheric Administration (NOAA) and private sources publish tide tables giving the dates, times, and height of high and low water levels at sites around North America. From this information and local observations, the speed and duration of tidal currents may be calculated.

Area *bathymetry* (i.e., depth and seabed contour) has a significant effect on the local tides. The speed of tidal currents depends on water depth and obstacles that may impede or channel the flow. A wide entrance to a bay leading to a narrow inland section may form a funnel that causes water to pile up at the bay's head. Canada's Bay of Fundy is the best example of this phenomenon. At its mouth in the Gulf of Maine, the Bay of Fundy is 62 miles (100 km) wide and between 400 and 700 ft (120 to 215 m) deep. At about 180 miles (290 km) inland, the Bay of Fundy is about 1.5 miles (2.4 km) wide and 45 ft (14 m) deep at low tide. The funnel shape results in an astounding 52-foot (16-m) tidal range at the most inland reaches of the bay.

Northern Norway's archipelago of large islands and narrow channels provides another example of tidal behavior influenced by bathymetry. Strong currents occur in the straits between the islands because the islands form natural dams that resist the ebb and flow of the tides. These straits are ideal locations for tidal power stations. Verdant Power's tidal current energy project in New York's East River takes advantage of the same phenomenon. Tidal flows between New York Harbor and Long Island Sound use the East River as a conduit and produce a strong current where Verdant Power has located its turbines.

A foundation for progress

The obvious similarity between terrestrial wind turbines and their marine counterparts might lead the casual observer to conclude that it is a relatively straightforward matter to take wind energy production offshore. The pioneering offshore oil and gas prospectors learned, however, that exploration and production offshore requires specific in-depth knowledge of the marine environment coupled with new and adapted technologies. Offshore renewable energy development faces those same challenges, but fortunately the offshore oil and gas experience and marine research during the 20th century provide a solid foundation for progress. The marine environment is undoubtedly energetic. The challenge lies in building machines that can capture this energy at reasonable capital and operating costs while surviving the storms that appear so often to test the best designs. The next chapter takes a closer look at these offshore renewable energy machines.

3 Marvelous Machines: Harvesting Nature's Offshore Energy Crop

Since my involvement with the construction of Limpet (Land Installed Marine Powered Energy Transformer) on Islay I've found myself a popular figure at social gatherings and am in receipt of the congratulations of relative strangers who uniformly desire the success of the emerging wave energy technology.

—*Tom Heath, Wavegen*

Renewable energy is the new frontier. President Barack Obama hopes to invest $15 billion annually over 10 years to advance renewable energy. Venture capital firms like Kleiner Perkins Caufield & Byers, which made millions for its investors in semiconductors, communications, and Internet technology, are aggressively pursuing renewable energy and other clean technologies. Ray Lane, a partner of the firm, believes that the sector will be "bigger than the Internet, by an order of magnitude."[1] Richard Branson, the founder of Virgin Atlantic Airways, has committed to invest $3 billion in the fight against global warming over 10 years. Seeking to mobilize a global brain trust to attack the problem, Branson has offered a prize of $25 million to the scientist who invents a method of extracting significant volumes of greenhouse gases from the atmosphere.[2] Entrepreneurs are searching for sustainable ways to provide an expanding population with food, water, sanitation, and energy. Opportunities to profit from effective solutions abound.

The consequences of failure to meet these needs are sobering for both developed and developing countries. The challenge is enormous, and the stress of a changing environment adds even more urgency to the quest for solutions. The Intergovernmental Panel on Climate Change (IPCC) has concluded that "warming of the climate system is unequivocal," global average sea levels are rising, and many natural systems are being affected by climate change. Droughts will strike agriculture, disease vectors will intensify, and increasingly powerful storms will batter low-lying coastal areas. Widespread species extinctions are predicted, and the world's coral reefs will probably die.[3] These changes are likely caused by anthropogenic greenhouse gas emissions.[4]

The IPCC models are not yet able to predict the timing and amount of eventual sea level rise attributable to the melting of ice fields such as those in Greenland and Antarctica. However, if sea level increases are significant, millions would be affected in numerous major coastal cities around the world. Facing threats of this magnitude, many feel largely at a loss for what to do.

An Antidote to Despair

Despair itself is part of the problem. Germany in the 1980s was faced with a host of environmental problems, including nuclear fallout from Chernobyl, heavy metal pollution in Germany's rivers, deforestation related to acid rain, and an expanding ozone hole. In an effort to raise the environmental consciousness of German schoolchildren and promote an earth-friendly ethic, educators revised the curriculum to expose students to the full smorgasbord of environmental degradation facing the earth. Follow-up studies conducted years later showed that the effort backfired. Children became discouraged by the enormity of the world's problems and resigned themselves to the belief that they could have no meaningful role in a solution.[5] The technologies discussed in this chapter are an antidote to despair. They signal the beginning of an era marked by the responsible harvesting of marine renewable energy and represent one slice of the multilayered strategy that must be implemented to control global warming.

Marine renewable energy technology falls into several categories. Wind turbine generators are the most prevalent installed technology at this time and will receive the most emphasis here. Wave energy converters are another promising technology that has sprouted many designs in various stages of testing. Tidal power can take the form of an impoundment that captures the tidal flow at high tide and releases the accumulated water slowly through turbines, much like a dam would. There are also no-impoundment, free-flow designs where turbines are submerged in a strait with strong tidal currents to generate power as the tide ebbs and flows. Open water currents such as the powerful Gulf Stream also can be harvested with turbines. Finally, the temperature differential between deep water and surface water can be harvested with an ocean thermal energy conversion plant. Each technology has its proponents, its niche within the ocean environment, and its unique technical challenges.

The Nantucket Sound Project

Horseshoe Shoal lies in Massachusetts' Nantucket Sound between Cape Cod, Martha's Vineyard, and Nantucket Island. The area is a mecca for summer vacationers and is known for superb yachting. Based solely on technical criteria, it also may be the most ideal place on the East Coast for an offshore wind farm. Nantucket Sound is sheltered from large ocean waves, the winds are strong (average wind speed of 19.75 mph, or 8.83 mps), and water depth over the shoal of 50 ft (15.2 m) or less allows for the economical installation of turbine foundations. The site's proximity to land also minimizes the cost of transmission cabling.

Cape Wind Associates, a developer, would like to place 130 GE 3.6-MW wind turbine generators on Horseshoe Shoal. With a total capacity of 454 MW, and an annual average capacity of 170 MW, the farm would produce approximately 1,491,000 MWh/yr. (The total installed capacity of 468 MW [3.6 MW × 130 turbines] is reduced to a stated project maximum generating capacity of approximately 454 MW, after allowance for energy losses within the system.) The GE turbines have a design wind velocity of 30 mph (13.4 mps) and a maximum operational velocity of 60 mph (27 mps). The design velocity is the wind speed at which the turbine produces its maximum rated output. A turbine will continue to produce at maximum output up to its cutout speed (i.e., the maximum operational velocity). At the cutout speed, the rotor blades tilt to increase their aerodynamic drag and eliminate the lift that spins them through the wind. In aircraft parlance this is referred to as a *stall*. The stall position protects the machine from destroying itself in very strong winds. Matching the power curve of a wind turbine to the expected wind resource at the site is a key aspect of project design. This selection determines the amount of wind energy captured by a turbine and the number of hours per year that the turbine will be generating electricity.

The GE 3.6-MW turbines to be used in the Nantucket Sound project represent the first generation of GE turbines designed specifically for marine applications. They were first used in 2004 in the Arklow Bank project on the Irish Sea and are similar in design to other wind turbine generators (WTGs) that are employed in Europe for offshore applications. The GE machines use a three-blade rotor, like an airplane propeller, that spins a horizontally oriented axle (or rotor shaft) attached to the turbine's transmission (or gearbox); hence the term *horizontal axis wind turbine (HAWT)*.

The main components of a HAWT are located in the nacelle that houses the transmission, generator, and associated electronics, such as the power converter and transformer and the turbine control system.

Other components include a brake on the rotor shaft that is used to fix the rotor in place during maintenance, and yaw and pitch control mechanisms. Modern HAWTs do not use a tail rudder to turn the rotor into the wind. Instead they employ instruments to continuously sample wind speed and direction and use a motor to rotate the nacelle to face the wind. This motion is referred to as *yaw* and is similar to the way a compass needle swings to constantly point north. Electronic yawing systems are preferred to a fixed rudder because they can be programmed to ignore brief wind gusts and keep the rotor pointed in the direction of the prevailing wind. The pitch control mechanism rotates the individual rotor blades, varying the angle at which they attack the wind. This allows the blades to produce maximum lift under a wide range of wind speeds and also to stall when the wind speed exceeds the maximum rated speed of the WTG. Variable pitch blades allow GE's machine to exploit winds as weak as 7.8 mph (3.5 mps) (the cut-in speed) and as strong as its 60 mph (27 mps) cutout speed.

The nacelle is mounted on a tubular steel tower of varying height depending on the site location. The Nantucket Sound project's towers will extend 247 ft from mean lower low water (MLLW) level to the hub. At their lowest point, the blades will pass 75 ft (22.9 m) above the surface of the water and, at their height, the blade tips will be 417 ft (127.1 m) above MLLW. Each tower will be grouted to a steel monopile foundation embedded approximately 80 ft (24.4 m) into the seabed. A *monopile*, or a hollow steel tube, is driven into the seabed with a pile driver. If the seabed is rock, a hole also may be drilled and the monopile cemented into place in what is known as a *rock socket*. Monopiles would vary in diameter from 16.75 to 18.0 ft (5.1 to 5.5 m), depending on water depth.

Cape Wind needs to acquire rights to approximately 22,000 acres (about 89 km^2) of the seabed from MMS to build the project, which would be arranged with the turbines spaced on a grid approximately 0.4 miles (629 m) apart in one direction and 0.6 miles (1,000 m) apart in the other direction. This leaves room for recreational boating and some kinds of fishing within the boundary of the farm. The spacing is designed to minimize wind shadowing. If the WTGs are spaced too closely, turbulent air in the wake of one turbine would reduce the wind energy available for downstream turbines and subject the downstream machines to increased fatigue-inducing vibration. Although the wind park is quite large, the seabed disturbance caused by the facilities is small. The total acreage occupied by the towers, electric service platform, and subsea cabling is about 10 acres (0.04 km^2).

A jet plow would be used to excavate trenches at least 6 ft (1.8 m) deep for electrical cable burial. The 33-kV interturbine cables would converge at an electric service platform that would house transformers to boost the collection system's voltage from 33 kV to 115 kV, allowing the farm's output to be efficiently transmitted to land in the form of high-voltage alternating current (HVAC). The platform also would hold spare parts required for maintenance and would include a helipad, two backup diesel generators (with fuel) to supply emergency power to the farm, and shelter in case maintenance workers are stranded offshore in bad weather. Transmission to shore is provided through dual cables about 12.5 miles (19.3 km) long. Redundancy in the cables should permit approximately 75% of the energy output of the farm to be transmitted to land in the event of a fault in one cable. At landfall, the cables would continue underground using rights-of-way under public roads until joining the existing transmission network at a substation owned by the local utility.

Substantial dock facilities are required to marshal the WTG components, complete partial assembly of components such as the rotor, and stage items for loading onto vessels for the trip out to the project site. The nacelle and rotor combined weigh about 290 metric tons, and the tower weighs about 190 metric tons. Due to the size, weight, and heights involved, specialized lifting and construction vessels are required to install the foundations and perform the lifting. Construction of the Nantucket Sound project would require approximately 18 months.

Marine Wind Turbines

The technology proposed for the Nantucket Sound project is typical of marine wind energy systems in use today. These WTGs look much like their terrestrial counterparts, except they are generally larger and have special design features that better adapt them to marine conditions. To improve maintenance accessibility and minimize downtime, marine WTGs often have provision for helicopter access and small cranes attached to the nacelle to facilitate lifting replacement parts.

Most offshore WTGs currently in use have employed three-bladed horizontal axis wind turbines. There are variations on this basic design. Some designs do without a gearbox, running the generator directly at the speed of the rotating hub. The design without a gearbox provides the advantage of a simpler machine, eliminating potential gearbox failure in a remote, difficult-to-repair location and certain routine maintenance, such as gear oil changes. A turbine may have only two blades, or even one blade offset with a counterbalancing weight. Variations in blade design may lower cost or make a machine quieter. For example, the triple-bladed rotor design allows for a slower rotor speed and lower noise than turbines with fewer blades. Since noise is less of an issue offshore, there may be more machines with two- and one-blade rotors to reduce blade costs. However, triple-bladed turbines seem to have an advantage in terms of balance. A WTG tower creates a wind disturbance that affects the blade as it passes the tower at the bottom of its circular path. Stronger wind speeds at higher altitudes apply more force against the blade at the top of its rotation than at the bottom. These two cyclical variations cause a vibration that is strongest in single- and double-blade rotor designs. To control the vibration, single- and double-bladed machines may employ an axle that swivels slightly at the hub during rotation to lessen wear on the axle bearings.

Wind turbulence also is a source of vibration and stress. Because water is smooth in comparison to land, ocean winds are significantly less turbulent than terrestrial winds. This has the dual benefit of a steadier power output and less stress on marine WTG components. Addressing the stresses on wind turbines that will eventually result in fatigue failures is an important component of designing for long service life. To illustrate the challenge, consider that the average automobile is designed for a service life of approximately 150,000 miles. At a speed of 45 miles per hour, a car would be worn out within fewer than 3,500 hours. Assuming a target service life of 20 years before a major overhaul, a wind turbine operating only 30% of the time should be designed to survive 52,560 hours of operation without a major failure. To attain long life, it is crucial to understand and counter mechanical enemies such as vibration and corrosion.

Foundations, towers, and cables

Most marine WTGs are supported on steel towers connected to tubular steel monopile foundations. Monopiles will typically have a diameter of 13 to 18 ft (4 to 6m) and be driven approximately 72 to 80 ft (22 to 24 m) below the seabed. Monopile foundations are generally limited to water depths no greater than about 100 ft (30 m). Beyond that depth, the foundation would need to be made much stronger to resist current and wave forces, at a significant additional cost. This reason, and the high cost of transmission lines, provides an incentive to locate offshore wind energy projects in shallow water close to shore.

A monopile foundation will not work for all seabed conditions. In some cases a concrete gravity base foundation set on a crushed stone pad can be used where the seabed provides adequate support. Steel caissons shaped like inverted cups also may be used as foundations on muddy bottoms. The water is pumped from the caisson and suction adheres the foundation to the bottom. In deeper water, tripod-type foundations also have been used. Rocks or artificial seaweed mats engineered to trap sediment are typically used as scour protection around the base of a foundation.

A transition piece sits on top of the monopile (or other foundation) to accept the tower. The tower, made in lengths of 65 to 100 ft (20 to 30 m), is connected to the transition piece with concrete grout or bolts. Grout is used to plumb the transition between the transition piece and the foundation because the monopile may not be exactly vertical. A small platform is installed on the transition piece to serve as a dock for maintenance boats. In northern locations, ice protection may be included in the form of a sloped collar at the water line that encourages ice to bend and break. Built-in ladders and/or an elevator in the tower provides access to the nacelle.

Power and communications cable is run from the foundation through a trench to a central collection point in the wind farm where a transformer increases the voltage. From there the main transmission cable proceeds to land, where it is connected to the high-voltage transmission grid. It is not unusual for each WTG to have its own Internet address allowing for individual turbine monitoring via the Internet and easy communication between maintenance staff and a central office.

Turbine downtime is a bigger issue with remote marine WTGs than it is with terrestrial machines. During stormy winter and fall months it may be a week, a month, or longer before seas are calm enough to permit a safe maintenance visit to an ailing turbine. To minimize unproductive time, engineers have given much thought to preventive maintenance, predicting component failures, and accessibility. Designing for corrosion resistance also is paramount. Steel components are triple coated with paint, and the nacelle and tower interiors are sealed to minimize the intrusion of moisture and salt. In contrast to air-cooled terrestrial turbines, marine turbines use heat exchangers to maintain separation between the environment inside and outside the nacelle. In addition, dehumidifiers maintain relative humidity below the level that would promote corrosion, and the main electrical components such as the generator and control systems may have standby heating systems to prevent condensation when temperatures change quickly.

Greater materials costs, higher offshore construction costs, and longer outage durations due to servicing difficulty during severe weather, among other factors, have contributed to electricity costs at European offshore wind power projects of between \$0.08/kWh and \$0.15/kWh. This is about twice the cost of onshore wind power.[6]

Future offshore wind turbines

With the first generation of offshore turbines in place and generating power, engineers and inventors are designing a variety of new machines. Future offshore wind turbines may float in place, much like floating oil production platforms. The Norwegian utility Norsk Hydro ASA has designed a spar buoy–mounted HAWT that would be anchored in water from 650 to 2,300 ft (200 to 700 m) deep. The spar buoy is a long, hollow, cylindrical concrete structure developed in the offshore oil and gas industry. The company is currently proving the concept by testing a 3-MW turbine with a hub height of 262 ft (80 m) above sea level and a rotor diameter of 295 ft (90 m). Designers expect that this combination will leave sufficient clearance for North Sea storm waves as high as 115 ft (35 m). The spar buoy would be attached to the seabed with cables pulled taught by the float's buoyancy and in this way resist the heaving action of waves and the twisting force of the rotating turbine.[7]

A number of designers also have considered *vertical axis wind turbines (VAWTs)* for offshore applications (fig. 3–1). As the name implies, the blades of a vertical axis machine rotate about a central vertically oriented axis driving a generator placed at the base of the machine. Vertical axis wind turbines are in some respects simpler than their horizontal axis cousins. They do not need a yaw mechanism to turn into the prevailing wind, and the location of the generator and other mechanicals low in the structure permits the use of a lighter and less costly tower. Simpler blade design, and the fact that the blade is usually connected to the tower at both the top and bottom, also may allow for a less expensive blade. Blade aerodynamic efficiency in vertical designs, however, appears to be less than the current generation of horizontal axis machines. In addition, problems with main bearing failure and in transmitting the torque generated at the blades to the generator through a long shaft have been difficult to overcome (table 3–1).[8] A design for a bearingless vertical axis offshore wind turbine that uses water as the bearing surface is a clever take at addressing the problem of bearing durability in very large machines. The massive floating turbine conceived by Florida Turbine Technologies, Inc. has rotors with a diameter of 1,500 ft (457 m), the length of the largest super tanker, and a design generating capacity of 90 MW (fig. 3–2). At this time, vertical axis turbines for offshore use remain largely experimental.

Fig. 3–1. NOVA (novel offshore vertical axis) floating wind turbine conceptual drawing
Courtesy of Wind Power Limited and Grimshaw Architects

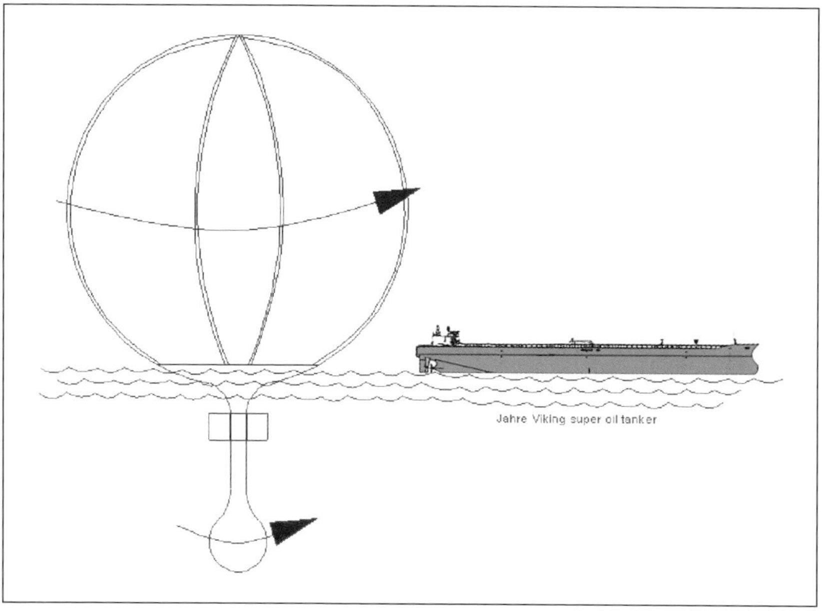

Fig. 3–2. Florida Turbine Technologies, Inc.'s bearingless floating vertical axis turbine design
Courtesy of Florida Turbine Technologies, Inc.

Table 3–1. Comparison of HAWTs and VAWTs

HAWTs	
Advantages	**Disadvantages**
• Are exposed to the slightly higher wind speeds that occur at these elevations. • Require less ground area since they usually use free standing towers. • Have constant aerodynamic loading with constant angle of attack.	• Require drive train to be placed atop a tower. • Require heavier towers—drive train located at tower top. • Require expensive and troublesome yaw drive systems. • Maintenance carried out high above ground often using expensive cranes. • Cantilevers used extensively: - high bending moments on blades; - high overturning moments on foundation. • Blade tips and drive train noise may propagate due to high tower heights.

VAWTs	
Advantages	**Disadvantages**
• Omnidirectional—no yaw equipment • Have drive train at ground level:- easier to operate and maintain. • Simpler design—no yaw gear, drop cables, etc. • Lower cost installation:- due to reduced overturning moments and lower tower top mass. • Have non-cantilevered blade supports. • May be quieter due to lack of blade tips and ground mounted drive train. • Unexplored technology may offer more potential for cost reductions.	• Have a low rotor height which reduces rotor wind speed. • Have lower tip-speed ratio for peak aerodynamic efficiency. Higher torque means more expensive gearboxes. • Require more area due to the guy wires. • Aerodynamic torque ripple provides cyclic loading on the drive train. • Longer blade length, about twice the length of HAWTs. Blades are major wind turbine cost item—low cost blades are essential to VAWTs.

Source: Paraschivoiu, Ion. 2002. Wind Turbine Design with Emphasis on Darrieus Concept: 378.

Offshore wind turbine design standards

Marine oil and gas platforms produce a concentrated resource with a high dollar value. People live and work on these platforms. Safety systems and extensive precautions to avoid oil spills are important. Oil and gas production will therefore continue to require elaborate and expensive custom-built platforms.

Ocean renewable energy projects, in contrast, collect a diffuse, low value resource. The potential for loss of life is low should a renewable energy system fail in operation, since people will be in, on, or around them infrequently. Environmental risks from structural failure also are low. WTGs contain lubricating oil and coolant but this amounts to approximately 200 gallons per machine, not the tens of thousands of gallons of heavy crude processed by oil rigs. Ocean renewable energy system design therefore places an emphasis on mass-produced components with broad applicability. Cost reduction also follows from designs that minimize weather delays and specialized equipment needed for offshore construction, by allowing much of the construction and assembly work to be performed onshore.

The design process generally starts with consideration of the likely extremes of wind, waves, currents, and perhaps ice loading at the chosen site over a 50-year period. The conservative designer then assumes as a starting point that the machine should be robust enough to withstand the maximum values of all forces occurring simultaneously and acting in the same direction on the machine and its structure. This, of course, requires access to reasonably accurate historical measurements of environmental conditions at or close to the prospective site for some time (ideally 25 years or more). Unless the designer is fortunate to have chosen a site with a long history of weather buoy observations, such data can be hard to find. Typically, a scientific observation station (e.g., meteorological tower or *met tower*) must be established at the proposed site to measure the variability of the site environment. Cape Wind, for example, erected a tower 196 ft (60 m) tall on Horseshoe Shoal and has collected three years of data such as wind speed and direction (at three altitudes), wave and tide height, current speed and direction, and water temperature.

In an effort to avoid overly conservative designs that are more costly to build than strictly necessary, designers are developing sophisticated models of the behavior of WTGs under multiple stresses. Engineers have found that the response of the tower structure to wave forces is partially dampened by the aerodynamic forces acting on the rotors.[9] This means that the structure need not be designed to resist the combined maximum wind and wave forces. Lessons like these suggest that offshore WTG capital costs will decrease over time as designs are refined. The onshore wind industry has reduced its capital costs on average 15% for every doubling of installed capacity.[10] This process is likely to occur in offshore wind energy as well. The increasing number of offshore wind energy projects in Europe provides experience that will drive down the cost on this side of the Atlantic. European manufacturing, engineering, and construction firms that have built offshore wind projects in the North Sea have been among the first to be called upon to build similar projects in the United States.

One sign of a maturing industry is when its players develop their own common language and expectations about the "right way" to conduct business. International offshore turbine design standards serve that function. They provide a baseline of expectations, promote trust and confidence among the many participants in a project, and help to speed project development. Several organizations, including the

International Electrotechnical Commission (IEC), Germanischer Lloyd (GL), and Det Norske Veritas (DNV), have contributed to evolving marine renewable energy standards.

MMS rules for offshore project construction require independent certification that a project has been built according to the designs submitted to MMS. DNV and GL review and document compliance with design standards and issue certificates evidencing compliance. This last step provides a tangible, impartial opinion regarding the build quality of a project that regulators, insurers, and lenders can rely upon. According to DNV, "Certification is a convenient process for the

Det Norske Veritas and Germanischer Lloyd

Det Norske Veritas (DNV) was established in Norway in 1864 as an independent foundation to safeguard life, property, and the environment by evaluating the technical condition of Norwegian merchant vessels. Since then, DNV has expanded into managing risk in other areas and has applied its experience with oil and gas exploration and production platforms to offshore renewable energy production. DNV researches, develops, and publishes (i) recommended practices for design and construction and (ii) offshore standards. DNV-RP-C203, for example, provides recommended practices for measuring and addressing fatigue in the design of offshore steel structures. Offshore standards are the technical requirements and acceptance criteria for various offshore structures. DNV's offshore standards OS-J101 and OS-J102, for example, provide principles and requirements for the design of offshore wind turbines, from foundations to nacelles, rotors, and blades. These compilations of recognized engineering practice provide a common language for project participants regarding the many aspects of offshore construction and operations.

Germanischer Lloyd (GL), based in Hamburg, Germany, is similar to DNV. GL provides consulting, testing, and certification services for onshore and offshore wind energy projects on behalf of turbine manufacturers, wind farm operators, suppliers, investors, and financial institutions. GL has its own standards, known as the "Guideline for the Certification of Offshore Wind Turbines," but GL also may certify a project in accordance with other relevant standards in the wind energy industry. GL claims that its guidelines are the only complete set of rules worldwide for the certification of offshore wind projects, covering not only the structure and machinery, but also condition monitoring and periodic testing and inspection.

developers to obtain an independent verification of their work and to provide evidence to financiers, partners, utility companies, insurers and the public that the energy converter will perform adequately within acceptable levels of safety, availability, reliability, asset integrity and environmental impact, complying with the Qualification Basis and, where applicable, to relevant standards."[11] Certification at the time of project commissioning also may be supplemented through regular inspections of facilities to ensure that the rules for ensuring safety for personnel, equipment, and the environment are followed.

The processes of setting guidelines and standards across all aspects of the performance of renewable energy systems, and then testing for compliance, are invaluable to developers and investors. Ultimately, the objective of all parties is to strike the right balance of risk avoidance, up-front capital expenses, and ongoing operational expenditures. Sound design, construction, and management practices increase the likelihood that the project will perform as expected.

Wave Energy Converters

As compared to wind power, where the horizontal-axis three-bladed turbine has become the design standard, wave energy machines are as varied in their appearance and technology as they are in their names—the Pelamis, Duck, Frog, OSPREY, Sea Clam, Power Buoy, and Limpet, to name but a few. Built by small teams of doggedly persistent engineers, each design seems to take a unique approach. A survey of patent activities indicates that several hundred wave energy converter (WEC) designs have been developed.[12] Today, there are more than 20 WEC manufacturers.[13]

WECs may float far offshore or be close in, even on land where the surf breaks. As waves move into shallow water, they slow and lose energy. Consequently, onshore or shallow water WECs usually are handicapped due to their lower energy environments. As explained in chapter 2, the energy in waves corresponds to their height, the distance between crest and trough, and their period or frequency (the time between successive wave crests or the number of waves passing during a given interval). All WECs have in common a means to capture wave energy, usually by resonating in harmony with the most prevalent wave frequency. That motion is then typically converted into mechanical, hydraulic, or pneumatic energy that is used to drive a turbine/generator unit.

WECs are generally grouped according to size and orientation to the wave front. The four principal categories of WECs are (1) point absorbers, (2) floating attenuators, (3) terminators, and (4) overtopping devices.

Point absorbers

Point absorbers are small compared to a typical wavelength. They are tethered to the bottom and bob up and down as waves pass. The AquaBuOY, made by the Canadian firm Finavera Renewables, uses the up-and-down motion of waves to extend a steel-reinforced rubber hose pump. In its extended position, the hose narrows and forces the water inside through a valve into a high-pressure tank. The pressurized water is then released to drive a Pelton turbine that is connected to a generator with a maximum output of 250 kW. The AquaBuOY's estimated average output is 46 kW, given a 28 kW/m average wave resource.

To minimize its environmental impact, the AquaBuOY uses a closed-loop design that recirculates freshwater, rather than using seawater. This prevents entrainment of small fish and other sea life and minimizes biological fouling of internal components. The exposed top of the buoy was flat in the initial prototype designs, but this was later changed to a conical shape to make it unattractive for bird roosting and to prevent its use as a resting site for seals and other marine mammals. The device's initial anchoring design was altered to substitute embedded anchors in place of concrete blocks and chains that could drag along the bottom and damage bottom habitat.

Finavera Renewables had received FERC authorization to install AquaBuOys in a pilot wave energy power plant located in Makah Bay, Washington. The Washington coast is known for rough seas, with wave heights up to 90 ft (27.4 m) having been measured on and beyond the continental shelf. In the project as conceived, four AquabuOYs would be arranged in an array 60 ft by 240 ft (18 m by 73 m) parallel to the prevailing wave front, about 3 miles (4.8 km) from shore in water approximately 150 ft (46 m) deep. Each AquaBuOY has a diameter of 19.5 ft (5.9 m) and a length of 98 ft (29.9 m). Power would be transmitted to shore via a DC submarine cable anchored to the seabed through most of its route, except close to shore where it would travel through a directionally drilled bore to a small building housing a rectifier, inverter, transformer, and switchgear. The site was chosen in part because of its proximity to an existing transmission line.

The mooring system is an important component of any floating wave energy system and can be surprisingly complex (fig. 3–3). Each AquaBuOY would be attached with cables to four surface floats. The floats in turn are attached to subsurface mooring buoys with a cable, and the mooring buoys are connected by chain to a *vertical load anchor (VLA)* fixed in the seabed. Developed for the offshore oil and gas industry, VLAs are specialized drag embedment anchors designed to withstand strong vertical and horizontal loads. They are installed with a cable that pulls at a shallow angle to the seabed allowing the *fluke* (the anchor body) to dig under the seabed like a plow. When in position, the anchor is attached to its load with a second cable that is oriented at right angles to the seabed. The weight of the sediment above the anchor helps to keep the anchor in place. The VLAs would be combined in opposing pairs in a square pattern to secure the AquaBuOY array on station with minimal disturbance to the seabed.

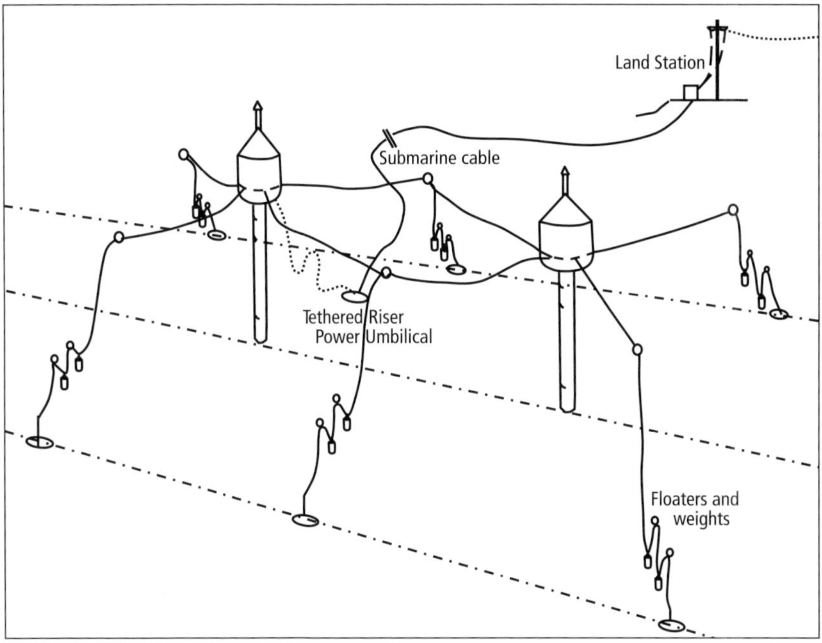

Fig. 3–3. AquaBuOY mooring system

In some locations, wave energy will inevitably compete with existing ocean uses. Unlike wind turbine generators installed on foundations fixed to the seabed, the AquaBuOY project and most likely other floating wave energy plants will, due to their mooring mechanisms, limit boat traffic through and fishing within a farm. Trawling, in which fishermen drag weighted nets along the seabed, for example, would be impossible within the farm's boundary.

Mooring also is a significant cost and reliability issue for WECs. Smaller WECs in particular face a difficult economic burden as the balance of plant costs such as moorings will generally represent a larger portion of overall project costs than would be the case with WECs with a larger output. An analysis by the Electric Power Research Institute (EPRI) indicates that maintaining a utility-scale power plant consisting of 1,000 AquaBuOY devices would be challenging given the length of mooring lines required. "The AquaBuOY mooring design requiring 2.5 mooring lines per device will require about 2500 mooring lines and 2500 anchors. In 50 m water depths, the mooring line is roughly 3x water depth or 150 m of chain per mooring. Total cable or chain-installed length is therefore: 150m × 2500 => 375,000 m or 350 km (280 miles). Even at a low failure rate, this will require a lot of intervention for O&M purposes and will critically affect the device's economic viability."[14]

EPRI's study shows that, though the ocean is vast, finding the ideal marine energy site can be difficult. To reduce the cost of mooring, it is likely that wave energy project developers will want to locate their wave farms in depths that provide access to waves of acceptable height and frequency, yet not so excessively deep as to unnecessarily increase mooring costs. Similarly, sites with the strongest waves are not necessarily the most desirable. Rather, it is the most consistent waves that fall within the design parameters of the particular device that are most sought after. The siting of the Makah Bay project was based on an analysis of wave energy, water depth, and effective distance from shore. In particular, Finavera required average wave energy of at least 20 kW/m, water at least 150 ft (45.7 m) deep, and a site within a few miles of shore and a nearby grid interconnection. One AquaBuOY sank in testing shortly after deployment due to a bilge pump failure, illustrating the value of field experience prior to commercialization.[15] Finavera Renewables has since abandoned its efforts to develop the Makah Bay project, citing the current economic climate and trouble raising the capital needed to pursue development of the early-stage, experimental technology.[16]

AWS Ocean Energy Ltd., based in Scotland, has built another kind of point absorber WEC. Based on a Dutch design, the Wave Swing consists of two nested cylinders that are suspended below the water surface (fig. 3–4). The lower cylinder is fixed to the seabed and the upper cylinder (the floater) moves up and down in response to the waves. Variations in water pressure as waves pass on the surface above the device move the upper cylinder in relation to the lower cylinder. The first generation of the Wave Swing was large (tower diameter of 10 m) and heavy (approximately 420 tons). Magnets attached to the upper cylinder moved past a coil attached to the lower cylinder to generate electricity. The two nested cylinders were held apart by a low-pressure air spring. Unfortunately, the tower was prone to flooding when submerged.

Back at the drawing board, attempts to address the flooding problem led to a better design. Designers determined to replace the air spring with high-pressure hydraulics, permitting a smaller and less expensive device. The ramming action sends pressurized hydraulic fluid to a turbine mounted on the ocean floor, which generates electricity.

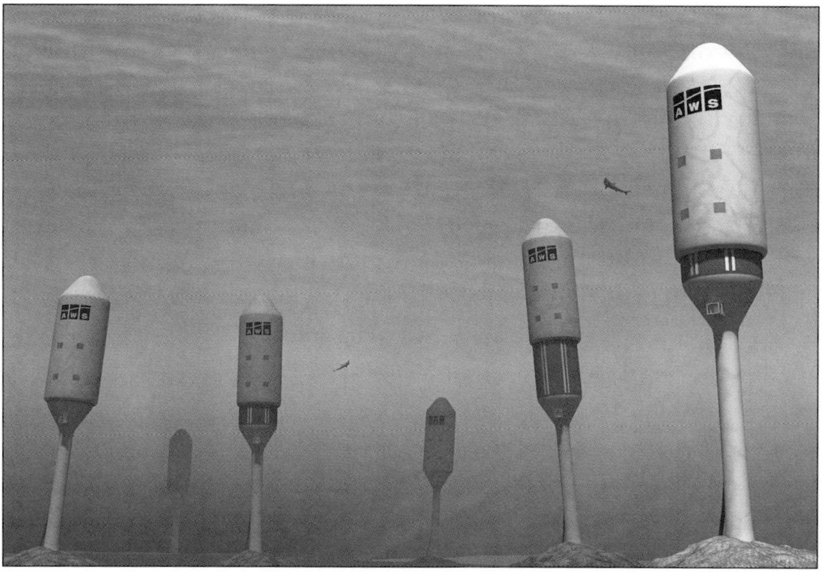

Fig. 3–4. Wave Swing
Courtesy of AWS Ocean Energy Ltd.

Unlike most point absorbers that float at the surface, the Wave Swing sits at least 20 ft (6 m) below the surface in water depths of 262 to 295 ft (80 to 90 m). In this position the device is acted upon by ocean swells, but it avoids extreme storm loading of the structure. As a result, a less expensive mooring or foundation can be used. According to AWS, its simple design with few moving parts and convenient maintenance improves reliability. AWS also believes that a remotely operated vehicle could be used for maintenance in most sea conditions to minimize downtime.

A measure of design efficiency is the amount of steel required to produce a given energy output. AWS hopes that the Wave Swing's relatively light weight and low maintenance requirements will make the Wave Swing an efficient, low-cost generator.[17] Output per unit of capital invested and maintenance costs are key metrics applicable to all WECs. If fewer tons of steel and other materials can be used to produce a given output of electricity, the unit cost of energy produced will be lower and more competitive with conventional electricity supplies. The weight of a WEC is an important variable because the cost of mass-produced items is largely dependent on materials cost and manufacturing complexity. Other costs such as research, design, and engineering fade in importance as production numbers increase.

In addition to capital costs, maintenance costs are a significant determinant of profitability for two reasons. First, it is expensive to pay for a maintenance crew to travel to a remote ocean site to fix a machine or tow it to a shore-based maintenance facility. Second, and most important, a machine that is out of service produces no revenue. Waters that are ideal for wave energy production because they have a strong wave environment are hazardous. During the winter season in some areas, it may be a month or longer before the sea is calm enough to permit maintenance of a disabled WEC. Maintenance and reliability are two sides of the same coin, and greatly affect a device's practical economic potential.

Floating attenuators

Floating attenuators may be as long as, or longer than, a typical wavelength. They are oriented perpendicular to the wave front and parallel to the direction of wave travel. An attenuator may have several sections, and as it rides atop a passing wave, the sections bend relative to one another. The bending motion may be used to pressurize fluid to turn a turbine generator.

The Pelamis, a creation of Pelamis Wave Power Ltd., is a floating attenuator. Named after a sea snake related to the cobra, the Pelamis is shaped like a tube, 394 ft (120 m) in length and 11.5 ft (3.5 m) in diameter (fig. 3–5). It has four articulated segments with three joints that flex as it rides along on a wave like its namesake. According its builders, "The wave-induced motion of these joints is resisted by hydraulic rams, which pump high-pressure oil through hydraulic motors via smoothing accumulators. The hydraulic motors drive electrical generators to produce electricity. Power from all the joints is fed down a single umbilical cable to a junction on the sea bed. Several devices can be connected together and linked to shore through a single seabed cable."[18]

Several Pelamis machines, each with a capacity of 750 kW, have been installed off the coast of northern Portugal in a 2.25-MW pilot project. The project is supported by a feed-in tariff of approximately 0.23 euros/kWh; indicative of the incentive required to encourage the commercialization of this new technology. A 3-MW project also is expected to be developed at the European Marine Energy Centre in Orkney, Scotland, and another project as large as 5 MW may be developed off England's southwest coast. According to Pelamis Wave Power, a wave farm occupying 1 km^2 of ocean could generate 30 MW of power. Designs call for mooring the Pelamis in water 50 to 60 m deep to provide access to large wave swells while minimizing the cost of the submarine cable needed to transmit the power to shore.

In its study of various WEC devices, EPRI found that the Pelamis could become commercially competitive with land-based wind generation in locations such as Hawaii and California once cumulative production volumes reach approximately 15,000 MW.[19] This projection is based on certain assumptions regarding the pace of product improvement and cost reduction corresponding to increasing production levels.

Fig. 3–5. Pelamis in tow.
Courtesy of Pelamis Wave Power Ltd.

EPRI concluded that the Pelamis and other WEC devices are worth developing for several reasons.

> *First, with proper siting, converting ocean wave energy to electricity is believed to be one of the most environmentally benign ways to generate electricity. Second, offshore wave energy offers a way to minimize the 'Not In My Backyard' (NIMBY) issues that plague many energy infrastructure projects, from nuclear to coal and to wind generation. Because these devices have a very low profile and are located at a distance from shore, they are generally not visible. Third, because wave energy is more predictable than solar and wind energy, it offers a better possibility than either solar or wind of being dispatchable and earning a capacity payment.*[20]

Terminators

A terminator may be just as long as an attenuator, but a terminator is oriented at a right angle to the direction of wave movement and acts to stop wave travel much like the shore does. Of the group of WECs classified as terminators, the *oscillating water column (OWC)* devices appear most promising. An Australian company, Oceanlinx, has developed an OWC device that would be installed along the shore in areas where fairly deep water meets land, such as at the base of a cliff or at a harbor breakwater. The company also has a floating version of its technology that can be used in water up to approximately 165 ft (50 m) deep. Oceanlinx's floating OWC machine is large, at approximately 450 tons, and its rated output varies from 500 kW to 2 MW, depending on the wave environment.

The heart of the device is a central chamber and a turbine. As the water column in the chamber rises and falls with passing waves, it displaces air, which is forced through a turbine to generate electricity (fig. 3–6). The air displaced by an incoming wave pushes through a small orifice to drive a Wells turbine. A vacuum is created in the chamber as the wave retreats and air rushes back across the turbine in the other direction. A Wells turbine rotates in the same direction regardless of the direction of airflow.

The turbine has variable pitch blades that increase average conversion efficiency to 60% as compared to approximately 30% with fixed-blade designs.[21] Variable pitch blades also make it possible to tune the device to the prevailing wave frequency.

As described in chapter 2, waves have mixed frequencies. Tunability expands the sea conditions under which wave energy may be collected. Devices that can draw power from a broad spectrum of wave frequencies are said to have good *bandwidth*, which not only gives a device greater performance but also allows developers more flexibility in siting.

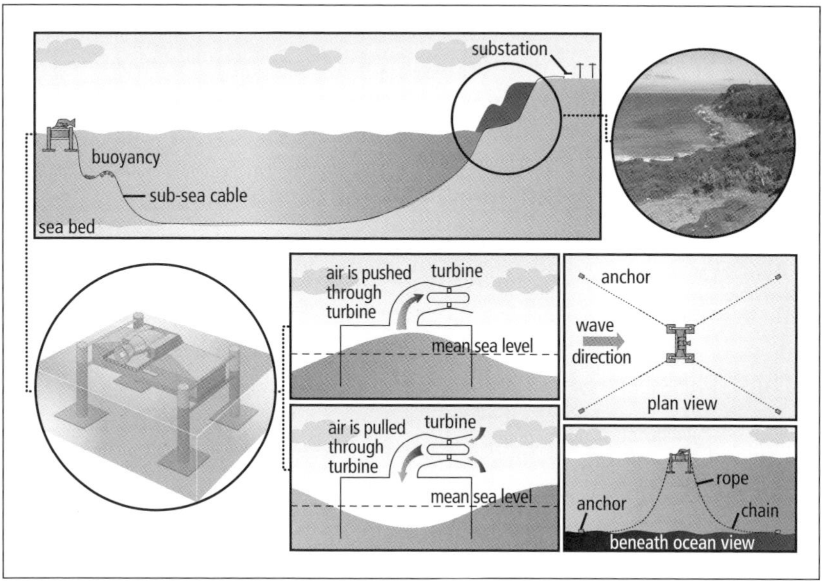

Fig. 3–6. Oceanlinx oscillating water column device
Courtesy of OceanLinx Limited

Tuning is accomplished by adjusting a device's size, shape, mass, stiffness, damping, or some combination of these attributes. Oceanlinx's variable pitch turbine blades adjust to the frequency of the air pulses in the oscillating water column chamber. Other designers have built WECs with multiple vertical capture chambers having different lengths, like the pipes of a large organ, to capture the energy in different wave frequencies. The air pressure from each chamber is combined and used to power a single generator. Orecon, a U.K. firm, is building a 1-MW floating OWC device with this design. The Wave Swing, which can achieve oscillation amplitudes of 23 ft (7 m) in a sea with wave amplitudes of 6.5 ft (2 m), is an example of a tunable device achieved by adjusting water ballast levels in its float chamber.[22] As with a child's swing, a series of small, well-timed pushes can result in a large swinging motion provided that drag is minimized.

Scientists at the University of Edinburgh have modeled the motions of the lamprey eel to better understand how it swims under different current conditions. The scientists think that this research will help them to improve the capture efficiency of the Pelamis wave energy converter by altering the flexibility of the Pelamis' joints under different wave conditions.[23] The ability to detune a WEC in heavy seas so that its response to waves is muted also is an important part of designing for survivability in extreme conditions.

Overtopping devices

Overtopping devices are the last broad category of wave energy converters. An overtopping device is similar to a low-head hydroelectric dam. The typical overtopping device uses a curved wall to focus incoming waves on a central ramp. The waves mount the ramp and spill over into a reservoir. Gravity then pulls the water through an outlet at the bottom of the reservoir and past one or more turbines to generate power.

Possibly the largest overtopping device currently under development is the Wave Dragon. Its wave collecting arms are as wide as 984 ft (300 m), and its reservoir can hold up to 8,000 m^3 of water—over 2 million gallons. The steel and concrete device has a rated power of 4 MW and weighs as much as 33,000 tons.

The Wave Dragon floats and swivels to face in the direction of prevailing waves. Air chambers may be flooded or emptied as wave conditions change to alter the Wave Dragon's profile above the water and improve the device's wave capture efficiency. The reservoir and multiple turbines in the Wave Dragon provide a smooth power output, much like a traditional hydroelectric dam. Field tests indicate, however, that fouling from marine organism growth and the accumulation of marine debris such as fishing nets, plastics, and floating containers could be a significant maintenance concern. Water intake screens used to keep the turbines free of debris will need regular cleaning.[24] Wave Dragon's developers argue that their design's basis in proven hydroelectric technology is a reliability advantage. Flow-through designs, however, raise the question whether marine organisms may be trapped against a screen or harmed by passing through a turbine.

Advancing wave energy technology

The shores of the United Kingdom experience some of the most energetic wave conditions in the world. The United Kingdom's Atlantic coast in particular enjoys wave energy densities on the order of 60 to 70 kW/m of wave front. In the United States, the waters off Alaska and the Pacific Northwest, with wave energy densities from 38 to 67 kW/m, are most comparable. Government support is one element of a plan to move the United Kingdom toward a low carbon economy. The British government has funded an independent entity, the Carbon Trust, to advance wave and tidal power technology. The Carbon Trust designed a technology "bake-off," called the Marine Energy Challenge, which has helped to reduce the cost of wave and tidal power through improved engineering and design. The Carbon Trust also sponsored the development by DNV of detailed guidelines on the design and operation of wave energy converters (wave energy guidelines) that is the leading resource in this field.[25] These guidelines cover the qualification of new technology, the selection of materials and equipment, reliability, construction, cable connections to shore, in-service operation and maintenance, and decommissioning, among other topics.

Device developers number in the hundreds, and the devices described in this chapter are only a sampling of the technologies in development. These examples illustrate common issues affecting the current generation of wave energy devices. Since the industry is still young, an unexpected technology quite possibly could emerge as the preferred method for wave energy conversion. Viable technologies may reside along a broad spectrum from simple, low cost, and relatively low efficiency machines to complex, high efficiency, and relatively high output machines. Each WEC design possesses a unique combination of initial capital cost, ongoing O&M costs, availability, and electrical output that produces a particular cost of energy.

As the spectrum of available technologies advances, there likely will be efforts to deploy a suite of energy production technologies at a single site. FPL Energy, the largest terrestrial wind energy developer in the United States, has observed, "Wind, wave and current power generation should also be capable of being co-located in the same general siting area. The ability to take advantage of a potential offshore substation and available power cable capacity could also assist in minimizing additional infrastructure needs and impacts."[26] Given that offshore cabling, foundations, mooring, and installation may amount to as much as one-half of project capital costs, collocation could make offshore renewable energy technologies that are not viable on their own practical in the context of a multidevice installation.

A suite of offshore renewable energy technologies also would ease the task of managing the electricity network to accommodate the intermittency of renewable energy. A study by Oxford University's Environmental Change Institute found that wave forecasting models could be used to predict the output of wave farms up to five days ahead. In addition, coordinated wind, wave, and tidal stream energy installations could improve the use of network capacity and reduce network balancing costs. Offshore areas that feature a combination of technologies should be better able to pay for the sea- and land-based transmission system upgrades that in many cases will be needed to bring marine renewable energy to the load. Lastly, lower variability in offshore energy farm output achieved by combining wave, tidal, and wind resources should earn project sponsors larger capacity payments to supplement energy revenues.[27] Currently, the intermittent nature of most renewables limits capacity payments. Chapter 10 describes further the challenge of integrating offshore renewable energy and the transmission grid.

Tidal and Ocean Current Energy Converters

San Francisco's Golden Gate Bridge spans a strait connecting the San Francisco Bay and the Pacific Ocean. A fall from the bridge's heights is usually fatal; furthermore, the Golden Gate Strait's treacherous currents, where peak velocities may exceed 5.6 mph (2.5 m/s), are also dangerous. San Francisco's Environment Department director has observed that "each day, nearly 400 billion gallons of water rush through the mouth of San Francisco Bay. If harnessed, the energy from this water could be an answer to the city's power needs."[28]

San Francisco has a history of efforts to harness ocean power. The Wave Power & Compressing Company, Hercules Wave Motor Company, Pacific Wave Motor Company, and other firms were active in San Francisco in the 1890s. In 1891, Henry P. Holland used the vertical oscillations of a large buoy to pump water up a cliff. From there the water was to fall through several waterwheels to generate electricity for local manufacturers. The machine did not live up to expectations and was abandoned a few years later. Some other wave energy projects of the time turned out to be less about technology than fleecing investors.[29]

More recently, HydroVenturi has proposed to build large concrete venturis with a diameter in excess of 15 ft (4.6 m) that it would place on the bottom of San Francisco Bay at a depth of about 190 ft (58 m). Water passing through the venturis would create a vacuum and draw air through a pipe extended to the surface. A turbine on the surface connected to an electric generator would be driven by the airflow in the pipe. HydroVenturi expected that its plant could run 16 hours each day and promised exceptional reliability since the device would have had no moving parts below water. The San Francisco Bay's tidal flow may support a 2,000-MW system —about twice the size of the typical nuclear generating plant. A separate project for the Bay proposed by Golden Gate Energy would use submerged turbines exposed to the free flow of the current. It is unclear whether either of the San Francisco projects will proceed beyond the concept stage.

Tidal power: The predictable renewable

Tidal power uses the ebb and flow of water, driven by the gravitational pull of the moon and sun, to generate electricity. James Ives is the chief executive of OpenHydro, which has installed a tidal stream generator in a channel off of the island of Eday in Orkney, United Kingdom. He states that "tidal energy sets itself apart from other forms of renewable energy in that it is completely predictable. It is the fastest growing emerging technology in the renewable energy sector and is set to make a major contribution to the security of energy supply and to carbon-free energy generation."[30]

Wind farm scheduling involves predicting the weather and ensuring that adequate traditional generating reserves are available in case the prediction proves wrong. Scheduling is especially tricky on a hot summer day when all generators are required to meet peak loads. Imperfect weather forecasting means that wind energy plants sacrifice significant value because other dispatchable generators must stand by to follow the load. Tidal power is attractive because it eliminates the forecasting challenge. The tides are regular and, given a plant's availability, dispatchers can predict with a high degree of confidence when a tidal energy plant will and will not be generating energy. Predictability reduces the system reliability implications associated with intermittent (and less predictable) resources and increases the value of tidal power.

Nevertheless, *predictable* is not the same as *uninterrupted.* A tidal energy plant will typically generate power for 6 to 12 hours each day; other resources need to be available during slack water periods. The tidal cycle is based on the period of rotation of the moon (24.8 hours), while electricity demand follows the cycle of the earth day. As a result, tidal energy production will not always be in phase with demand, occurring during peak usage times during some parts of the month and off peak at other times. Some dual-basin tidal energy designs attempt to address this limitation and are discussed further below.

Tidal range technology

There are several ways to extract energy from the tides. In the past, following the example of large hydroelectric projects, estuaries and lagoons have been impounded with a barrage. A typical barrage is a dam built with large concrete blocks (*caissons*) and gates (*sluices*) that permit the incoming tidal flow to enter a reservoir behind the dam. The sluices are closed when the tide peaks. During the ebb tide, while the sea recedes, power is produced as the water impounded by the barrage is released through turbines located in the caissons. On France's Rance River a plant with an installed peak capacity of 240 MW, and an average output of 68 MW, has been operating in this way since 1967.

Tidal range devices are low-head hydroelectric machines. This means that the hydraulic head, or the vertical distance between the water level in the reservoir and the sea, is small compared to conventional river-sited hydroelectric dams. In addition to the head, the electrical output of a tidal range device depends on the water area impounded behind the reservoir and the generating capacity of the turbines in the caissons. These factors determine the water volume and pressure available to generate electricity.

Other factors such as the design of the barrage and whether it generates power solely on the ebb tide, or on both the ebb and flood tides, play a lesser role. Robert Clark offers a comprehensive discussion of this technology in the *Encylopedia of Energy Technology and the Environment*.[31]

The natural tidal range at the plant site determines the hydraulic head that can be captured by a barrage. Cobscook Bay, Maine, with an average tidal range of 17.7 ft (5.4 m) and Cook Inlet, Alaska, with an average ranging from 11.4 to 30.3 ft (3.5 to 9.2 m), depending on location, provide two of the best tidal energy resources in the United States. It is not a coincidence that the estuaries with the greatest tidal ranges, the Bay of Fundy, the Severn, Cook Inlet, and the Gulf of California, to name a few, have long, slender funnel shapes. Like a resonance tube, these bays amplify the tidal pulses. Placing a barrier across an estuary can change its resonance and alter the tidal range that developers hope to exploit. Accordingly, mathematical and physical models of the estuary should be built to predict the effect of a barrage on the tidal range in an estuary. Understanding the interaction between the barrage and estuary is necessary to predict both environmental impacts and the potential energy output of the structure.

Cobscook Bay and nearby Passamaquoddy Bay in New Brunswick, Canada have intermittently attracted the interest of tidal energy developers since the early 1900s. In 1921, in what was known as the Cooper Project, it was proposed to link several islands along the Gulf of Maine with a series of barrages to form an upper basin out of Passamaquoddy Bay and a lower basin out of Cobscook Bay. Water would pass from the upper basin through a powerhouse to the lower basin. In this manner a head could be maintained and used to generate electricity independent of the timing of the tides. Though the project was estimated to be capable of producing 3 TWh (3 billion kWh) at a capital cost of about $100 million, it was not built due to a lack of local energy demand and the inability to transmit the power to more distant loads such as Portland and Boston.

During the Great Depression the project was later revived, in U.S.-only guise, as a single-basin project in Cobscook Bay, coordinated with a pumped storage reservoir located 12 miles (19 km) away. This project was sponsored by the Public Works Administration and employed as many as 5,000 workers. Unfortunately, this iteration of the project also failed to advance very far after an economic analysis by the Federal Power Commission revealed that the energy produced would cost more than twice the U.S. average cost of energy, and significantly more than traditional river hydroelectric power. According to L.B. Bernshtein's *Tidal Energy for Electric Power Plants*, the project was assailed as an "expensive whim" by the president of the American Society of Civil

Engineers and an "economic folly" by the Edison Electrical Institute.[32] Bernshtein argues that the project output was unduly expensive not because tidal power technology is costly, but because the engineers imposed an artificial output restriction on the plant since there was not enough demand in rural Maine. Cobscook Bay remains a focus of tidal energy developers to this day, and several companies have competed at the Federal Energy Regulatory Commission to obtain priority of right to develop a project in the bay.

A tidal barrage proposed for the Severn Estuary in the United Kingdom clearly aims for maximum output. The project would feature a barrage 10 miles (16.1 km) in length, a generating capacity of 8.64 GW, and a projected annual average electrical output of 17 TWh (about 4% of U.K. electricity supply in 2006). Reasonably maintained, the project could deliver power over a 120-year life. Concerns about the project's cost and environmental impact have kept it at the conceptual stage for the present. Like a traditional hydroelectric dam, a tidal barrage placed across an estuary will restrict the free movement of water through the estuary, causing several undesirable effects. The tidal range within the confined basin is usually lessened (higher low tides and lower high tides), with the result that the ecologically vibrant intertidal zone is narrowed. A decreased tidal range leads to weaker currents within the estuary, inhibiting the flushing action of the tides. Sediment flows will change, possibly leading to increased siltation in the basin and greater erosion seaward of the barrage. Reduced interchange between the sea and basin also will affect water salinity. Large marine mammals would tend to be excluded from the estuary or, should they find their way in through the sluices and locks, could find themselves trapped within the basin.

Experience at the Rance River project indicates that fish will likely pass through the large-diameter turbines unharmed. The report of the Sustainable Development Commission (SDC) reviewing the Severn project, however, cites the experience of a pilot barrage scheme on the Bay of Fundy that "resulted in a very high impact on migratory fisheries" and concludes that "actual injury rates would be sensitive to precise turbine designs, rotation speeds, fish pass size and location and patterns of generation."[33] Wildlife and habitats will be affected by the many changes brought about by a barrage in ways that are complex and often difficult to predict. For example, the Severn Estuary's muddy water gives rise to a food chain based more on detritus than photosynthesis. A barrage will slow currents, the water will clear, and a greater degree of photosynthetic activity will occur, altering the mix of organisms participating in the food chain, perhaps all the way through to the birds that use the Severn on their annual migrations. The SDC report stated, "It is clear that a barrage would cause major changes in the physical and biological marine environment, upstream and downstream of the development, and that these changes would also affect coastal land."[34] On the positive side, a tidal barrage will contribute to achieving the United Kingdom's climate change goals, may provide coastal defense from storm surge flooding, and may double as a bridge upon which to build a road or railway connection.

Tidal range devices, in the form of a tidal barrage strung across the mouth of an estuary, may change the estuarine environment so profoundly that the ecological cost is considered too high. Before dismissing tidal range technology entirely, however, it would be wise to take a closer look at the novel Tidal Wing, developed by Halcyon Marine Hydroelectric and its founder, Ramez Atiya. In the spirit of full disclosure, one of the authors, Markian Melnyk, is involved with Halcyon in project development activities. The Tidal Wing appears to answer many of the objections to traditional tidal barrages while promising large, utility-scale capacity at a cost that the company states is competitive with new coal-fired power plants. The device would be sited along the shores of an estuary, rather than across the opening, which preserves free passage for marine mammals, fish, and shipping. In addition, it minimizes interference with the tidal resonance in the bay (fig. 3–7). The reservoir enclosed by the Tidal Wing would be sized to fill and empty with the flow of the tide. By using the Tidal Wing's turbines to pump during some parts of the cycle, the full extent of the natural tidal swing within the reservoir could be maintained. This should preserve the preexisting habitat of the intertidal zone.

A Tidal Wing is a modular assembly of numerous steel-reinforced concrete "minicaissons" about 3 m wide, connecting two points on shore with a wide arc to enclose a basin (fig. 3–8). Large turbine/generator caissons are positioned at the deepest point at the center of the arc. Electricity would be produced from the head created by the tidal range at the site. The gates of the turbine/generator caissons would remain closed as the tide rises. When the maximum head is established, the gates would be opened, and the turbines would generate electricity as water fills the basin. The turbines would be used as pumps towards the end of the filling cycle to bring the basin level to the high tide level existing at the site before the installation of the facility. As the tide recedes a head is reestablished, and the flow of water through the turbines would begin again, allowing the basin to empty. Pumping out of the basin would be used at the end of the emptying cycle to recreate the correct low tide level within the basin. The Tidal Wing design also includes vertical lift gates that can be left open, generally during slack water periods, to allow fish and marine mammals to pass between the basin and the rest of the estuary.

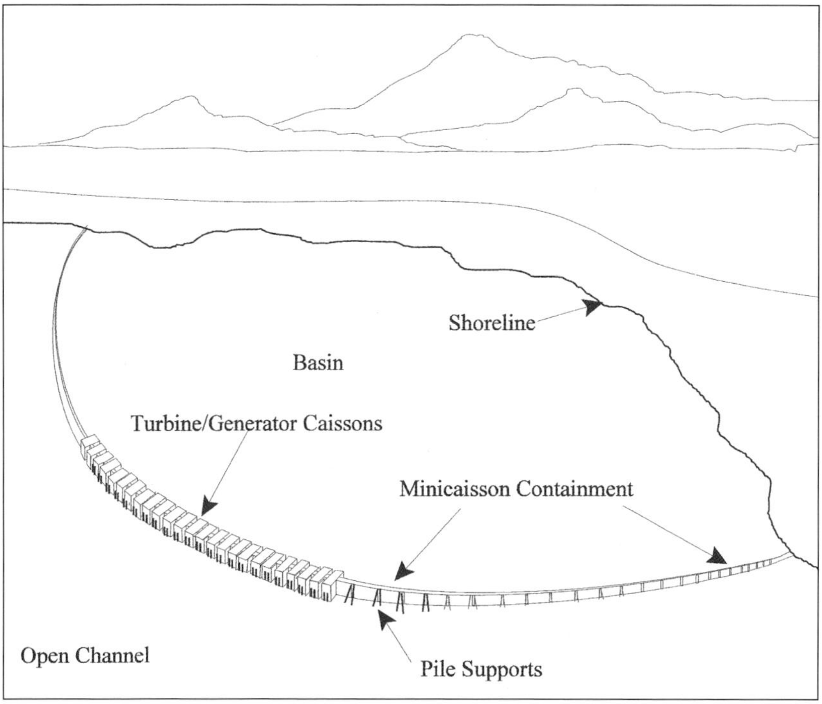

Fig. 3–7. Halcyon Marine Hydroelectric's Tidal Wing
Courtesy of Halcyon Marine Hydroelectric

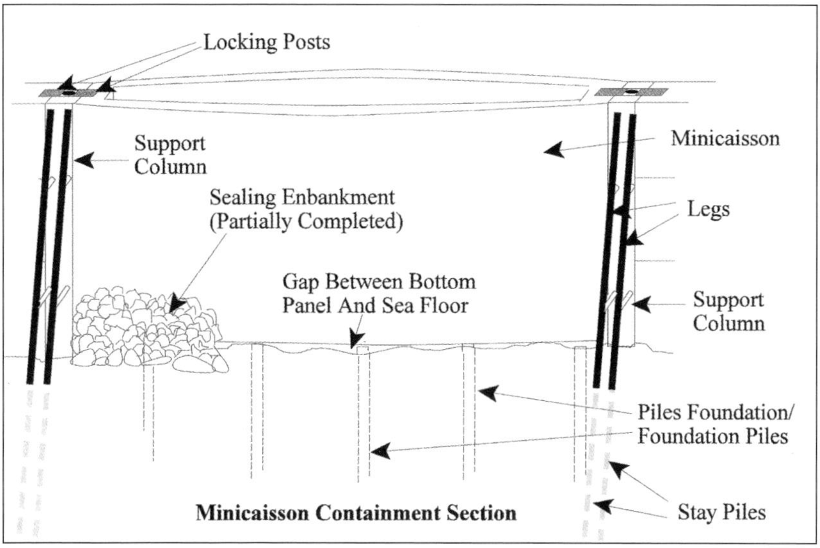

Fig. 3–8. Major structural components for the pile-supported/modular barrier construction used in the Tidal Wing
Courtesy of Halcyon Marine Hydroelectric

The Tidal Wing design expands the potential scope of tidal power because it does not need to be sited across a highly indented shoreline. River outfalls and their heavy sediment loads can be avoided, and conflicts with shipping, recreational boating, and fishing can be minimized, all of which expand siting options. Civil engineering costs are a large component of traditional barrage construction. The minicaissons used in the Tidal Wing, in contrast, are supported by piles driven into the seabed, and therefore the structure can be narrower and lighter, using much less concrete than a traditional containment structure. Little or no dredging is required to place the minicaissons on the seafloor. The modular/pile-supported construction technique is expected to allow for the construction of a Tidal Wing with a perimeter length of 25.5 km, enclosing a basin of 104 km^2, in as little as 21 months.

According to Halcyon Marine Hydroelectric, a Tidal Wing project proposed for the Bay of Fundy in Canada's New Brunswick Province, using hydroelectric turbine generators with an aggregate capacity of 3,740 MW, operating at a 35% capacity factor, would have an annual output of 10 TWh (10 billion kWh). Halcyon is exploring opportunities to run such a project in coordination with the traditional river hydroelectric resources of the region. The output of river hydropower plants could be regulated to fill the gaps in tidal power production. A significant portion of the province's energy needs could be supplied by reliable, dispatchable renewable energy sources.

Lastly, it is notable that the turbines used by tidal range devices, including the Tidal Wing, are derivative of those employed in low-head, river-sited hydroelectric projects. This mature technology has known costs, reliability, and durability. Similarly, the engineering of reinforced concrete structures built on piles is well understood and backed by extensive experience with bridge building and offshore oil and gas facilities. Tidal range technology therefore has an important advantage over other marine hydrokinetic devices that still require many years of development, testing, and improvement before they will be ready for utility-scale applications.

Free-flow tidal power turbines

While tidal range devices impound water and use the stored potential energy contained in the hydraulic head to produce power, free-flow tidal turbines use the kinetic energy in moving tidal currents. Free-flow designs will typically have power densities on the order of 4.9 W/m^2, while tidal energy devices designed to produce a head may have a power density of as much as 80 W/m^2 at the turbines.[35] The free-flow analog to the tidal barrage is the tidal fence, which uses a row of vertical axis turbines placed across a channel to produce electricity on the ebb and flood tides (fig. 3–9). Though they are not impoundments, tidal fence designs also are not entirely without critics, since they may disrupt shipping and block the passage of large marine mammals.

Fig. 3–9. Side view of a tidal fence of Davis Hydro turbines linked in series across a channel
Courtesy of Blue Energy Canada

Other free-flow designs dispense with fences and typically feature freestanding turbines submerged in a channel to capture energy as the tides flow in both directions. Shrouds may be used to concentrate flow on the turbines to increase power. OpenHydro has an interesting design that integrates a shroud with a ring-type generator (figs. 3–10a and 3–10b). The company has been testing its prototypes at the European Marine Energy Centre.

In 2003, a modern grid-connected tidal energy plant commenced operation in Norway's Kvalsund Channel. The 300-kW pilot plant generates an estimated 700 MWh annually under a maximum current speed of 5.6 mph (2.5 mps). The turbine is mounted to the seabed on a 66-ft (20-m) tripod. Its blades are rounded at their leading edges and spin at only seven revolutions per minute, minimizing adverse effects on fish and marine mammals. To ease maintenance, the turbines have a modular design that allows critical components to be lifted out of the water in one operation for maintenance and repair.[36]

Fig. 3–10a and 3–10b. (a) Open-Centre Turbine awaiting installation and (b) in conceptual drawing

Courtesy of OpenHydro Group Limited

Verdant Power's New York City tidal power plant provides another example. Driven by the tides, water follows the channel of the East River to pass between the western portion of Long Island Sound and New York Harbor. Verdant's plant would consist of approximately 60 turbines located at various depths in a narrow part of the East River that experiences the strongest currents. Verdant Power estimates that there are 120 commercially exploitable tidal energy locations in the United States, with a total potential of approximately 1,200 MW, and that their technology may be adapted to run-of-the-river locations. The resulting electricity generating potential could be as high as 170,000 MW.

Design challenges with submerged turbines include the avoidance of *cavitation*, which occurs when the pressure of a liquid passing through a turbine drops so low that the liquid ruptures and forms a void or bubble. As the bubbles collapse, they create noise and a shock wave that may damage turbine blades, cause vibrations, and reduce efficiency. The collapse of cavities along the surface of a propeller or turbine also involves the release of energy at high temperatures that causes erosion, pitting, and eventual failure of the part. Other design challenges include preventing marine growth buildup and controlling corrosion. Fishing nets and other debris also will invariably foul turbines and cause maintenance issues.

Most fish and small marine mammals should be unharmed by slowly rotating submerged turbines. Protecting larger marine mammals such as whales may require the construction of protective fences or equipping turbines with sonar-activated brakes. As with tidal barrages, it also is possible that numerous free-flow turbines sited at the entrance to estuaries may slow the ebb and flow of the tides sufficiently that they alter the natural flushing action of the estuary, affecting sedimentation and water turbidity. Slower tidal ingress and egress also could change the salinity of an estuary by altering the mixing of salt and fresh water. These and other impacts on the marine environment and activities such as commercial and recreational fishing, shipping, and recreational diving will need to be considered.

Ocean current energy

The principle behind free-flow tidal current power also can be applied to currents in the open ocean. The Gulf Stream alone is said to have a flow of water that is 50 times the flow of all the world's freshwater rivers. Although ocean currents are slow compared to wind, water is approximately 835 times denser than air. Ocean currents also are less variable. A 12-mph (5-mps) current, for example, provides the equivalent power per unit area of a 110-mph (49.2-mps) hurricane force wind. The total power available in ocean currents is estimated at 5,000 GW, and power densities as high as 15 kW/m^2 are not uncommon.[37]

Ocean currents are generally extracted with devices that are similar in appearance to wind turbines. They can be vertical or horizontal axis designs and may extract power through drag, like an old-fashioned waterwheel, or through hydrodynamic lift. Like wind turbines, the rotors would be connected to an electrical generator, and cables would transmit the power to shore. The devices could be secured on site with a solid foundation, or they could float and be secured with a mooring scheme. Given that the turbines are submerged, reliability is paramount to minimize the expense of recovering turbines from the water for maintenance.

Ocean Thermal Energy Conversion Devices

A discussion of marine renewable energy would be incomplete without at least a passing reference to ocean thermal energy conversion (OTEC). OTEC is a unique method of electricity production that was considered quite promising, especially in the 1970s, but failed to gain commercial momentum. OTEC systems generally have a low efficiency (3% or less) due to the small difference in input and output temperatures and large pumping loads. Interest in tapping thermal differences within the ocean and other large bodies of water continues, nevertheless, and methods of harvesting this energy for district cooling, desalination, aquaculture, and even agriculture have been proposed.

Ocean water is stratified by temperature and salinity. Cold and salty water is dense and sinks to the bottom, while warmer, less saline water floats. Water is at its most dense at 39°F (4°C). Below this temperature, water's molecular structure shifts as it freezes, causing water to increase in volume, become lighter, and float to the surface. In an open-loop OTEC system, warm surface water is pumped into a vacuum chamber, where it boils and forms steam to drive a turbine. Cold water from deep in the ocean is pumped to condensers on the output side of the turbine to cool the steam down, lowering its pressure. Like other steam turbine generators, the pressure differential between the input and output sides powers the turbine. A closed-loop OTEC system pumps warm surface water through a heat exchanger filled with a working fluid such as ammonia or Freon. The heated working fluid gasifies, drives the turbine, and is condensed back to its liquid state in a condenser chilled by the deep cold water.

The key requirement for an OTEC system is a temperature difference of at least 36°F (20°C) between surface water and deep water. Latitudes between 25°N and 25°S, where the temperature difference between the surface and depth averages 40°F (22°C), are the most attractive areas to locate an OTEC system.[38] A system typically draws cold water from a depth of 2,000 to 3,300 ft (600 to 1,000 m) and pumps large quantities of surface water, since only a small fraction (less than 0.5%) of the warm surface water used in an open-loop system is turned into steam. The pumping energy required to circulate the fluids through the system consumes a large portion of the energy extracted by the plant.

Although OTEC systems provide clean, renewable power, there is some concern about the environmental impact of discharging large quantities of cold, nutrient-rich water near the surface of tropical oceans. The nutrients in the discharged water encourage the growth of marine plants that may interfere with existing tropical ocean ecology. Entrainment and impingement of aquatic organisms also is an issue with any technology that circulates large volumes of ocean water.

Open-loop OTEC systems may be built more for the fresh water they produce than for electricity. At the Natural Energy Laboratory of Hawaii at Keahole Point, an OTEC plant generates 100 kW per day of energy and 7,000 gallons (26,000 liters) per day of fresh water from condensed steam. The cold water pumped from below also is used for air conditioning.

Air conditioning is a significant consumer of expensive peak-load electricity. Cold waters from the depths of the Great Lakes have been tapped to make air conditioning systems in large building complexes more efficient. Enwave Energy Corp. of Toronto has constructed a deep lake water cooling (DLWC) system that withdraws water from 272 ft (83 m) below the surface of Lake Ontario, where the water maintains a constant temperature of 39°F (4°C). The water is pumped through a filtration and treatment plant and then proceeds to another plant, where it passes through heat exchangers connected to a closed-loop chilled water system serving numerous downtown Toronto office buildings. The treated lake water exits the heat exchangers at approximately 55°F (13°C) and enters Toronto's drinking water supply. The water in the closed-loop system is cooled by the lake water to 42°F (5.6°C) and then is further cooled through steam-powered chillers to the 38°F (3.4°C) design temperature of the system. Both Enwave and the city of Toronto benefit from combining the district cooling and municipal water supply systems. The city pays the cost of pumping the water from the depths of Lake Ontario but benefits from access to cleaner water provided by the new intake pipes financed by Enwave. The DLWC system is reported to provide a 90% reduction in the electricity consumption of a conventional chiller system.[39]

OTEC's future may unfold as part of an integrated solution to several problems. John Craven, who founded the Natural Energy Laboratory, has a new enterprise called Common Heritage Corp. The company is planning a development on the island of Saipan that would use OTEC for power and freshwater production in connection with a real estate development. After running through the OTEC plant, the cool ocean water would be circulated through soil in a vineyard to make it possible to grow table grapes and other northern climate fruits in the tropics. As Craven observed, the combination of cool soil and warm air spurs plant growth and can shorten the time plants need to mature, permitting multiple harvests each year.[40] The cool water output of an OTEC plant also can be used to support aquaculture of cold water species such as salmon and lobster. It is possible that projects such as Craven's that use an OTEC plant to earn revenues from multiple sources may succeed where previous OTEC projects have not.

It is fitting that OTEC, a technology that once attracted great attention but has largely slipped out of favor, concludes this chapter. It is a caution to take the long view and avoid overly optimistic assessments of any particular technology's potential. Writ large, ocean renewable energy technologies show great promise because the ocean is so vast and energetic. Some of that energy can even be found in relatively concentrated form, as it is in ocean swells and fast currents. Yet each technology has individual strengths and weaknesses, and all face the very considerable hurdle of long-term survival in a harsh and extremely variable environment.

Chapter 2 included a summary of George Hagerman's findings about wave energy sites off the New England coast. In an exposed site north of Boston, Hagerman found that storm forces could impact a device with 100 times the energy present under normal operating conditions. That observation, in a nutshell, captures the dimensions of the marvelous challenge that engineers designing marine renewable energy machines must confront. It is similar to asking an architect to draw plans for a cottage that can withstand a nuclear blast. Will the designers succeed? Judged solely by the number of creative solutions already advanced, there is good reason for optimism. The next chapter takes a closer look at the surprisingly challenging process of identifying the best offshore sites for renewable energy machines.

4 Carving Out a Piece of the Shelf: Project Siting Decisions

If you have built castles in the air, your work need not be lost.

That is where they should be. Now put the foundations under them.

—*Henry David Thoreau*

It is rush hour on the continental shelf. A visitor to the shore on a summer day may see passing fishing boats, freighters, pleasure boats, surfers, and perhaps a slow flying, single-engine plane pulling an advertising banner past the sunbathers. The number of users of, and uses for, the marine environment has grown. Oil, gas, sand, and gravel are extracted from the seabed. The military monitors the U.S. coastal borders for threats and conducts training exercises. Air traffic has dedicated flyways and airport approaches. Even archaeologists have laid a claim to shipwrecks and the submerged remains of prehistoric settlements from a time when sea levels were lower. Some parts of the seabed are used as dump sites, and other areas are crossed by communications cables, sewer pipes, and oil and gas pipelines. Pressure on the offshore environment grows as energy security concerns drive the expansion of drilling beyond the Gulf of Mexico to elsewhere along the U.S. coast. Aquaculture also seeks a foothold offshore, while the centuries-old tradition of fishing for a wild catch struggles in an environment of competition and broadly declining fish stocks.

These are just the human uses of the marine environment. Birds, fish, sea mammals, and other sea life migrate through marine environments, as well as use them for breeding, feeding, and sheltering purposes. Threading through these multiple competing uses to situate a renewable energy project is no small feat.

Positioning a project both geographically and in the public's perception requires a well-considered siting plan. Disregarding important criteria may lead to problems during permitting, design, construction, or operation. Siting affects project economics as well as a developer's success in enlisting stakeholders as allies and in resisting other stakeholders who perceive their interests to be threatened. Developers should consult with fishermen early, for example, to learn about and hopefully avoid their most valuable fishing grounds. Avoiding battles and other pitfalls, without sacrificing fundamental project value, is the object of the siting process.

A good plan integrates commercial, economic, technical, engineering, regulatory, land/water use, and environmental factors. The plan should

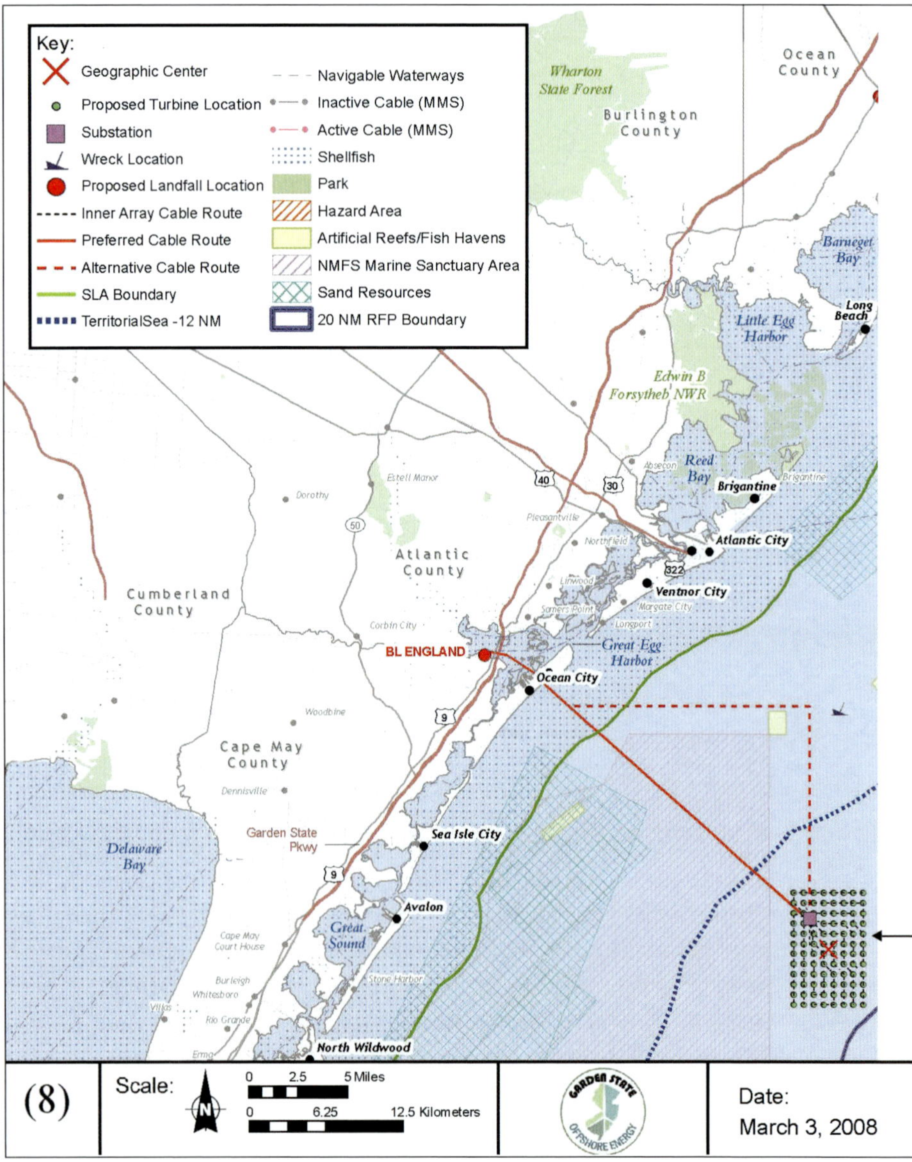

Fig. 4–1. Siting map created with GIS software
Courtesy of Ecology and Environment, Inc.

balance a project's technical aspects, such as water depth, wind speed, and distance to the grid interconnection, with qualitative aspects, such as the impact on tourism. Because many types of spatial information from diverse sources are gathered together in a siting analysis, a geographical information system (GIS) helps to synthesize the information, focus the analysis, and facilitate presentations to stakeholders (fig. 4–1).

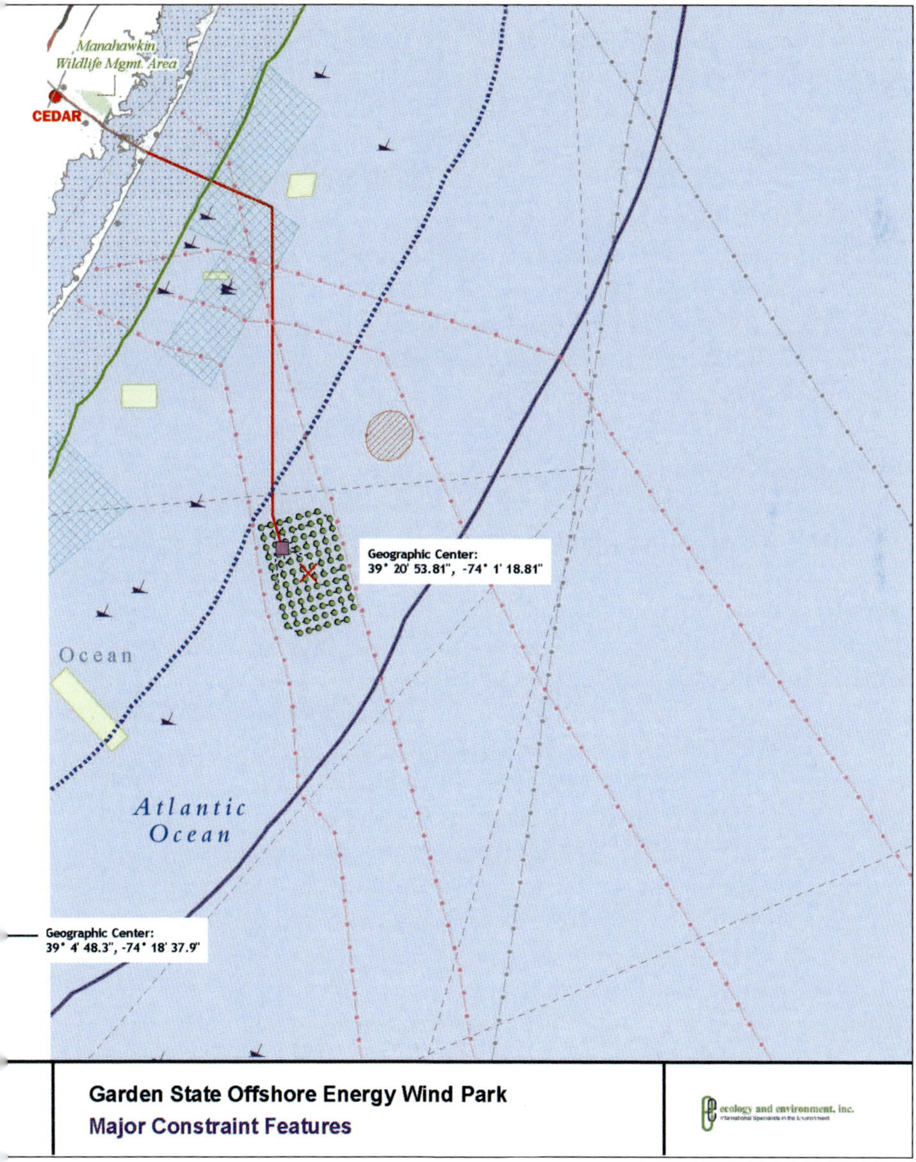

Garden State Offshore Energy Wind Park
Major Constraint Features

ecology and environment, inc.

Organizing the Siting Analysis

Markets where renewable power sells at a premium

Strange as it may sound, finding the right offshore renewable energy project site begins on land. The first order of business should be to identify the markets where renewable energy would bring the highest price. Hawaii relies almost exclusively (94.5%) on imported fuel oil and coal for generation. Due to seabed geography, there also is limited opportunity for electrical grid interconnection among the islands. For these reasons, electricity prices are high. According to the U.S. Energy Information Administration, in August 2008, the average retail price of electricity sold to all customer classes (residential, commercial, and industrial) in Hawaii was $0.3265/kWh.[1]

The abundant sun, wind, wave, and geothermal resources of Hawaii invite the development of distributed renewable energy sources. Recognizing this opportunity, the state of Hawaii has identified potential sites for renewable energy. The state's excellent study encourages and directs development efforts to particular areas based on site characteristics such as the availability of transmission, access to oceanfront land, and conflicts with tourism.[2]

New York City and Long Island illustrate a different offshore renewable energy siting opportunity. These heavily populated areas feature large electrical loads that are constrained by limited electrical import capability and insufficient local generating capacity. Electrical engineers call the region a *load pocket.* New York City's peak electrical demand in 2003 was approximately 11,000 MW. Transmission lines leading to the city have the capacity to import approximately 3,700 MW from Westchester County to the north, 1,000 MW from New Jersey to the west, and 300 MW from Long Island to the east.[3] Power importers pay a premium for this transmission service, which is reflected in the local market price.[4] The remaining 6,000 MW of demand must be supplied by in-city power plants, a large number of which are old and inefficient or fueled by expensive natural gas. The average retail price of electricity in the state of New York in 2008 was $0.19/kWh, made possible in part by low-cost hydroelectric power from Niagara Falls and coal-fired and nuclear power plants located in the state's rural areas. New York City, however, has limited access to these resources, and the price of power is much higher. Locally sited renewable generation, like Verdant Power's East River tidal current energy project, may be competitive in this market.

Also relevant to the competitive market environment are state incentives. Marine renewable energy projects may earn renewable energy credits (RECs) under a state-imposed renewable portfolio standard (RPS). The New York State Energy Research and Development Authority (NYSERDA) purchases RECs under long-term contracts, and this reliable cash flow becomes part of the revenue stream that can support long-term project debt. NYSERDA purchases the RECs with funds provided by a "systems benefit charge" imposed on New York's investor-owned utilities. In 2005, for example, NYSERDA entered into seven REC purchase contracts, with terms up to 10 years, primarily for new terrestrial wind power.[5]

Massachusetts, with its high energy prices ($0.17/kWh), also is an attractive location for renewable energy developers. Massachusetts REC prices, in excess of $50/MWh, are the highest in the nation. Contrast these possible locations with the southern states of Georgia and South Carolina, where electricity prices are in the range of $0.085/kWh to $0.098/kWh, and RPS incentives are unavailable.

States with RPS programs are likely to be generally receptive to renewable energy projects. On the East Coast, the RPS states include Maine, Massachusetts, Rhode Island, Connecticut, New York, New Jersey, Delaware, Maryland, and North Carolina. Along the Gulf Coast, Texas is the only state with an RPS program. All the states bordered by the Pacific, except Alaska, have RPS programs.

Nevertheless, general receptivity towards renewables did not keep Cape Wind's project from becoming a lightning rod. Stiff opposition from citizens groups, town governments, and prominent politicians, including Democratic Senator Edward Kennedy and former Republican Governor Mitt Romney, arose in Massachusetts even though citizen support for renewable energy on a conceptual level is strong. Proving that project timing also is important, Massachusetts' current governor, Deval Patrick, has reversed Romney's position and is supporting the project.

Actions by many states seem to signal an evolving favorable climate. To cite just one example, in 2004 Richard Codey, acting governor of New Jersey, imposed a 15-month moratorium on state authorizations for wind projects off of the New Jersey coastline. Governor Codey called on a blue ribbon panel to report on the impact of offshore wind on New Jersey's environment and coastal tourism–dependent businesses. Fast-forward to 2008, and much has changed. New Jersey's current governor, Jon Corzine, released an energy master plan that, among other initiatives, calls for the construction of 3,000 MW of offshore wind capacity by 2020. The state's first offshore wind energy project is now underway with the help of a grant for siting and permitting work.

Protecting economies propelled by coastal tourism is often a main concern of state governments. Just four oceanfront counties generated 64.3% of New Jersey's $32.2 billion in total tourism revenue in 2004.[6] Yet most coastal tourists will likely continue to visit and spend money if wind farms are built offshore. A survey by University of Delaware researchers indicated that 78.8% of Delaware residents would support an offshore wind farm such as that proposed by Cape Wind for Nantucket Sound. Even residents living within one-half mile of the Delaware shore were strongly supportive. According to the survey, a wind farm's effect on beach tourism would be minor, with 88.6% of respondents indicating that they would continue to visit the same beach they last went to in Delaware even if a large wind farm were constructed offshore.[7] Studies conducted in Denmark show that popular support for wind energy projects increases after they are constructed and begin operating. In New Jersey, the blue ribbon panel's report recommended that the state proceed with an initial offshore wind farm of up to 350 MW or 80 turbines and then evaluate the impact of such development before issuing additional permits.

Rhode Island and Oregon are in the midst of zoning offshore areas for renewable energy development. Offshore zoning benefits developers by reducing uncertainty in the siting process and benefits the public by providing a forum for determining the use of public lands. Oregon's plan is focused on wave energy facility siting and addressing conflicts with other ocean users, particularly fishermen. The Rhode Island Coastal Resources Management Council is developing an ocean special area management plan that would determine where in state waters wind and wave farms could be sited. Some proponents of the plan called for a moratorium on offshore development until the plan could be developed, but Rhode Island's Governor Donald Carcieri resisted, arguing that a moratorium would chill developer interest in building projects in the state. South Carolina Governor Mark Sanford called for a moratorium in his state pending the development of a clear national policy on ocean-based development. Imposing moratoria on project development until impacts are fully understood and regulations and policies are fully developed would hamstring offshore renewable energy development. New Jersey's support for the construction of a reasonably-sized project that will be closely evaluated before more projects are permitted is a reasonable way to proceed in the face of uncertainty.

Existing transmission facilities

There is no shortage of windy sites offshore. The challenge is finding suitable sites close to existing high-voltage transmission lines. A planning study of the offshore project proposed by the Long Island Power Authority (LIPA) found that the most important siting parameters included the following:

- The proximity of an onshore transmission interconnection with the capacity to accept the output of the proposed wind farm without expensive upgrades
- Distance from shore
- Water depth
- Bird activity[8]

There is very little existing undersea transmission cable infrastructure in the United States to which a new offshore renewable energy project could connect, and existing terrestrial high voltage conduits generally do not run along the coast. Building new transmission lines is time-consuming and expensive. The farther a project is from an onshore high-voltage transmission interconnection point, the greater the environmental impacts, permitting obstacles, and overall expense associated with the project. A developer should try to minimize onshore construction, particularly of overhead lines in coastal areas that are likely to be densely populated, in order to avoid significant project delays.

Roadways, bridges, causeways, and sewer pipes may provide rights-of-way for the construction of underground transmission facilities. Plans that incorporate these structures, or areas near them, may reduce disturbances to sensitive natural areas. Collocating subsea or subterranean transmission cables with existing linear facilities may reduce the environmental impact, but collocation may raise concerns about the potential adverse effects of electromagnetic fields on other infrastructure, such as pipelines. Opportunities for collocation should be considered in consultation with the owners of these existing facilities to determine the technical viability of locating within the same right-of-way.

In the Mid-Atlantic region, PJM Interconnection controls the transmission system along the coast from New Jersey to northern North Carolina. PJM offers maps of the transmission infrastructure from 69 kV to 765 kV that can be used for preliminary site analysis. The transmission system operator also can provide information about transmission line congestion in the area of interest. As explained in more detail in chapter 10, congestion in the area of a proposed interconnection affects project cost and timing. In congested areas, a transmission study evaluating the impact of adding the renewable energy supply to the grid and options for increasing grid capacity will likely be required. Congestion also may point to a queue of unresolved transmission access requests and a backlog of required transmission upgrades that could delay the interconnection of an offshore project for years. Developers will need the assistance of transmission system engineers to identify the best interconnection options in an area of interest.

Land-based staging areas

Land-based facilities also are necessary as a staging area for offshore construction activities. Dock facilities and a large open area for storing and marshalling foundation components, towers, turbines, blades, cranes, cables, and other equipment are required to build an offshore renewable energy project. The distance from the dock to the project site will determine transit time, the number of trips per day, the number of ships required, and the exposure of workers and equipment to dangerous weather. These factors will have a substantial effect on the construction timeline and costs.

Offshore energy resource strength

Having preliminarily qualified certain zones of interest on land, the next step is to look offshore from those areas to the conditions at sea. Factors such as water depth, distance from shore, average wave height, average wind speed, probable maximum wave and wind conditions during storm events, and seabed characteristics are important site discriminators. The importance of each factor will depend largely on which renewable energy technology is used. Depth may be the determining factor for wind turbines due to the cost of foundations, while average wave energy, storm wave energy, and distance to shore may be key factors for a wave energy plant.

Average wind speeds are typically 25% greater over water than nearby land. Since the energy in the wind increases by the cube of its speed, the energy production potential over water is thus generally much superior to land.[9] Evaluating the resource can be challenging because historical offshore wind and wave data is not as plentiful as terrestrial weather data. Furthermore, when data is available, it often requires adjustment to make it useful for project planning. Offshore weather buoys, for example, typically record wind speeds at 16 to 33 ft (5 to 10 m) above water level. Wind close to the surface is slowed by friction with the water, although not by as much as wind is slowed by friction over land.[10] (For example, in densely forested areas, the effective ground level is the height of the tree canopy, which as a rule of thumb is two-thirds of the height of the trees. In a forest where the average tree height is 45 ft [15 m], ground level is assumed to be at 30 ft [10 m]. A tower must be built that much higher to reach winds of equivalent speed over open terrain.) Weather buoy data may be adjusted to reflect the wind speeds that will be present at wind turbine hub height, typically 245 to 295 ft (75 to 90 m) for large offshore turbines.[11] At that height, wind force will be considerably greater than it is just above the water.

Resource consultants can help. They use mathematical weather models to predict wind speeds and wave heights in particular locations from measurements taken in other areas. Wind mapping tools use global weather patterns, geographic databases, and atmospheric simulation models to forecast wind speeds. These tools may identify promising offshore sites when actual measured historical wind data is scarce. An initial wind survey using a wind mapping tool should produce a high resolution image indicating which possible sites deserve closer consideration (figs. 4–2a and 4–2b).[12] Increasingly sophisticated, the models can even accommodate complex wind conditions such as the effects of terrain and land-sea breezes.[13] One such wind map resource is the Wind Resource Explorer, which features interactive wind maps that developers may query for wind speeds at heights from 30 m to 100 m, wind power density in W/m^2, and surface features such as transmission lines, highways, cities, vegetation, and terrain. AWS Truewind claims that its wind projections using the MesoMap System mathematical weather model are accurate within 5% to 7% of measured wind speeds at the height of a typical wind turbine's hub. It is interesting that the model's dominant source of error comes from variations in topography and land cover that are not accurately reflected in the database. Topographic variations should not be as significant offshore, and models of offshore winds should therefore have greater accuracy. Realistically, it is unlikely that a decision to build a wind farm would be based solely on the output of a weather model. However, the preliminary information from a model can be used to determine where to site a meteorological tower to measure actual offshore winds, waves, air density, and other data.

Understanding typical weather is important for predicting a project's likely energy output. Understanding potential weather extremes allows a developer to design for survivability and estimate the capital costs of fortifying a project against nature's worst storms. Knowledge of historical patterns of storm severity and frequency helps to differentiate between alternate sites. Storm track records and other data associated with major hurricanes are maintained by NOAA.[14]

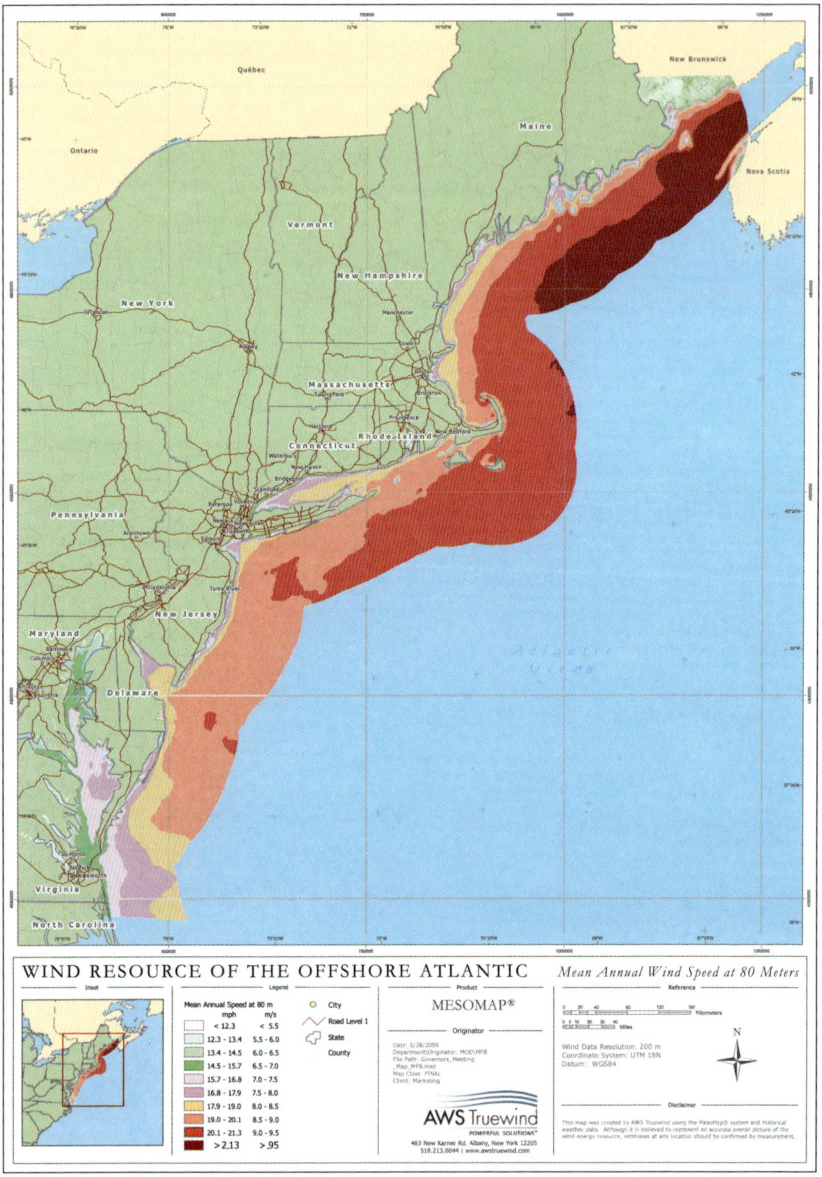

Fig. 4–2a. New England offshore wind map
Courtesy of AWS Truewind, LLC

The south shore of Long Island, for example, is exposed to tropical storms tracking northeast along the coast. Over the last 150 years, more than 85 tropical storms have passed within 115 miles (185 km) of Long Island, and nine hurricanes have made landfall. Of those nine storms, the strongest five were category 3 on the Saffir/Simpson scale, capable of winds up to 130 mph (58 mps) and a storm surge of 9 to 12 ft (2.7 to

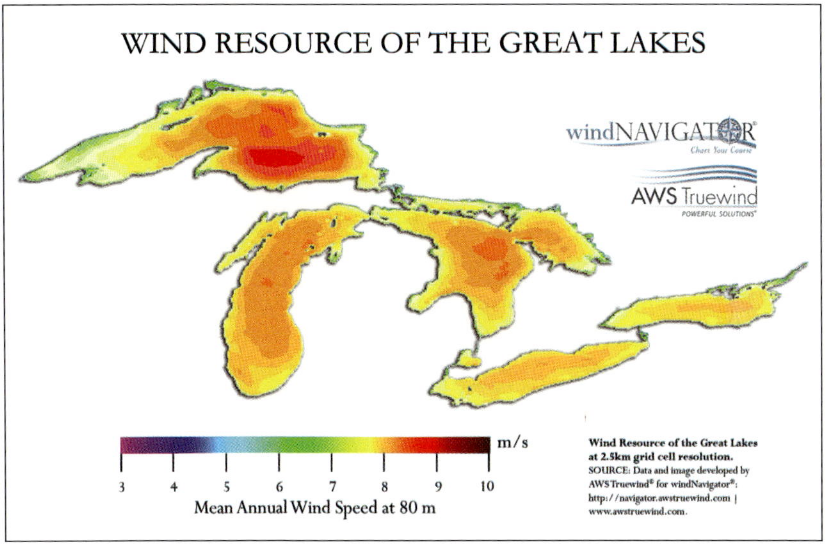

Fig. 4–2b. Great Lakes offshore wind map
Courtesy of AWS Truewind, LLC

3.7 m).[15] Yet even along Long Island's south shore, the potentially severe effects of a hurricane may differ. The eastern portion of Long Island is more exposed to wind-driven waves that can build over a broad fetch of ocean driven by counterclockwise hurricane winds. In contrast, the western parts of the Long Island shore are less vulnerable due to the sheltering effect of the nearby New Jersey coast. To properly evaluate the vulnerability of a potential offshore renewable energy site, engineers, meteorologists, and oceanographers working together should estimate the wind speed, wave height, current, and storm surge that would be associated with the most severe storm event that could affect the area of interest and its probability of occurrence. This information will affect foundation design, project insurability, and other cost components.

Shipping lanes and safety zones

Existing offshore activities will further limit potential renewable energy project sites. Shipping lanes and fairways are clearly an offshore area that should be avoided for safety reasons. The U.S. Coast Guard is the primary agency with responsibility for domestic shipping and navigation regulations. Near busy ports and navigational choke points, the Coast Guard has established traffic lanes as part of a scheme to separate shipping traffic traveling in opposing directions.[16] Coast Guard regulations also establish safety fairways in the Gulf of Mexico to allow for safe passage of shipping through oil and gas drilling platforms.[16]

Drilling rigs must be located outside of fairways, and associated anchor lines may not cross a fairway at a depth of less than 125 ft (38 m). Certain fairway ship anchoring areas also are specified, and structures built within the ship anchoring areas are generally required to be spaced more than 2.3 miles (3.7 km) apart. Other safety fairways have been designated off of the Atlantic coast and the coasts of California and Alaska.[18] Nautical charts produced by NOAA show these areas.[19] MMS will not authorize alternate energy leases that encroach on shipping fairways.[20]

The Coast Guard also may establish safety zones around vessels and facilities such as oil production platforms and deep water ports.[21] (The establishment of safety and security zones in harbors, around vessels carrying hazardous cargoes [including LNG] in specified areas, and at other places or vessels is at the discretion of the commander of the port.) The safety zones restrict vessel traffic in the adjacent waters. The authority to establish a safety zone is quite broad, covering measures necessary to safeguard life and property and "to protect the living resources of the sea from harmful agents."[22] Safety zones around existing offshore structures may limit the area available for offshore renewable energy facilities. Alternatively, the Coast Guard may determine that an offshore renewable energy facility requires a safety zone, which would restrict other activities within that zone. A safety zone may extend as much as 1,640 ft (500 m) from the outer edge of the facility, but it may not interfere with recognized shipping lanes "essential to navigation."[23] A safety zone may increase the perceived footprint of a marine renewable energy facility and the chance that other ocean users will be adversely affected. Recreational and small fishing vessels, however, are less affected by safety zones, which generally are not applicable to vessels of less than 100 ft (30.5 m) in length. As a rule of thumb, offshore renewable energy facilities should be set back from shipping lanes sufficiently to allow for the maximum safety zone should the Coast Guard determine one to be necessary.

Opponents of offshore renewable energy projects may seek to influence the Coast Guard's safety determinations. One particularly bold effort to derail the Nantucket Sound project involved an amendment offered by Alaska's Rep. Don Young to a Coast Guard budget reauthorization bill. The amendment would have excluded wind turbines within 1.5 nm (2.8 km) of a shipping lane.[24] The result would have been to remove much of the wind farm project's site on Horseshoe Shoal from development, and with fewer turbines, the project would have been economically infeasible. The Coast Guard and supporters of offshore wind energy opposed the amendment, and it was defeated.

In nautical parlance, an *allision* occurs when a vessel strikes a fixed object. The risk of an allision between a ship and a wind turbine tower or other marine renewable energy facility cannot be entirely eliminated, even with careful siting and safety zones. The Maritime Research Institute Netherlands has developed a mathematical model that uses the level of vessel traffic in an area and vessel types to calculate the risk of allisions between ships and wind farms, the increase in the frequency of a mishap due to a wind farm, and the expected spill size resulting from an allision.[25] Allisions involving relatively small fishing vessels, for example, may be a more frequent occurrence but present relatively low environmental consequences. An allision with an oil tanker, however, is likely to be infrequent but has potentially disastrous environmental consequences.

What would happen if an oil tanker ran off course and allided with a wind farm foundation or an offshore transformer substation? Several safety features should minimize the likelihood of a disastrous oil spill. Oil tankers operating in U.S. waters must by 2015 have a double hull design that reduces the opportunity for oil spills in case a tanker runs aground, collides with another ship, or, in this example, hits a turbine tower. Turbine tower manufacturers also generally design their structures to cause minimal damage to colliding ships. Towers located in northern waters must be able to resist the pressure of sea ice colliding with and pushing past a foundation, which dynamically is not dissimilar to an allision with a ship. A tower is generally built to give or flex somewhat until the moving ice breaks or bends. Tower designers also can deflect the impact of a ship or ice flow on a tower by incorporating a conical section or fender at the tower's water line.[26] Obstructions in the water also must have markings, lights, and warning horns. These requirements are described further in chapter 8.

The consequences of an oil spill depend greatly on the type of oil spilled. Diesel oil, such as is used to fuel smaller ships and the emergency generators on a wind farm's substation, is comparatively light, and it evaporates quickly. Bunker oil and crude oil, on the other hand, are heavy, tarlike substances that can coat marine life and beaches in a thick, oily mat.

Although the risk of an allision and spill in a particular area may be low, a developer must still demonstrate that it is prepared for such events. Each project should have emergency procedures and a spill response plan that identifies the most sensitive human and ecological features in the vicinity of the project and gives these locations priority for protection.[27] (The MMS requires lessees to "prepare waste management plans, hazardous material plans, and oil spill prevention plans, as appropriate, for the facility."[28]) The project emergency plans should complement the emergency response plans that ships also must have under international treaty (MARPOL 73/78) and the Oil Pollution Act of 1990 in case they become involved in a spill.

Renewable energy facilities sited in shallow water provide the benefit of marking a navigation hazard, much as a lighthouse or warning buoy would. On average, one ship a year is grounded on the Middelgrunden shoal outside Copenhagen's harbor. Developers of a wind farm on the shoal posited that their project would reduce the risk of ship groundings and consequent oil pollution as compared to the no-farm case.

Avian uses of the marine environment

After more than two decades of commercial wind farming, the wind power industry now has extensive experience with the impact of wind farms on birds.[29] Unfortunately, much of the public's perception of wind power's effect on bird life has been shaped by the problems with an older generation of wind turbines located in California's Altamont Pass. The placement of the Altamont Pass wind farms on a bird migration route has caused high-profile deaths of golden eagles and other raptors that frequent the area year-round. Given this history, the effect of an offshore renewable energy facility on bird life will be an important part of the project's environmental analysis and should be considered early in the site selection process.

Recent avian studies of terrestrial wind farms have generally shown no significant impact on bird populations. The consensus is that modern turbine design, featuring slower rotating blades and tubular towers (instead of truss-type towers, which provided roosting places), has significantly lessened bird strikes. Collisions with building windows, vehicle strikes, pesticides, feral cats, and numerous other causes each kill significantly more birds than wind turbines.

As Altamont Pass demonstrated, siting also has a role in reducing adverse effects on birds from wind farms. A preliminary project siting analysis should identify and avoid wildlife preserves and other areas where birds congregate, wetlands sanctuaries protecting nesting areas, migration stopover points, and other critical habit. Developers should be aware of the location of important bird habitat and migration pathways and seek to avoid siting in these areas. This is particularly important when species that are listed as endangered or threatened under the Endangered Species Act, the Migratory Bird Treaty Act, or similar state laws would frequent those places.

The U.S. Fish and Wildlife Service (FWS) has issued wind power siting guidelines for operating wind turbines to minimize avian and other wildlife impacts. Although the guidelines were developed for terrestrial wind power sites, they provide a useful reference.

> *Avoid locating turbines in known local bird migration pathways or in areas where birds are highly concentrated, unless mortality risk is low (e.g., birds present rarely enter the rotor-swept area). Examples of high concentration areas for birds are wetlands, State or Federal refuges, private duck clubs, staging areas, rookeries, leks, roosts, riparian areas along streams, and landfills. Avoid known daily movement flyways (e.g., between roosting and feeding areas) and areas with a high incidence of fog, mist, low cloud ceilings, and low visibility.*[30]

The fact that many birds tend to migrate in corridors along the coasts and inland waterways presents a special challenge for marine renewable energy projects (fig. 4–3). Although little is known about bird migrations over water, it is thought that many bird species stay close to the coast, which provides marshes and other areas to rest, shelter, and feed. Feeding opportunities in shallow, nearshore waters draw birds; lower bird concentrations can be expected over deeper waters. The further offshore a wind farm is located, the less likely it is to impact birds.

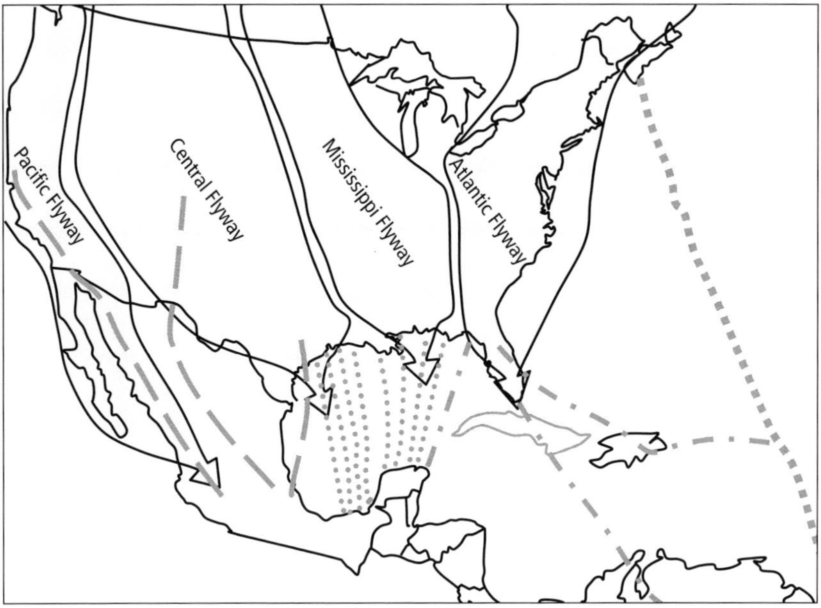

Fig. 4–3. Map of principal North American bird migration corridors

That is not to say that some species do not make long, uninterrupted migrations over open water or that there is not tremendous variation in the behavior of species.

> *With our present knowledge of bird migration, recognizing distinct broad belts of migration down the North American continent encompassing groups of distinct populations or species is not realistic. About all we can say for sure now is that birds travel between certain breeding areas in the North and certain wintering areas in the South; that a few heavily traveled corridors are used by certain species; and that more generalized are routes followed by other species.*[31]

Lightweight radio transmitters that can be attached to birds and radar tracking have added to biologists' understanding of migration patterns. "One interesting fact to come out of current radar work is the discovery of relatively large movements of warblers and other small land birds migrating over oceans rather than along coastlines and in directions about which ground-based observers were completely unaware."[32] More such research clearly is needed to help determine the potential impact that large numbers of offshore renewable energy facilities may have on birds. According to the FWS,

> *The potential for significant offshore turbine impacts on waterbirds is great, virtually no research has been conducted in the United States to quell these concerns, and finding carcasses at sea is very challenging... While habitat loss is not believed to be a serious concern, its impacts continue to be assessed. Disturbance may be problematic since some species such as Common Eiders avoid wind farms and may not return to a coastal area for several years (Guillemette and Larsen 2002). Disturbance may lead to displacement, and turbines may serve as barriers to seaduck movements. Only a few studies have been conducted in Denmark, the Netherlands, and Sweden, so further research is needed.*[33]

One study by Danish researchers involved observations of migrating ducks and geese with radar and a thermal animal detection system (TADS) (a heat-activated infrared video camera) mounted on a wind turbine tower. The researchers found that birds generally altered their path to skirt the wind farm. When the birds did enter the farm, they usually flew safely between turbine rows to avoid the spinning blades. The birds were observed during the day and night. "During more than 2,400 hours of monitoring...[TADS] spotted only fifteen birds and bats and one moth flying near the turbine, and it recorded one collision involving a small bird or bat."[34] In another study, Danish researchers used decoys to attract ducks to a wind farm. The ducks, however, were reluctant to come within 328 ft (100 m) of the turbines. The researchers also switched the turbines on and off to determine whether rotating blades and turbine noise would drive feeding birds away from the area. The birds appeared to be indifferent to turbine operating status. These promising findings should be supplemented with additional study to determine whether birds habituate to turbines over time, the risk of collision during low visibility conditions, such as in fog, and the behavior of other bird species.[35]

An avian impact study will typically be required to evaluate bird abundance, migration, and use patterns at a proposed wind farm site. In connection with the Long Island offshore project, planners considered and eliminated a significant area along Long Island's south shore so as not to interfere with important bird habitat. The project siting assessment notes,

> *All waters in the vicinity of Montauk, including an area extending roughly 10 n-mi around Montauk Point, were excluded from consideration because of the high concentrations and diversity of birds. More so than any other portion of Long Island's south shore, this area supports an abundance of nesting, wintering, and migrating birds. In addition to Montauk, the waters within 3 n-mi of all south shore inlets were excluded from consideration because they have much greater avian use than the waters between such inlets.*[36]

NOAA's National Ocean Service prepares environmental sensitivity index maps that are useful for preliminary site analysis.[37] Intended mainly as a reference for oil and chemical spill responders, the maps show a wealth of wildlife information and provide a quick view of the biological diversity in an area of interest (fig. 4–4). The maps show shoreline sensitivity to oiling and human structures, such as water intake pipes and swimming beaches. Of perhaps more interest to the offshore renewable energy developer, the maps also show the location and extent of sensitive biological resources, such as seabird colonies and marine mammal haul-out grounds. Additional data include the nesting, laying, hatching, and fledging schedules of bird species present in the area and the level of concern for a species' population health. Times for spawning, migration, calving, and other key life stages also are provided for fish, shellfish, reptile, and sea mammal species. Understanding the timing of species life stages may help a developer to anticipate possible conflicts between wildlife and a proposed project construction schedule.

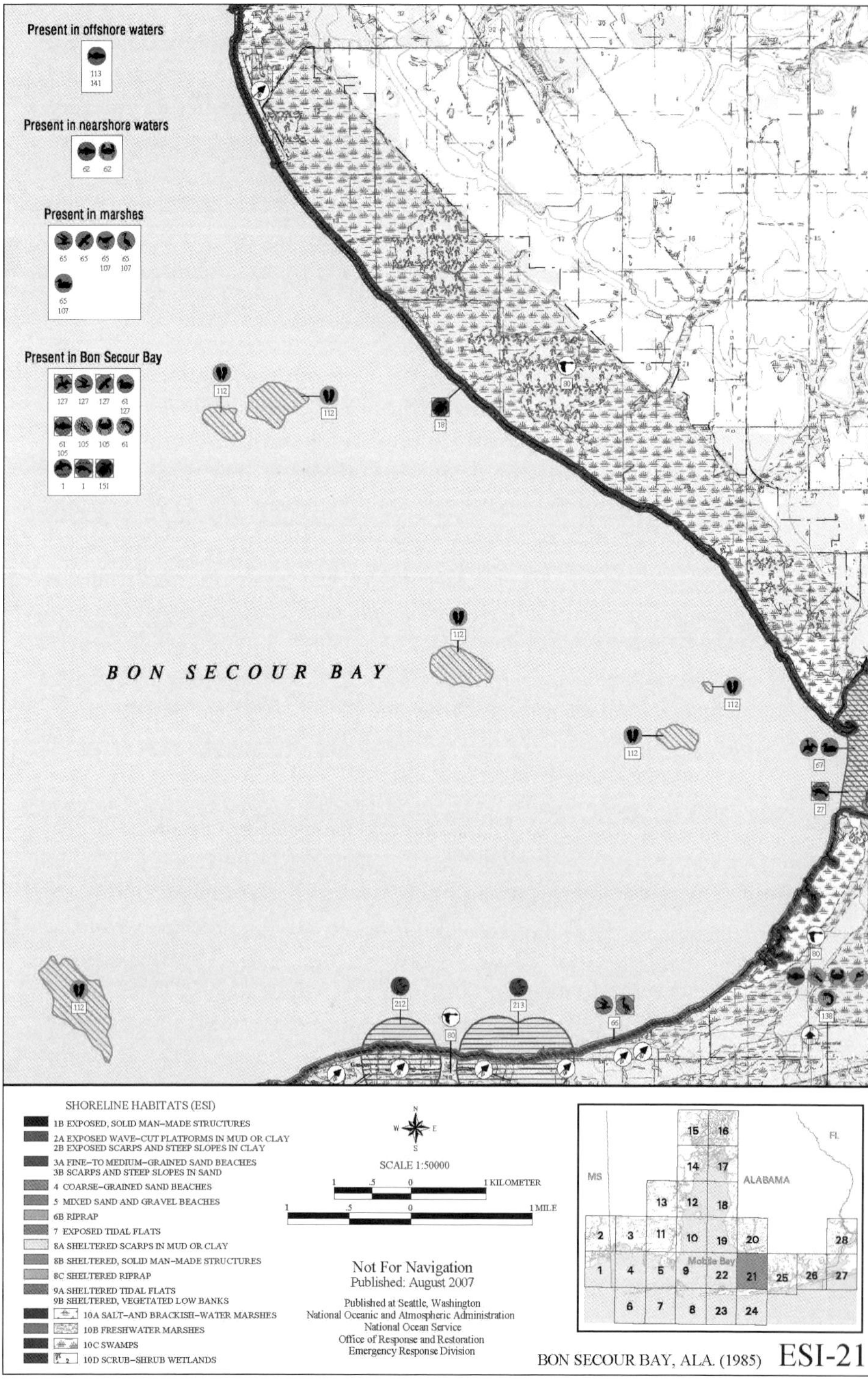

Fig. 4–4. Environmental sensitivity index map for Bon Secour Bay, Alabama
Source: National Oceanic and Atmospheric Administration

ALABAMA

SHORELINE HABITAT RANKINGS

- 1B) EXPOSED, SOLID MAN-MADE STRUCTURES
- 2A) EXPOSED WAVE-CUT PLATFORMS IN MUD OR CLAY
- 2B) EXPOSED SCARPS AND STEEP SLOPES IN CLAY
- 3A) FINE- TO MEDIUM-GRAINED SAND BEACHES
- 3B) SCARPS AND STEEP SLOPES IN SAND
- 4) COARSE-GRAINED SAND BEACHES
- 5) MIXED SAND AND GRAVEL BEACHES
- 6B) RIPRAP
- 7) EXPOSED TIDAL FLATS
- 8A) SHELTERED SCARPS IN MUD OR CLAY
- 8B) SHELTERED, SOLID MAN-MADE STRUCTURES
- 8C) SHELTERED RIPRAP
- 9A) SHELTERED TIDAL FLATS
- 9B) SHELTERED, VEGETATED LOW BANKS
- 10A) SALT- AND BRACKISH-WATER MARSHES
- 10B) FRESHWATER MARSHES
- 10C) SWAMPS
- 10D) SCRUB-SHRUB WETLANDS

HUMAN-USE FEATURES

- AIRPORT
- ARCHAEOLOGICAL SITE
- BOAT RAMP
- CH CRITICAL HABITAT
- HISTORICAL SITE
- MANAGEMENT AREA
- MARINA
- PARK
- RECREATIONAL BEACH
- WILDLIFE REFUGE
- 123 HUMAN-USE NUMBER
- BRIDGE
- MANAGEMENT AREA BOUNDARY
- STATE BOUNDARY

SENSITIVE BIOLOGICAL RESOURCES

BIRD

- DIVING BIRD
- GULL / TERN
- PASSERINE BIRD
- PELAGIC BIRD
- RAPTOR
- SHOREBIRD
- WADING BIRD
- WATERFOWL
- NESTING SITE

FISH

- FISH

MARINE MAMMAL

- DOLPHIN
- MANATEE

TERRESTRIAL MAMMAL

- SMALL MAMMAL

REPTILE

- TURTLE
- OTHER REPTILE

INVERTEBRATE

- BIVALVE
- CEPHALOPOD
- CRAB

INVERTEBRATE (cont)

- ECHINODERM
- GASTROPOD
- SHRIMP

HABITAT

- FLOATING AQUATIC VEGETATION
- PLANT
- SUBMERGED AQUATIC VEGETATION

MULTI-GROUP

123 **RAR NUMBER**

THREATENED / ENDANGERED / SPECIES OF SPECIAL CONCERN

Marine mammals and other ocean dwellers

Areas where marine mammals and other ocean dwellers congregate, or through which they pass during migration, also present project siting issues. These range from concerns about the effect of transmission cable electromagnetic fields on the navigation and feeding habits of marine mammals, turtles, fish, and sharks, to potential harm associated with construction noise and possible entanglement in the mooring systems used to secure floating marine energy systems.

Transmission cables are shielded to minimize the emissions of electromagnetic fields, and they are usually buried, which further reduces field strength. (MMS requires shielding of transmission cables where necessary to control the intensity of electromagnetic fields.) A review of the scientific literature conducted in support of the Cape Wind project concludes that adverse effects are unlikely.

> *Buried underwater cables would not appear to be capable of interfering with the navigation of marine animals because of the limited area over which they would be sensed (in the current project, about 60 mG at the seafloor at maximum loading, but dropping to 3 mG beyond 30 ft from the cable centerline) and the fact that the 60 Hz time variation causes the magnetic field from the cables to alternately add to and subtract from the Earth's magnetic field, resulting in a time-averaged deviation of zero.*[39]

Underwater noise interference with echolocation and whether noise may drive mammals from their habitat is another issue. Construction noise and normal operational noise are separate issues. The effect of the former on fish and mammals may be mitigated by scheduling construction for times when sensitive species are not present. To prevent marine mammal exposure to harmful noise levels, observers should be posted to maintain a lookout for marine mammals in the vicinity of the construction site.[40]

Pile driving is likely to be the most significant noise concern, and seismic sounding to characterize the seabed for wind turbine foundation engineering also may disturb sea mammals. Regarding MMS regulations relating to seismic testing and marine mammals, it is notable that "the consensus of the data is that virtually all marine-mammal species are potentially impacted by sound sources with a frequency above 500 Hz. However, relatively few species are likely to receive significant impact from frequencies well below 500 Hz, which is where the main energy from airgun arrays [used in seismic sampling] resides."[41] A study of offshore wind farm noise by COWRIE Ltd. (Collaborative Offshore Windfarm Research into the Environment), an independent corporation established in the United Kingdom by government and industry, found that the broadband peak sound pressure associated with pile driving was measured at 228 dB at a distance of 3.3 ft (1 m) and 189 dB at 1,312 ft (400 m). COWRIE found that "the noise created during pile-driving operations involves sound pressure levels that are high enough to impair the hearing system of marine mammals near the source and disrupt their behavior at considerable distances from the construction site."[42]

Certain marine mammals such as seals and porpoises could hear pile driving above normal ocean background noise more than 50 miles (80 km) from the source, according to COWRIE. These mammals would exhibit behavioral responses to the noise over a broad range. Whale behavior, for example, would be affected over a 37-mile (60-km) radius from the origin of the noise. A Swedish study noted that pile driving has been observed to stun fish temporarily, causing them to drift in the water as if they were dead. This corresponds to findings in the COWRIE study that "physical effects, like internal or external injuries or deafness [temporary or permanent] up to cases of mortality, are possible in the close vicinity to pile-driving."[43]

Other issues include whether mooring cables may present a possible entanglement danger for whales and whether a renewable energy platform or buoy may attract roosting birds or serve as an artificial haul-out site and thereby alter local habitat, nesting, and feeding patterns. Anticipating these and other issues at the project siting stage or, even better, in the technology development phase, may help to develop mitigation strategies and prevent serious obstacles later.

Ocean dump sites

The oceans were once thought to be so vast that they could absorb all of man's transgressions. Men hauled in nets heavy with sea life from one part of the ocean and emptied waste-laden barges in another part. Prior to 1972, U.S. waters were widely used for the dumping of sewage sludge, industrial wastes, construction debris, scuttled vessels, and much more. Recordkeeping was thin or nonexistent, and the content of many dumps is anyone's guess. Massachusetts Bay, for example, had at least four disposal sites for radioactive waste during the 1940s and 1950s.[44] Assorted toxic and hazardous wastes continued to be disposed of at sites in Massachusetts Bay during the 1970s.

Heavy industrial activity in the coastal zone also may be responsible for significant dispersed sediment contamination offshore. The seabed outside of New York harbor, known as the New York Bight, is particularly polluted. The Bight has been the subject of a long-term study by the U.S. Geological Survey to determine the extent of sediment contamination and its dispersal. The study found that lead and other heavy metals have migrated throughout the New York Bight over time. Contaminated sediments are continually redistributed by currents and biological processes.

Setting foundations for an ocean renewable energy facility or digging trenches for transmission cables could disturb contaminated sediments. When the sediments become suspended in the water column, they are made available to enter the food chain. For this reason, it is prudent not only to avoid dump sites but to know about the nature of sediment contamination generally in any area under consideration.

Denmark's Middelgrunden wind farm was built on contaminated material dredged from Copenhagen's harbor over decades. Testing for heavy metal contaminants was required at each turbine foundation location, and dredged material that proved heavily contaminated was subject to special handling. A similar procedure would apply in the United States. Under the Marine Protection, Research, and Sanctuaries Act of 1972, and the Clean Water Act, disposal of material that could degrade public waters or endanger human health or the marine environment is generally prohibited.[45] (Title I of the Marine Protection, Research, and Sanctuaries Act is generally known as the Ocean Dumping Act.) The U.S. Environmental Protection Agency (EPA) and the U.S. Army Corps of Engineers (Corps) have joint oversight over ocean dumping, which is generally restricted to material derived from dredging as long as the sediments are not too contaminated.[46] Contaminated dredge spoil requires special disposal, such as transportation to an onshore site or treatment to destroy or isolate the contaminants. Confined aquatic disposal is another alternative in which a deep pit is dug in the seabed, sediments are deposited in the cell, and then a 3-ft layer of clean sand is used to cap the contaminants. Old dump sites present particular challenges, since construction personnel and marine life are more likely to be exposed to higher levels of toxins than present in dredge spoil. Disposing of uncovered hazardous material will be costly.

Military waste dumps in the oceans are less common than commercial and industrial waste dumps, but they can be much more hazardous. In 1946, for example, 30 mustard gas bombs, each weighing 1,102 pounds (500 kg), were dumped 20 miles (32 km) off the coast of Mobile, Alabama in 200 to 600 ft (61 to 183 m) of water. One bomb was later recovered after it floated ashore.[47] The other bombs may or may not be where they were dumped.

An investigation by the *Daily Press* (Newport News, VA) in 2005 reported that the U.S. military had disposed of thousands of tons of unexploded ordinance, chemical weapons, and radioactive waste in the ocean.[48] The weapons include rocket-propelled missiles holding lethal

VX gas, a nerve agent that has a long persistence in the environment and is capable of causing death in extremely small doses. Although much of this weaponry was probably sunk in deep water, it is possible that munitions may be found in the shallow nearshore waters where the first generation of offshore renewable energy facilities is likely to be sited. In one recent discovery, a clam trawler operating off the New Jersey shore unintentionally retrieved a mustard-gas-filled artillery shell.

Many ocean dump sites are indicated on the nautical charts published by NOAA, but military dump sites are generally unidentified. The U.S. Army has put only 1 of its 26 known chemical weapons dumps on nautical charts. MMS also has only a "vague idea" where chemical weapons were thrown into the ocean.[49] Fishermen, who have an interest in identifying and avoiding sites that could foul or tear their nets, may provide valuable informal information about dumps.

In addition to consulting charts and collecting anecdotal information, various technologies and techniques may be used to survey an area. Seafloor surveys (sometimes referred to as *shallow hazards surveys* or *geohazards surveys*) are standard practice in siting and designing offshore infrastructure. The surveys are done with the aid of geophysical remote sensing equipment such as sidescan sonar, multibeam sonar, subbottom profilers, and magnetometers. The results serve a variety of purposes, including dump identification, cultural resources interpretation, and determining the suitability of bottom conditions for various trenching/plowing technologies. Surveys also help to determine mooring/anchoring methods and aid in identifying biologically sensitive habitat (hard bottom, reefs, etc.).

Sand and gravel mines

Offshore sand and gravel mining of the seabed is incompatible with marine renewable energy facilities or electrical transmission cables. Beach replenishment involves dredging sand from offshore borrow pits and placing it on shore, where after a while, it usually erodes once again. Offshore gravel deposits may be mined for road and other construction projects. Developers siting a project close to a borrow pit should consider whether the alteration of the seabed in the area of the mine could cause local currents and scouring around foundations, moorings, and cables.

Erosion control may be an unintended benefit of wave energy facilities. Because they extract energy from wave motion, wave energy projects located close to shore (i.e., within 2 to 2.5 miles [3 to 4 km]) could reduce erosion on shore. The amount of the reduction will depend on the wave energy device and the array configuration. One study by the Electric Power Research Institute estimated a 12%

reduction in wave height caused by a three-row arrangement of Pelamis floating wave machines. Projects located farther from shore would not reduce the strength of waves impacting shore significantly, since waves that pass around the plant will mix with the weaker waves to the lee of the plant.[50]

Offshore communications cables, pipelines, and sewer outfalls

Undersea communications cables, pipelines, and sewer outfalls should be identified in the siting process; these items also are indicated on NOAA's nautical charts. As mentioned previously, these facilities and associated rights-of-way may provide opportunities for collocation of renewable energy facilities if they are determined to be compatible. Developers should consult with the owners of such facilities to explore collocation options and obtain an engineering analysis to ensure compatibility. Whether the electromagnetic field emitted by transmission cabling might promote pipeline corrosion, and how a collocated cable would be installed and accessed for maintenance, are issues to consider.

The LIPA offshore siting analysis proposed using sewer outfalls as a potential pathway to shore for the wind facility's marine cable.[51] MMS requires developers to minimize impacts to wetlands and to "avoid hard-bottom habitats, including seagrass communities and kelp beds, where practicable, and restore any damage to these communities."[52] Collocation with existing linear facilities may help to reduce disturbance to important natural habitat areas near shore and onshore.

Offshore sites of historical interest

It may seem unlikely that a marine renewable energy project could disturb important historical artifacts, but underwater archaeological sites do exist, and developers should take care in the planning stage to identify them. Finding historical material during the construction phase may tie a project in a knot while the archaeological significance of the site is evaluated and necessary project adjustments are made. One useful reference for conducting archaeological evaluations in the context of offshore renewable energy development is the *Guidance for Assessment of Cumulative Impacts on the Historic Environment from Offshore Renewable Energy*.[53] Although written with U.K. history and regulations in mind, it provides useful general guidance for U.S. projects.

Shipwrecks are the principal source of archeological material, but underwater archaeology also includes the mundane objects used by prior inhabitants and the remnants of their settlements, which due to historically lower sea levels may be located in water depths as great as

250 ft (76 m). Evidence of prior coastal peoples may include fish traps, salt-making sites, oyster shell piles, and other remains of inundated or eroded settlements. NOAA's Automated Wreck and Obstruction Information System provides shipwreck locations, the type of wreck, and the date of the loss.[54] A developer may use this information to distinguish archaeologically significant wrecks from mere junk.

The locations of significant wrecks may be kept confidential to lessen plundering by unauthorized divers. A marine renewable energy developer should look beyond NOAA's database to identify and avoid potential archaeological sites. State historical preservation agencies may release information on sensitive archeological sites to project planners on a confidential basis.[55] MMS also requires site lessees to conduct magnetometer surveys in areas where there is a high potential for shipwrecks.[56] A magnetometer survey of a wreck site can locate buried metal objects and differentiate a ship with a metal hull from a possibly more significant wooden hull vessel. Divers or remote-operated vehicles also may be used for a closer inspection of suspected wrecks. Identifying archaeological sites early in the project planning stage increases a developer's options and helps to avoid expensive delays during the project construction phase.

Seabed geology

The gently sloping East Coast continental shelf is characterized by generally thick sedimentary deposits of clay, sand, and gravel. The Gulf Coast is similar. The West Coast continental margin is characterized by a steeper slope, rugged outcroppings of exposed bedrock, and extensive rock bottom habitat. Maps showing the geology of the seabed are available from the U.S. Geological Survey (USGS) and NOAA's National Geophysical Data Center. The USGS and NOAA maps show bottom topography, sediment types, and likely conditions below the surface of the seabed. This information should be supplemented by seabed sampling and seismic profiling in areas of interest during a project's engineering design phase to determine the composition of the seabed for foundation design.

Shoals, canyons, valleys, rock outcrops, and other bottom features may influence site selection. A shoal can provide a relatively shallow area suitable for locating wind turbine foundations in the midst of deeper water, permitting access to strong winds with minimal penalty in terms of increased foundation costs. Underwater canyons and valleys may date to a time when the sea level was lower and rivers carved their way farther out on the continental shelf than the present day shoreline. The historic flows of the Hudson River created a valley in the continental shelf east of

New Jersey's coast. Canyons and valleys present the hazard of turbidity currents that may damage foundations and sever transmission cabling. Rock bottoms present different challenges. Foundations or moorings may require drilling, increasing construction costs. Rocky areas also tend to be productive habitat for marine life. Building there may displace or disturb a greater number of species than in sandy or mud-bottomed areas. MMS requires developers to "avoid locating facilities near known sensitive seafloor habitats, such as coral reefs, hard-bottom areas, and chemosynthetic communities."[58] (*Chemosynthetic organisms* exist at great depths and extract energy from dissolved gases, rather than through photosynthetic processes used by organisms populating the upper portion of the water column, such as phytoplankton.) The agency also requires developers to survey proposed sites to identify sensitive and unique marine habitats in the vicinity of a project. Project plans should minimize adverse impacts to these habitats.[59]

The relative scarcity of hard bottom areas along the eastern seaboard and Gulf Coast has been addressed by the creation of artificial reefs. Constructed with riprap, scuttled vessels, and even old subway cars, these reefs provide a hard substrate on which mussels, sponges, anemones, corals, and other sea life attach. The colonizers in turn attract mobile sea life that feeds on the reef dwellers and may shelter on the reef. The hard surfaces of oil and gas platform foundations have served as the basis for artificial reef communities that have become popular recreational fishing and diving areas. Wind turbine foundations and other marine renewable energy structures can be expected to provide similar benefits.

Fishing grounds

Wind turbine foundations and other marine energy device moorings can pose an obstacle to trawling, net fishing, and longline gear. Project developers should evaluate the extent and type of commercial fishing conducted in an area of interest to understand how an offshore renewable energy facility may impact commercial fishing. Activities such as clam dredging and lobster trapping, for example, which employ maneuverable equipment, should be relatively unaffected by an ocean energy installation. What is the mix of commercial and recreational fishing in the project area? Which species are targeted, and with what type of gear? How healthy and productive are the fisheries? Is fishing income a principal component of the local economy or merely one of many sources of income in a diverse seashore community? There is no substitute for meeting with fisher organizations early in the project design stage to learn firsthand their concerns and discuss how renewable energy and fishing may coexist. Engaging fishers in the information collection program improves the likelihood that the results will be perceived as fair.

All fishers may not oppose an offshore renewable energy facility. Excluding the most aggressive fishing methods from a project site may improve the health and economics of a regional fishery outside the site boundary. Marine sanctuaries offer a promising example. A study of 100 no-take marine reserves around the world showed that, as compared to unprotected outside areas, reserves increase the following:

- Fish population density (on average 91% higher)
- Total biomass (192% higher)
- The average size of sea life (31% larger)
- Species diversity (21% greater)[60]

Thirty years of experience with marine reserves in New Zealand revealed that commercial fishermen who initially opposed the creation of these no-fishing zones later became the reserves' staunchest advocates. "Spillover and larval export—the drifting of millions of eggs and larvae beyond the reserve—have become central concepts of marine conservation. Reserves where fishing is banned are now seen as potential stud farms and fish hatcheries, replenishing the surrounding seas."[61]

Wild fish catches around the world are peaking or declining as increasingly sophisticated fishing gear, including satellite locators and airplane spotters, is employed to chase fewer fish. Between 1970 and 1990, the capacity of the world fishing fleet doubled, increasing the pressure on fish stocks and making fishing less profitable.[62] According to the U.S. Commission on Ocean Policy, despite attempts at scientific management, the current management techniques have "allowed overexploitation of many fish stocks, degradation of habitats, and negative impacts on many ecosystems and fishing communities."[62] Commercial fishers are under pressure from declining stocks, catch limits, shorter seasons that create a "race for the fish," overcapitalized fishing fleets, and increasing competition for fishing grounds from alternate uses.

Some fishing communities may welcome viable alternatives to the diminishing returns provided through fishing. By creating de facto marine reserves, offshore renewable energy projects can improve fishing and at the same time improve the economies of shore communities through activities such a project construction and maintenance. Project developers should be sensitive to opportunities that allow fisheries and fishers to benefit from offshore renewable energy.

Airport flight paths and military training areas

Air travel over water is another use of the offshore environment that could be affected by a wind farm. Obstructions located near an airport attract the interest of the Federal Aviation Administration (FAA) if they penetrate (or would when built) an imaginary cone of airspace extending upward from the border of the airport's runways. The cone is intended to provide an unobstructed area for takeoffs and landings. Objects that penetrate the cone may cause the FAA to issue a "Determination of Hazard to Air Navigation" that may result in other agencies denying necessary project permits. The FAA also requires objects that extend more than 200 ft (61 m) above the surrounding terrain or water level to be marked and lighted in a specific manner. The agency's rules are described in greater detail in chapter 8.

Military use of large offshore areas for training and testing may also cause a siting issue. Encroachment on terrestrial military ranges caused by environmental restrictions and development near range boundaries has affected military operations. Encroachment concerns also extend offshore. Large portions of the eastern Gulf of Mexico and the Virginia capes are designated training areas.[64] The Department of Defense (DOD) has resisted offshore drilling proposals in these waters and likely would respond similarly to offshore wind farms. Subsea hydrokinetic energy projects also could attract DOD opposition. The U.S. Navy operates the South Florida Testing Facility, an electromagnetic and acoustic testing range near Ft. Lauderdale, Florida. This range is a siting constraint for Gulf Stream current energy project development.

Radar interference

The effect of wind farms on radar stations is the latest roadblock for wind farms in the United States, both onshore and offshore. The concern is twofold. First, the height of the turbines may block radar from "seeing" behind the wind turbine towers. Second, interference caused by rotating blades may affect the ability of radar to discriminate a wind turbine from aircraft. Utility-scale wind turbine blades are quite large—similar in diameter to the wingspan of the largest passenger jets. Wind turbine blade tips also can move at speeds comparable to aircraft speeds. This similarity can make differentiating wind farm clutter from real aircraft a challenge.

The DOD is concerned that wind farms may lower the effectiveness of air defense and missile warning radars. Rather than imposing a complete ban, the DOD believes that each proposed wind farm should be evaluated on a case-by-case basis for its impact on military readiness.[65]

Radar interference usually arises when a proposed wind farm is in the "line of sight" of a radar facility. The closer a wind farm is to a radar facility, the greater the potential interference. A nearby wind farm will obscure a larger sector of airspace behind it from radar than will a distant farm. (In addition, vibrations from a wind farm also may create seismic noise interference affecting nearby monitoring sites for nuclear test explosions under the Comprehensive Test Ban Treaty.)

Understanding how interference issues may be addressed requires a little background in radar physics. Radar is a sensing system that uses a transmitter of radio frequency waves (3 MHz to 100 GHz), an antenna, a receiver, and a processor. The transmitter emits pulses of waves, some of which encounter a target and other objects such as trees, buildings, and precipitation (e.g., rain, snow, or hail). The radio waves bounce back from all of these objects. Together with radiation from natural and man-made sources, the radio waves travel to the receiving antenna at various levels of strength and with varying frequency, depending, for instance, on the motion of the target. The processor analyzes and compares the signals received by the antenna and discriminates targets from noise or "clutter."

The radio frequency waves emitted from the radar device are concentrated in one central direction, the *primary lobe*; but there also are *secondary* and *tertiary lobes* of radar power extending generally 60° to each side of the main lobe. Consequently, clutter and interference can come from reflections of the primary, secondary, and tertiary lobes over a fairly broad angle in front of a radar facility.

How far a radar facility can see depends on many factors, such as its height above the ground, intervening obstacles that prevent propagation of the radio waves, and the power of the radio waves emitting from the device. The distance also depends on the sensitivity of the device receiving and interpreting the reflected waves and the curvature of the earth. Radio waves are refracted, or bent, by the atmosphere, with the result that radar actually can detect objects that are over the real, or geometric, horizon. As a rule of thumb, to account for refraction, a radar facility's line of sight is calculated using a "4/3 earth model," in which the radius of the earth is assumed to be 5,279 miles (8,496 km) (i.e., four-thirds of earth's actual radius). For example, assume a radar facility situated 50 ft (15.2 m) above the surrounding terrain and a "bald" earth with no obstacles such as mountains between the radar and a wind turbine. A turbine with its blade tips extending no more than 300 ft (91.4 m) above the surface would need to be approximately 34.5 miles (55.5 km) away from the radar facility to be out of sight, or below the range of the radar.[66] The earth, of course, is not bald, and terrain between the radar facility and the turbine towers may mask the towers from the radar facility's line of sight. For this reason, evaluating whether a proposed wind farm will impact a radar station's line of sight requires a case-by-case analysis.

Currently, the principal method of mitigating the effect of a wind turbine on military radar effectiveness is to site the turbines out of radar line of sight. The layout of a wind farm also may be changed to minimize the *shadowing effect*, whereby wind turbine towers block the propagation of radio waves behind the farm. The DOD also is investigating software solutions that might enable the radar signal processors to better recognize and suppress clutter from wind farms. Another solution under review involves coordinating two or more radar facilities that have different, but overlapping, views of the wind farm. One radar facility would fill in the shadow behind the wind farm that the other radar facility cannot illuminate. It also may be possible to install a small radar station within the wind farm to fill in the shadow and coordinate its input with the main air defense system. These approaches remain experimental.[67]

The FAA administers air traffic control radars for civil aviation purposes, and the National Weather Service (NWS) uses radar for weather forecasting. Air traffic control radar may be less affected by wind farms than military air defense radar. In addition to passive radio wave reflections from objects within the radar's scope of vision, air traffic control radar also uses an active system that sends out an interrogatory "beacon" to aircraft. Aircraft then use a transponder to respond with a unique signal back to the radar. The active system should lessen concerns regarding clutter and shadowing.

A study of the effect of wind farms on weather radar concludes that, thus far, forecasting ability has not suffered a "major negative impact." The study warns, however, that as more wind farms are built—particularly in close proximity to radar stations—adverse effects are expected.[68] More studies of the effects of wind farms on weather radar are currently being conducted by the Atmospheric Radar Research Center at the University of Oklahoma.[69]

A wind farm developer should consult with the FAA and NWS to determine whether a proposed wind farm will adversely affect nonmilitary radar facilities. The DOD's Long Range Radar Joint Program Office will provide informal guidance to wind farm developers to assist them in avoiding military radar line of sight concerns. Low profile wave, tide, and ocean current energy facilities have the obvious advantage of avoiding these issues.

Visual and other community impacts

Recognizing the special cultural and resource value of certain areas, Congress prohibited MMS from issuing offshore leases or easements for projects within the boundaries of any national park system, national wildlife refuge system, national marine sanctuary system, or any national monument. MMS also requires projects to employ design elements such as visual uniformity, the use of tubular towers, and other aesthetic features to minimize the visual impact of offshore renewable energy facilities.[70]

Despite efforts to minimize aesthetic objections, some people will nevertheless argue that so-called industrial wind generation is out of place in sight of pleasant seaside communities. Wind farms, in their opinion, are inconsistent with the character and harmony of the landscape. Presumably, the wind farm opponents would prefer to see wind farms sited near coastal areas where industry is already the dominant theme. Environmental justice—the concept that residents of a particular community should not be unfairly burdened by industrial nuisances because they belong to a lower socioeconomic class or a minority group—takes on an interesting twist when the topic is offshore renewable energy. Assuming that the sight of a wind farm is reasonably considered a detriment, is it fair to burden the residents of the industrialized coastal community with yet another "industrial" facility? Or is it more just if those living in the bucolic seaside town assume some of the burden of generating the energy consumed in their homes and businesses by learning to live with the site of an offshore wind farm?

Aesthetic values are highly personal and not easily quantified. Beyond the visual, a community's culture, history, and sense of place are aspects of project planning. These intangible elements are difficult to measure or locate on a map, yet it would be foolish to ignore them. The intangible factors strongly influence the reaction of the public to a proposed project and the capacity of community members to organize in support of or opposition to a project. Project developers err when they focus solely on the purported visual impact of a proposed project and ignore the larger context, the psychological environment at the root of a community's emotional attachment to a place.

A focus on visual impact leads a developer to reason that if the visual impact of an offshore renewable energy project can be shown to be small, then the intangible aspects of the siting analysis should be dismissed in favor of objective factors that argue for approving the project (e.g., environmental improvement, jobs, or energy security). A developer may understandably try to find a way to objectively dismiss the visual impact concerns, for example, with simulated photographs showing very small turbines extending only an inch or so above a distant horizon. Unfortunately, this approach may not succeed, because visual impact is in the "mind's eye," not the biological eye. A developer looking at a visual simulation of a proposed offshore wind project sees a group of turbines painted battleship gray. In the haze of a summer day, they are almost invisible. The structures would likely escape the notice of the average passerby.

Show the same photograph to a group of community residents, however, and they may see something quite different. The owner of a local seaside motel may feel uncertain whether tourists will choose to vacation elsewhere. The surfcaster recalls tranquil days at the shore escaping the stress of the workday and wonders whether his personal refuge will be the same. The resident of a seaside cottage recalls evenings sitting on the back deck contemplating the dark horizon with a glass of wine and thinks about how red aircraft warning lights will look as they blink in the distance. The commercial fisherman, town council member, and local ice cream shop proprietor each have a different personal and economic connection to the water that affects how they will interpret the photograph.

In one recent case involving a proposed high-voltage electric transmission line, Dominion Virginia Power, which serves rapidly growing northern Virginia, argued that demand for power required the construction of a coal-fired plant in western Virginia and a new 240-mile (386 km) transmission line. The line, which would cross picturesque rolling hills and Civil War battlefields, among other important historical and cultural landmarks, ran into a firestorm of opposition from local property owners and politicians. They argued that conservation and locally sited generation were superior ways to meet growing power demands and that rural counties along the way should not be burdened to provide a benefit to another part of the state. Opponents also speculated that the real motivation for building the line was to bolster the utility's profits from shipping cheap power out of state.

According to a local realtor, "When you're out in the country and you're selling property, what you're selling is the open space and the bucolic views and the history...Running power lines through an area like this is just devastating."[71] The utility is now advancing a less controversial, and less direct, transmission path along existing transmission rights-of-way.

Permitting agencies will consider socioeconomic and cultural impacts, aesthetics, and environmental justice issues as part of a comprehensive analysis of alternatives to a proposed project. A developer's failure to understand and adjust to the intangible aspects of project siting can lead to a difficult and lengthy project permitting experience. Appreciating these issues may help a developer to anticipate public reaction and proactively mitigate impacts to the extent possible.

The efforts of FPL Energy to integrate a proposed Long Island offshore wind project with local popular sentiment illustrate the point. The company influenced community attitudes by sponsoring community events and arranging for a native Long Island celebrity to endorse the project, among other public relations techniques. The developer's approach is described further in chapter 9. Although some residents nevertheless opposed the project, a significant number supported and advocated for it. Many Long Islanders came to view the project as theirs and welcomed the opportunity for Long Island to enter the vanguard of the clean energy movement.

Merging hard and soft data

Extensive science and engineering-driven comparisons among potential offshore sites are needed to decide where to place a project. In many cases simple adjustments to a project's layout may be all that is required to avoid disturbing sensitive areas. An appreciation of local culture and attitudes also is indispensable. A successful site evaluation merges hard scientific, engineering, and economic data, the alchemy of emotion, and the art of public relations.

Offshore renewable energy projects, particularly wind farms visible from shore, may attract powerful opposition. If the opposition does not completely stop a project, it certainly may extend project timelines and increase costs. Avoiding this outcome by identifying the best project locations, educating local opinion leaders, and engaging the community in dialogue is time well spent. A developer that has identified a promising offshore site will learn in the next chapter how to secure the legal right to build a project in that space.

5 Acquiring Rights to Project Sites: MMS and Offshore Renewables; FERC and Hydrokinetics

Verdant is near to entering its sixth year of regulatory process since first proposing to install 10 MW of tidal power in the East River... Like other start up companies, our ability to survive without revenue is limited.

—*Gilbert P. Sperling, Verdant Power, Inc.*

Grays Harbor Ocean Energy Company has a combination wind/wave energy platform concept. The Y-shaped platform, with three retractable legs, is mated to a conventional tower and wind turbine at dockside and loaded onto a barge. Once towed on site, the platform's legs are deployed like a tripod, and the platform lifts above the water. Each leg is hollow and built with openings below the waterline. As waves rise and fall, air pulses through the legs and drives small oscillating water column turbines mounted on the platform, which supplement the wind turbine's output. The interesting bit, at least for purposes of this discussion, is that Grays Harbor tried to use the wave energy aspect of its wind/wave combination platform to obtain a preliminary permit from the Federal Energy Regulatory Commission (FERC). The company believed that FERC's process, rather than the still-developing Minerals Management Service (MMS) process, would give it priority of right to develop the site. According to Grays Harbor, "the Company has obtained the first-mover advantage and expects it will not have to compete for the sites because that is not required by the FERC as long as wave power is produced."[1]

Unfortunately for Grays Harbor, the company's regulatory strategy fizzled recently when MMS and FERC sat down to agree on how to license renewable energy projects on the outer continental shelf (OCS). Under a Memorandum of Understanding (MOU) between the agencies, FERC agreed not to issue preliminary permits for hydrokinetic projects on the OCS.[2] The agencies coordinated their jurisdiction over hydrokinetic projects and closed a loophole that offshore wind energy developers thought might lead to unfair site banking. This chapter describes the FERC and MMS processes for obtaining control of offshore renewable energy project sites and how, in the area of hydrokinetics, the two agencies propose to work together.

MMS Leads the Leasing Process

The Energy Policy Act of 2005 granted the secretary of the U.S. Department of the Interior authority to lease sites on the outer continental shelf for renewable energy development. The secretary delegated this authority to MMS, which has since analyzed the environmental impacts of offshore development, issued guidelines for data collection and test facilities, and issued regulations for its offshore renewable energy program.[3] The Energy Policy Act was intended to resolve project delays caused by confusion over which federal agency could convey offshore property interests. That aim was immediately frustrated, however, when FERC asserted authority over offshore hydrokinetic energy projects, leaving only offshore wind projects in federal waters clearly within MMS jurisdiction.

MMS issued proposed rules, entertained public comments, and delivered its final rules to the Office of Management and Budget (OMB) in 2008, but the rules sat stalled at OMB through the close of the Bush administration. Immediately after President Obama took the oath of office, Rahm Emanuel, the president's chief of staff, issued a government-wide order freezing all new regulations pending review by the new administration. Final rules for offshore leasing, project development and operations were at last issued on Earth Day, April 29, 2009, more than three-and-a-half years after Congress opened the door to offshore renewable energy leasing. Secretary of the Interior Ken Salazar commented, "It is fitting that on Earth Day President Obama is taking this bold step towards opening America's oceans and new energy frontier, so that we can wisely build a clean energy economy that will create millions of new jobs across the country. This new framework will enhance our energy security and create the foundation for a new offshore energy sector that will employ Americans developing clean and renewable energy."[4]

Experience leasing federal offshore lands for oil and gas development makes MMS a sensible choice to lead the offshore renewable energy program. As it has done for oil and gas interests on the OCS, MMS will evaluate the suitability of areas for leasing, determine interest in the sites among potential lessees, and issue leases on a competitive or noncompetitive basis. MMS will coordinate the environmental reviews conducted by both federal and state agencies necessary to authorize a project. In addition, MMS regulations establish royalties, fees, and other payments and impose safety and environmental standards on offshore renewable energy operations.

The agency's role as steward of the OCS derives from section 8 of the Outer Continental Shelf Lands Act (the OCS Lands Act), which requires MMS to ensure that renewable energy development on the OCS achieves the following:

- Is safe
- Protects the environment
- Prevents waste
- Conserves natural resources
- Is coordinated with other agencies
- Protects national security
- Considers and protects the rights of others to use the OCS for purposes such as fishing or navigation
- Provides a fair return to the United States
- Reflects public participation on leasing decisions
- Is properly monitored[5]

Under this legal standard, MMS expects developers to design projects and conduct operations with high regard for safety and in a manner that minimizes adverse effects to the coastal and marine environments.

A precondition to investing significant sums on resource evaluation, site assessment, project design, and permitting is the developer's expectation that it will have a reasonable opportunity to acquire rights to build on the chosen project site. This issue lies at the heart of the MMS leasing process, because lease sites are subject to competitive bidding. Developers that invest in preliminary resource and siting analysis, and have prepared concrete, credible plans for an offshore renewable energy project, are no more entitled to a particular offshore site than a speculative bidder. Managing the site acquisition process, therefore, is essential to planning any offshore project, especially one that requires an MMS lease.

Interim policy for the collection of resource data and technology testing

Pending the adoption of final rules in 2009, MMS implemented an interim policy for permitting precommercial facilities on the OCS. The interim policy allows the installation of meteorological towers for the collection of resource measurements and facilities to test renewable energy generating technologies. MMS will issue such research leases (*limited leases* in MMS parlance) generally for a five-year term, with the possibility of extension. By law, leases must be issued on a competitive basis unless, after issuing a notice, MMS determines that there is no competitive interest in the site.[7] Research leases do not confer a priority right to develop a site commercially and may permit only limited sales of power. Nor do research leases provide a shortcut around applicable laws and regulations. MMS will subject each lease to review under the National Environmental Policy Act (NEPA) and the Coastal Zone Management Act (CZMA), and lessees must comply with numerous other federal laws and state requirements designed to protect the environment and public safety. These requirements are examined in detail in chapters 6, 7, and 8.

Numerous developers expressed interest in acquiring research leases when MMS issued a notice soliciting competitive interest in 16 offshore areas in New Jersey, Delaware, Georgia, Florida, and California. As it happens, the majority of the individual sites drew no competitive interest, and MMS is leasing them on a noncompetitive basis. Several sites in Florida and one site in California, however, did attract interest from competing developers. Rather than offer those sites at auction, MMS asked the competing developers to collaborate in the development of resource data under leases issued noncompetitively. Collaboration reduces each developer's exposure to the cost of gathering site data and provides some symmetry in the information that each developer will eventually use in bidding on a commercial lease. Now that MMS has adopted final regulations for offshore leasing and operations, project developers have the option of acquiring long-term commercial leases and benefit from greater clarity with respect to the project development process.

Acquiring an OCS alternative energy lease under MMS regulations

MMS's newly-issued final rules for the offshore renewable energy program build on the previously-described interim policy and expand the program to commercial projects. This discussion will proceed by explaining how a developer could acquire access to the OCS and obtain permission to construct offshore facilities under the new rules.

Leasing options. There are several paths to a lease. The process may start with a developer identifying an appropriate site and communicating to MMS an interest in leasing it from the government. Alternatively, a developer could wait for the MMS to issue a *request for interest*—an invitation to the public to suggest sites to be evaluated for leasing. A third approach could be to negotiate with a party that has already acquired an OCS site lease from the MMS, which the prospective developer could acquire through an assignment subject to MMS authorization. Finally, a developer might find an existing facility on the OCS, perhaps an oil production platform that is sitting on a depleted field, and negotiate with the platform owner and lessee for the rights to occupy the platform and use it for an alternate use. The alternate use could be renewable energy production or another activity, provided that the activity is not otherwise authorized under other federal law. In response to opposition from environmental and fishermen's organizations, MMS has clarified that it would not authorize aquaculture operations.[8]

MMS offers several lease types under its rules that are appropriate for different purposes. Limited leases may be good for technology testing, but they are not ideal for a developer interested in building a full-scale project for commercial energy sales, which would require a commercial lease.

Rights-of-way and *rights-of-use and easements* are other interests (called *grants* by MMS) that might be useful in special circumstances. A right-of-way (ROW) grant entitles the holder to install cables, pipelines, and associated facilities on the OCS for the transport of renewable energy (or an energy product, such as hydrogen, generated from renewable sources) that has **not** been produced by a single project located on the OCS. The ROW grant is puzzling unless one first understands that the lease for a renewable energy project located on the OCS would include the required cable or pipeline rights-of-way in the lease, obviating the need for a separate ROW grant. A ROW grant would be necessary, for example, if a renewable energy project located on state submerged lands needed to transmit power across the OCS to reach an interconnection point located elsewhere on state lands.

A right-of-use and easement (RUE) grant is even more specialized. The RUE grant authorizes facilities on the OCS that support the production, transportation, or transmission of electricity (or other alternative energy product) from any renewable energy resource. However, the actual generation of electricity (or production of the alternative energy product) cannot occur on an alternative energy lease issued by MMS. Consider, for example, a pumped-storage hydroelectric project that uses wave-powered pumps located on the OCS to force water through a pipeline and up a cliff to an elevated reservoir. The facility generates electricity as the impounded water is released and falls to generators located at the base of the cliff. This project would need a RUE grant for the facilities located on the OCS. Since ROW and RUE grants are specialized instruments, they will not be covered in depth here. Nevertheless, the concepts discussed in this chapter are generally applicable to ROW and RUE grants as well.

MMS OCS leases are purpose specific. The same tract could conceivably be leased to several different developers for renewable energy projects that would not conflict. "A lease issued for one type of alternative energy activity would not necessarily result in prohibition of other types of activities in that same area, which could be authorized by separate leases issued subsequently."[10] For example, wind and wave energy projects that share the same general area and common transmission interconnection could be recipients of separate leases.

Commercial and limited leases. Commercial leases are likely to be the most common OCS interests. A commercial lease permits the lessee to install and operate facilities and sell power for a relatively long operations term (generally 25 years; potentially much longer upon renewal). Renewals are subject to MMS discretion and possible adjustment of lease payments.[11] Depending on the condition of the foundations, cables, and other infrastructure, it may be possible to extend the life of a wind farm considerably past the 20 to 25 years of useful life of the original turbines. Where feasible, the lessee can renew the lease, replace the turbines, and use the existing infrastructure for another 20 to 25 years. Commercial leases also include an additional period prior to the initial operations term of up to 5 years for the purpose of evaluating the lease site. This *site assessment term* allows the developer to assess the resource, conduct geophysical surveys, perform biological assessments, and otherwise evaluate site suitability. Unless the developer finds the site unsuitable, the site assessment process will typically lead to the developer preparing a *construction and operations plan (COP)* for submission to MMS. The operations term begins on the date MMS approves the COP.

Developers seeking to engage in precommercial activities, such as technology testing or resource evaluation, also may choose a limited lease. The limited lease lasts for a term of five years, with the option of extension. Notably, limited leases do not permit commercial use of the site, although MMS may permit lessees under limited leases to sell electricity produced in connection with technology testing within limits specified in the lease.[12] Another difference when compared to commercial leases is that a limited lease does not provide a path to obtaining a subsequent commercial lease should the resource evaluation indicate that the site would be suitable for a project. A limited lease does "not convey any preferential rights to obtain a commercial lease to develop the leased area."[13]

While it may not seem that a limited lease has many advantages over a commercial lease, the limited lease option should reduce the expense and burden of obtaining rights to the OCS. Since a limited lease does not convey long-term development rights, such leases should be less costly to acquire. Because installations would be temporary and of a smaller scale than those at a commercial development site, MMS expects an easier regulatory approval process for facilities built under limited leases as compared to those built under commercial leases.

Having to make a strategic choice between a commercial or limited lease will likely lead project developers to place a premium on early efforts to evaluate the renewable energy resources at various sites of interest. Using environmental models, developers will seek to develop a short list of preferred sites. Developers may then pursue commercial leases for the short-listed sites to secure long-term development rights. After acquiring a commercial lease, actual resource measurements taken during the lease's site assessment term may indicate that development would not be profitable. At that time, the developer may relinquish the lease back to MMS. Under this strategy, bonus and rental fees paid by the lessee would be lost. However, this expense should be weighed against the risk of competition. A developer that has proven the adequacy of a renewable energy resource at a site where it holds a limited lease may subsequently face interest from other bidders for a commercial lease.

Competitive leasing process. To ensure that use of the OCS provides a fair return to the U.S. government, MMS regulations establish a competitive process for issuing leases and grants. If only one party is interested in a particular site, however, MMS may issue a lease on a noncompetitive basis. The leasing process follows a defined path that includes several notices and environmental evaluations (fig. 5–1). The process may begin in two ways: through an unsolicited submission by a developer to MMS that expresses interest in leasing certain sites, or through an MMS-issued *call for information and nominations* or call. When MMS receives an unsolicited lease request it must determine whether there is competitive interest in the sites proposed by the developer. MMS would issue a *request for interest in the Federal Register* to make that determination. If it appears that there is competitive interest in certain sites, MMS would issue a call in the *Federal Register* seeking comment on areas which should receive special consideration and analysis (including geological, environmental and other conditions), and suggesting areas to be considered for leasing.[14]

When responding to a request or a call, the developer should provide MMS with certain information about the site, the developer's project plan, and the developer's qualifications as a prospective lessee.[15] MMS regulations require a respondent to disclose "available and pertinent data and information concerning renewable energy and environmental conditions in the area of interest, including energy and resource data and information used to evaluate the area of interest."[16] Clearly, this may be sensitive information with competitive value, and developers may request that this information be kept confidential.

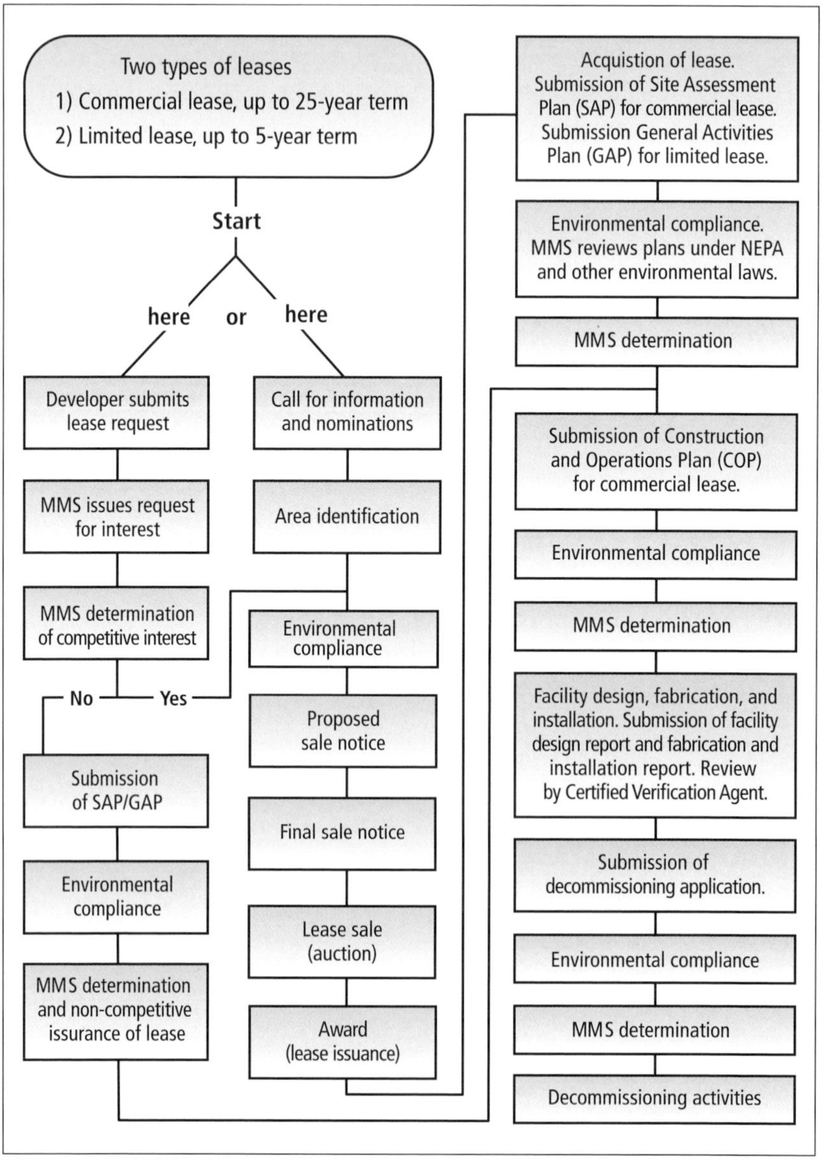

Fig. 5–1. MMS leasing process

MMS should protect sensitive information from disclosure to the extent allowed by law. The Freedom of Information Act (FOIA) requires a government agency to provide public access to its records unless an exception from the disclosure rules applies. Two relevant exceptions are:

1. "Trade secrets and commercial or financial information obtained from a person and privileged or confidential"
2. "Geological and geophysical information and data, including maps, concerning wells"[17]

Energy and resource data should be considered protected commercial or financial information, provided that the developer was careful to maintain its privileged and confidential status by restricting its availability to other persons. It is less clear whether the exception for "geophysical information...concerning wells" would be read broadly enough to apply to similar information describing renewable energy resources. During the MMS processing of expressions of interest in the 16 OCS limited lease sites mentioned previously, developers designated information they had amassed on the value and potential of the site as confidential business information, and MMS honored those designations.

If a potential bidder asserts the confidentiality of any sensitive resource data that it provides to MMS, the availability of the FOIA exceptions should protect MMS from having to release the information to the bidder's competitors. MMS, however, may use the information in estimating a site's value for the purpose of setting the reserve price or other terms of a lease.[18] Whether this works to the advantage or disadvantage of the prospective lessee depends on whether MMS will discount lease rates on suboptimal lease sites in addition to raising the rates on the most resource-rich areas.

Based on the response to a call, MMS would consider the nominated sites (and any others that MMS determines may be appropriate for leasing) for further evaluation in a process called *area identification*.[19] This phase entails consideration of the potential effects of leasing the sites of interest on the natural and human environment, along with possible related mitigation measures. MMS may hold public hearings and consult with appropriate federal, state, and local agencies to assist with this review. MMS will use the information gathered for the required environmental assessment under NEPA to support the decision to offer the leases. MMS also would consider whether the particular sites identified may raise issues under any affected state's coastal zone management plan. If the leases would not violate the state plans, MMS would provide a consistency determination under the Coastal Zone Management Act.

The public learns which nominated sites have survived the area identification review in a proposed sale notice published in the *Federal Register.* The notice invites public comment on the terms and conditions of the lease sale. In addition to a description of the areas available for leasing, the notice will include details such as the following:

1. Lease size
2. Lease term
3. Payment and bond requirements
4. Performance requirements
5. Site-specific lease stipulations

The notice also provides information about the auction process, including the following:

1. Bidding procedures and systems
2. Minimum bid
3. Deposit amount
4. The place and time for filing and opening bids
5. The lease award method (including the criteria used to evaluate competing bids)
6. Bidding or application instructions

Finally, the notice also will include or reference the official MMS lease form that the winner of the auction would be expected to execute.[20]

After MMS has considered the comments in response to the proposed sale notice, it may issue a final sale notice that, following the same format as the proposed sale notice, describes the agency's final decision regarding the areas to be offered, as well as the lease and auction terms.

Auction methods. Table 5–1 identifies the auction methods and bid variables that MMS may use.

The simplest method is *sealed bidding*, which involves a single round of bidding by each lease sale participant. Bidders typically would submit their bids by post or e-mail. The bid variable is the focus of the competition between bidders and takes the form of an upfront payment, called a *cash bonus*, or a recurring charge, known as the *operating fee rate*. When the bids are in, MMS opens and ranks them within each tract offered for lease, as well as among all of the tracts offered. MMS then determines which bids to accept and notifies the winners in writing, typically within 90 days.

A more complex bidding method is *ascending bidding*, which features multiple bidding rounds. Bidders would typically attend the auction in person and submit increasing bids based on the bid variable (either cash bonus or operating fee rate) up to the close of bidding. MMS is most likely to employ this format if there is strong interest in the leases and several bidders have good information about the tracts and their potential value.

Two-stage bidding is a hybrid of the other auction formats. Such an auction typically would begin with ascending bidding until only two bidders remain, or at least two bidders have offered the maximum bid amount specified by MMS. The first-round survivors would then participate in a second round of bidding using a sealed bid format. Second round bids would be made based on a bid variable that was not employed in the first round.

MMS also may employ a fourth auction format that considers multiple factors relating to proposed OCS renewable energy projects. Under this format MMS may consider both financial bid variables (such as cash bonus, operating fee and rental rate) and nonmonetary variables (such as technical merit, timeliness, financing and economics, the environment, public benefits, and consistency with state and local needs and requirements). "A method of assessing multiple factor bids may be employed to rank proposals, resulting in a lease award to the bidder making what MMS perceives as the best offer."[21] MMS acknowledged that this format "is likely to be less transparent to the public and more susceptible to favoritism and manipulation by selected parties than other auctioning formats."[22]

Table 5–1. MMS auction methods

Type of Auction	Bid Variable	Bidding Process
Sealed bidding	Cash bonus or operating fee rate	One sealed bid per company per lease or packaged unit
Ascending bidding	Cash bonus or operating fee rate	Continuous bidding per lease
Two-stage bidding (combination of ascending and sealed bidding)	An operating fee rate in one, both or neither stage and a cash bonus in one, both or neither stage.	Ascending or sealed bidding until: (i) only two bidders remain, or (ii) More than one bidder offers to pay the maximum bid amount. Stage two sealed or ascending bidding commences at some pre-determined time after the end of stage one bidding
Multiple-factor bidding	Factors may include, but are not limited to: technical merit, timeliness, financing and economics, environmental considerations, public benefits, compatibility with state and local needs, cash bonus, rental rate, and an operating fee rate.	One proposal per company per lease or packaged bidding unit.

Source: 30 C.F.R. § 285.220.

The term "package bidding" refers to an auction method that allows developers to identify possible synergies between tracts that would then be grouped into a single lease area. The developer and others could then bid for a "package" of tracts based on the development potential of the overall proposed project. In an auction where packaged bidding is an option, it appears that competing developers might each designate separate packages that may or may not overlap. MMS would determine the winning packages using a software algorithm that maximizes the sum of the package bids submitted in successive rounds. MMS also may hold a bidder's choice ascending auction in which the high bidder in each round earns the right to choose one tract, or multiple tracts to form a logical development unit from all tracts offered.

After an auction closes, MMS reviews the bids and must decide whether to award the leases. This decision is not as simple as picking the highest qualified bid for each tract. MMS regulations state, "We reserve the right to reject any and all high bids, regardless of the amount offered or bidding system used."[24] The agency implements *intertract competition* through its bid ranking process in an attempt to home in on the true market value of a tract when competition for individual tracts is weak. "In addition to enhancing competition, the object of intertract competition would be to provide signals through the bids which serve to assist us in leasing only the most valuable sources of energy needed to meet the expected demand."[25] MMS may, for example, reject the high bid for a tract A because, in the context of the high bids cast for tracts B, C, and D, the bid for tract A appears too low.

A bid that does not meet the minimum bid levels set by MMS in the final sale notice will not be accepted. Assuming that hurdle is cleared, the highest bid also must generally clear the unstated reservation price set by MMS. Reservation prices reflect MMS estimates of the value of individual tracts, whether for renewable energy production or other incompatible uses, such as sand and gravel mining. MMS relies on its estimate of a tract's value particularly when competition for that tract has been weak. Because it represents the lowest value at which MMS may award a lease, MMS will not disclose the reservation price.

Lease payments. The bidding system specified by MMS affects how the bid variable may change during the auction (table 5–2). A sliding operating fee rate variable is one notable option that lessens a project's royalty expenses in the early years. Such an expense deferral may help developers to better manage cash flow constraints during the years when debt service burdens weigh most heavily on project finances. Operating fees do not apply to limited leases, which by definition cannot be used for commercial-scale energy production, and therefore the cash bonus would be the only bid variable in a limited lease auction.

Table 5–2. MMS bidding systems for commercial leases

Bid System	Bid Variable
(1) Cash bonus with constant operating fee rate.	Cash bonus.
(2) Constant operating fee rate with fixed cash bonus.	A fee rate used in the formula below to set the operating fee per year during the operations term of the lease.
(3) Sliding operating fee rate with a fixed cash bonus.	A fee rate used in the formula below to set the operating fee for the first year of the operations term of the lease. The fee rate for subsequent years would change according to a mathematical function specified by MMS in the Final Sale Notice.
(4) Cash bonus and constant operating fee rate.	Cash bonus as in (1) above, and operating fee rate as in (2) above. Applicable to the two-stage auction format only.
(5) Cash bonus and sliding operating fee rate.	Cash bonus as in (1) above, and operating fee rate as in (3) above. Applicable to the two-stage auction format only.
(6) Multiple-factor combination of nonmonetary and monetary factors	MMS will identify bidding variables in the Final Sale Notice. Variables may include: (i) Nonmonetary (e.g., technical merit) factors and (ii) Monetary (e.g., cash bonus, rental rate, fee rate) factors.

Fee Formula:

F =	M *	H *	c *	P *	r
(annual operating fee) ($)	(installed capacity) (MW)	(hours per year) (8,760)	(capacity factor) (the anticipated efficiency of the facility's operation expressed as a decimal between zero and one)	(power price) (based on prior year's average retail power price in the state where the project's transmission cable lands, in $/MWh)	(operating fee rate) (expressed as a decimal between zero and one)

Source: 30 C.F.R. §§ 285.221; 285.506.

The operating fee formula deserves further discussion. Note that, unlike a royalty rate, the operating fee is based on MMS estimates of what the lease should produce, not actual production. MMS chose this method so that it could avoid tracking and auditing actual renewable energy production levels. The fee rate would either be set by MMS or, if used as a bid variable, be determined by the outcome of bidding at auction. Unless MMS specifies otherwise, the operating fee rate will start at 2% when the project begins commercial operation. MMS may set lower fees, or waive the fee rate for new technologies or when necessary to encourage continued or additional activities.[26]

The installed capacity and capacity factor elements of the operating fee formula are related. The first factor is based on the nameplate capacity of the equipment installed at the site. MMS will estimate the capacity factor based on comparisons with other projects operating in similar conditions.[27] Offshore wind capacity factors might be expected to be in the 35% to 40% range. MMS may subsequently adjust the capacity factor based on actual performance of the renewable energy machines installed on the leasehold. Maximizing energy output and revenues by selecting the optimal turbine for the expected wind environment makes more sense than does minimizing installed capacity or capacity factors in an effort to reduce the operating fee. In reality the operating fee is just a fraction of the expected energy revenues. Capacity factors and generator capacity are trade-offs made during project design. Under a given wind regime, for example, a 3.0-MW turbine may operate at a higher capacity factor yet have a lower total energy output than a 3.6-MW turbine that operates at a lower capacity factor. The 3.6-MW machine may have a large energy output at high wind speeds, while the 3.0-MW unit may exhibit more consistent performance, operating for more hours per year during the most prevalent, lower wind speed conditions.

The prior year's average wholesale power price in the state in which the project's transmission cables make landfall also is part of the operating fee formula. MMS will refer to wholesale sales prices as published by the Department of Energy's Energy Information Administration or other publicly-available wholesale power price indices.

Noncompetitive leasing. MMS may lease sites on a noncompetitive basis if the response to a call indicates that interest in any site is limited to just one potential bidder. Alternatively, comments responding to notice of an unsolicited lease request (designated a *request for interest*) will be reviewed by MMS to determine whether other persons may be interested in bidding on the site. An unsolicited request must include essentially the same information provided in response to a call, including energy and resource data, to permit MMS to evaluate the suitability of the area for leasing and its potential value.[28]

Upon determining that there is no competitive interest, MMS will publish its finding in the *Federal Register*, beginning a 60-day period during which the prospective commercial lessee must submit a *site assessment plan (SAP)*. This is just the first of several plans required by MMS that are described further below. Once MMS reviews and approves the SAP, it may offer the applicant a noncompetitive lease.

The fees paid in connection with a noncompetitive lease largely mirror those for a competitive lease. The noncompetitive lease differs, however, in that an acquisition fee, typically \$0.25/acre, must be paid when the lease request is submitted to MMS. A prospective lessee that subsequently refuses to acquire the lease will forfeit this fee. Competitive leases differ in that a bidder may walk away from a high-priced lease without losing its deposit, provided that it is not the winning bidder. All competitive lease sale bidders are required to submit a bid deposit of 20% of the bid amount, or another amount specified in the final sale notice. The winning bidder's deposit is applied to the bonus payment, and the winning bidder must pay the balance of the bonus amount shortly after being notified by MMS that it has been awarded the lease. Losing bidders receive their deposits back. A winning bidder forfeits its deposit if it declines to accept the lease or otherwise fails to comply with the terms of the final sale notice or MMS regulations.

With the exception of these initial payments, competitive and noncompetitive lease holders would pay the same rents and fees (table 5–3). During the site assessment term of a commercial lease, the annual rent of $3.00/acre must be paid in advance. Acreage used by easements acquired under a lease accrues annual rent charges of the greater of $5.00/acre or $450/year, applicable for the entire lease term, not just the site assessment term. Once MMS approves a plan authorizing construction under a commercial lease and commercial generation of electricity begins, rent payments (other than easement rent) stop, and operating fees begin. The operating fees are calculated based on the formula set out previously and may feature a fixed or variable fee rate.

Table 5–3. Commercial lease cash flows

Payment type	Amount	Timing of payment
Deposit (competitive lease)	20% of bid amount, or as specified	Deposit is paid when the bid is submitted.
Acquisition fee (noncompetitive lease)	$0.25 per acre	Acquisition fee is due when submitting a request for a noncompetitive lease.
Balance of bonus	As specified in bid award	Due within 10 days of receipt of MMS bid acceptance notice.
Financial assurance (e.g. bond)	$100,000 minimum	Due within 10 days of receipt of MMS bid acceptance notice. Additional financial insurance must be provided before the approval of each of the Site Assessment Plan and the Construction and Operations Plan.
Rent	$3 per acre for the main lease area (greater of $5 per acre per year or $450 per year for easement acreage)	First six months rent due within 10 days of receipt of MMS bid acceptance notice. Subsequent rent payments due at the beginning of each subsequent one year period.
Cost recovery fees	Case specific, depending on cost of preparing environmental impact statements, etc.	Periodic billing based on actual costs incurred during document processing.
Operating fee	Unless the fee rate is a bid variable, generally an operating fee rate of 2% would apply during the operations term.	Begins when COP is approved. Payments due per the schedule specified in the lease or Final Sale Notice.

Submitting project plans to MMS. MMS conducts a preliminary environmental review with respect to the tracts it proposes to lease prior to commencing the competitive commercial leasing process.

That preliminary review is avoided, however, when MMS proposes to issue a noncompetitive lease. Instead, the prospective lessee under a noncompetitive lease has 60 days from the time MMS issues its determination that there is no competitive interest to submit its SAP. This document describes the lessee's plans to evaluate the site and renewable energy resource for full-scale commercial development, including specific site characterization studies and plans for observation and measurement facilities. The SAP, along with the information the prospective lessee submitted earlier in the process, provide the basis for MMS to evaluate the environmental impact of the noncompetitive lease sale for the particular purpose intended by the lessee.

The competitive leasing process follows a longer timeline, as it would be unfair to ask all bidders to submit a SAP before they learn whether they have been awarded a lease. Competitive leaseholders have a six-month preliminary term at the inception of the lease to prepare and submit an SAP. Limited leases follow a similar pattern with a plan called the *general activities plan (GAP)*. Lessees under a competitively issued limited lease must submit a GAP within an initial six-month preliminary term, while those seeking a noncompetitive limited lease must submit the GAP prior to the MMS decision to issue the lease (fig. 5–2).

The SAP (for commercial leases) and GAP (for limited leases) are similar. In both documents, lease applicants must describe in detail their planned resource assessment and site characterization activities. When facilities such as meteorological towers or bird observation posts must be constructed, those also would be described. A developer need not wait for an approved SAP, however, to conduct geophysical and geological surveys, hazard surveys, archaeological surveys, and baseline collection studies such as biological surveys that may provide data useful for preparing a SAP. These kinds of surveys may be permitted in many cases under the Nationwide Permit Program administered by the U.S. Army Corps of Engineers (ACOE). MMS encourages developers that want to conduct preliminary surveys under the Nationwide Permit Program to contact MMS and ACOE before commencing any surveys to ensure that the activity is covered under the Nationwide Permit.

MMS requires each plan to demonstrate the applicant's intention of using the best and safest technology, management practices, and properly trained personnel available. The plans should demonstrate conformity with all applicable laws and explain why the planned activities would not unreasonably interfere with other uses of the OCS, such as fishing or military training. The plans also should show that the activities would not cause undue harm or damage to natural or cultural resources.[29]

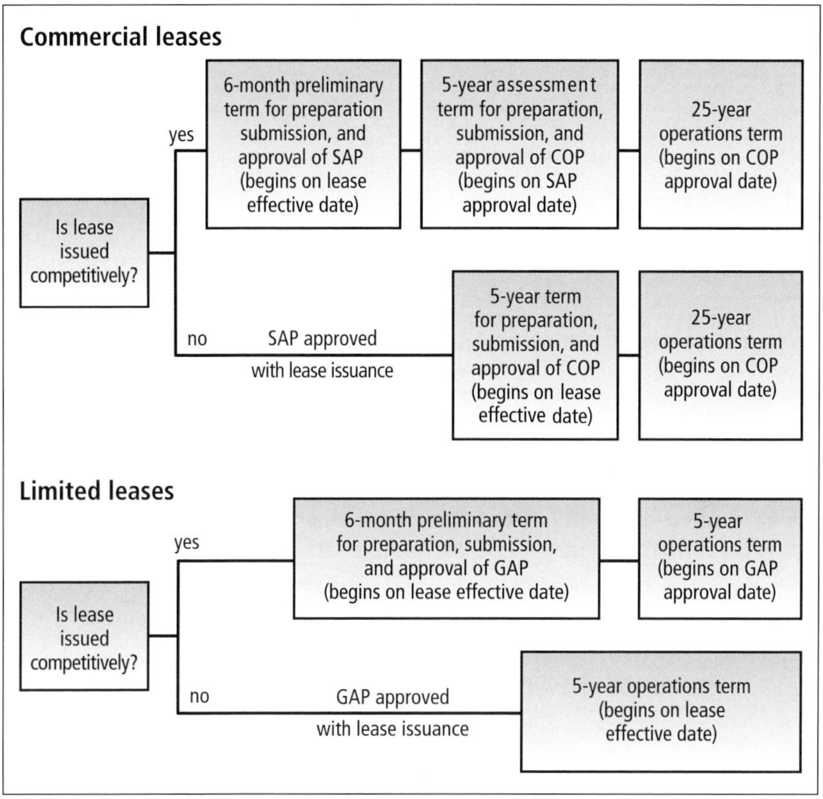

Fig. 5–2. Lease and plan timing

MMS may impose conditions on activities conducted under a SAP or GAP. In consultation with wildlife agencies, for example, MMS may incorporate as conditions certain agency recommendations for protecting essential fish habitat and marine mammals or minimizing the incidental taking of endangered species. Upon MMS approval of the plan, the lessee can begin all site evaluation activities with the exception of the construction of facilities that MMS deems "complex or significant."[30] Before construction of complex or significant structures may begin, the lessee must comply with the facility design, fabrication, and installation rules in subpart G of the MMS regulations, and submit a description of its Safety Management System.

A carefully drafted SAP or GAP anticipates problems that may arise during the site assessment term and provides the developer with the flexibility to respond. Should the need for activities not described in the plan arise, the lessee is required to inform MMS in writing of the change and obtain authorization before proceeding. Revisions that result in a "significant change in impacts previously identified and evaluated; require any additional Federal authorizations; or involve activities not previously identified and evaluated" may require the filing of an amended plan and additional NEPA and CZMA reviews.[31] A well-drawn plan may avoid such delays.

A commercial leaseholder must submit and receive authorization of a third plan before project build out may begin. The COP, mentioned briefly in the previous text, describes the facilities required for the full commercial project, both offshore and onshore, and the construction, operations, and decommissioning processes. The COP should include design drawings, fabrication, and installation information, and details of the related transmission infrastructure.

The COP also should describe any easements required for the project, which MMS would typically approve in connection with its authorization of the COP and include in the lease by means of an amendment. Generally, easements are limited to a width of 200 ft (61 m). Identifying alternate easement routes is a good idea if there is any doubt as to the feasibility of the preferred route. Remember that significant changes to the project after the COP has been approved may require additional environmental and legal review. Having an alternate easement route described in the COP in the first instance can prevent the delay and expense of an additional environmental review.

The SAP expires once MMS approves the COP. It might be desirable, however, to continue site assessment activities on the lease during the operations term—for instance, under a strategy of developing a large leasehold in phases. In such cases, the COP should describe the continuing site assessment work. There could be several reasons to choose to build a project in phases. It may be cost-effective to build a large capacity transmission tie line to shore, but due to market conditions, it may be advisable to install new generating capacity more gradually over time. The lessee also may seek to pace its development of the lease with the evolving capacity of the supply chain for offshore turbines and other necessary equipment. A COP should explain the lessee's phased development plan, if any, and describe those portions of the lease to be developed currently and those that will be reserved for the future.

In the normal course of events, a lessee would proceed first to have its SAP approved and then undertake site assessment research and planning. The SAP work informs the conceptual development of the project as described in the COP, which would be filed and approved later. The lessee could attempt to compress the project review timeline by submitting the SAP and COP together, which would eliminate one level of environmental compliance review.

> *If the applicant decides to submit the COP and SAP simultaneously, then sufficient data and information must be submitted with the COP for MMS to conduct needed technical, NEPA, and other required reviews. If new information becomes available after the applicant completes the site assessment activities, then the COP will require revision. Furthermore, MMS may need to conduct additional reviews, including NEPA, on any new information.*[32]

Clearly, the success of a combined SAP/COP approach will depend on the developer's confidence in the quality of its site information. Realistically, the first project built in a particular area may not be the ideal candidate for a combined SAP/COP filing. However, it is possible that subsequent projects in the same general area might use such a strategy to accelerate the approval timeline.

MMS considers each SAP, GAP, and COP to be a plan subject to the requirements of the CZMA. At the time a lessee submits a plan, it must also submit a consistency certification affirming that "each activity which is described in detail in such plan complies with the enforceable policies of such state's approved [coastal zone] management program and will be carried out in a manner consistent with such program."[33] The state may concur, object, or remain silent with respect to the lessee's consistency certification. If the state has not responded within three months of the receipt of the certification and supporting information, it is presumed to have concurred. If the state objects, the lessee may submit a new or amended plan, and the three-month period would begin again. The secretary of the interior may override a state's objection to a plan if the secretary finds that the proposed activities are consistent with the objectives of the CZMA or are otherwise necessary in the interest of national security. A state's ability to delay a project through CZMA review illustrates the importance of developer efforts to lay a strong foundation for the project through full consultation with state and local officials, a process described in greater detail in chapters 8 and 9.

One of the criticisms of the MMS proposed rules was that they required as many as three separate NEPA and CZMA reviews: at the time of the lease sale, SAP review, and COP review. These assessments are time consuming and costly and, in response to comments from developers, MMS has reduced the number of CZMA and NEPA reviews by combining the lease sale and site assessment reviews into one review. A non-competitive limited lease, which only has a GAP, would receive just one round of CZMA and NEPA reviews (table 5–4).

Table 5–4. Frequency of NEPA/CZMA reviews based on instrument held

Instrument Held	MMS Process	NEPA Documentation and CZMA Review
Competitive Commercial Lease	Conduct lease sale and issue decision on plans	1. Lease Sale and SAP 2. COP
Noncompetitive Commercial Lease	Negotiate and issue lease	1. Lease Issuance and SAP 2. COP
Competitive Limited Lease	Conduct lease sale and issue decision on plan	1. Lease Sale 2. GAP
Noncompetitive Limited Lease	Negotiate and issue lease	1. Lease Issuance and GAP
Competitive ROW, RUE Grant	Conduct ROW, RUE sale and issue decision on plan	1. ROW, RUE Sale 2. GAP
Noncompetitive ROW, RUE Grant	Negotiate and issue ROW, RUE grant	1. ROW, RUE issuance and GAP

Source: MMS Final Rules. 194, Table 2.

Designing, Fabricating, and Installing a Facility

A COP includes general structural, project design, and operational information. It is not surprising, given the experience that MMS has with the hazards of offshore oil and gas production, that accident prevention is a significant component of a COP. MMS requires inclusion of plans for safety management, inspections, maintenance, and monitoring systems. The COP also should demonstrate that activities would not harm or damage natural resources, life, property, or objects of historical or archaeological significance. To ensure that the lessee builds the facilities according to the design and specifications represented in the COP, the lessee must either nominate a *certified verification agent (CVA)* for approval by MMS or request an exemption, in which case the project engineer would perform the same duties and responsibilities as the CVA. The CVA is an independent pair of eyes that observes critical stages of construction, fabrication, and testing to confirm that the developer follows both industry standards and the facility design as represented to MMS.

After the COP is approved, but before installation begins, the lessee must file with MMS two separate reports detailing the facility's design, construction, and installation plans. The *facility design report* provides MMS with details regarding the project's facilities, including cable and transmission infrastructure—plans that previously had only been outlined in more general terms in the SAP and COP. Among other items, the report should include the following:

1. A location plat and detailed drawings.
2. A summary of the engineering, geotechnical, and design data used in the design or analysis of the facility.
3. Design calculations, including a description of the loads imposed on the facility.
4. A statement certifying that "the design of this structure has been certified by a MMS approved Certified Verification Agent to be in accordance with accepted engineering practices and the approved SAP, GAP, or COP as appropriate. The certified design and as-built plans and specifications will be on file at (given location)."[34]

MMS regulations also provide that "the CVA must certify in the Facility Design Report to MMS that the facility is designed to withstand the environmental and functional load conditions appropriate for the intended service life at the proposed location."[35] This is a more specific statement than the certification noted in item 4 above, which merely provides that the facility was designed according to accepted engineering practices. MMS will review the facility design report to confirm that the project has been designed to ensure safety and minimize the adverse effects to the coastal and marine environment to the extent practicable.

The second document is the *fabrication and installation report*, which describes how the lessee will proceed from the design plan to a completed facility. This report should include the following:

1. A schedule
2. Information about the industry standards that will be used to ensure that fabrication quality conforms to the design
3. A detailed description of the installation process, including the equipment and materials that will be used, anchoring and mooring patterns, and the design and installation of project cabling
4. Information about construction-related waste and pollution, such as water discharge and waste disposal
5. Evidence of the appropriate permits related to regulated discharges, such as those that may be required for air emissions and dredged soil disposal.[36]

Due to the relative infancy of the marine renewable energy industry, MMS has not required that projects comply with specific industry design and construction standards. This decision provides leeway for new designs to emerge, although those designs must still be sufficiently sound to earn certification from a CVA.

The CVA serves as an honest intermediary between the lessee and MMS. MMS may periodically visit the site and conduct inspections, but the agency places significant reliance on an independent CVA to thoroughly review the project work. Project lenders would generally expect to receive an independent engineering firm's certification of construction quality, so the CVA's opinions will often serve a dual purpose. The lessee nominates a CVA for MMS approval by submitting a qualification statement, which describes the nominee's previous experience in third-party verification and technical capabilities.[37] A key qualifying point is the CVA's lack of a conflict of interest. It would therefore be unacceptable for the CVA to have a financial interest in the developer or the project itself.

The CVA should have considerable experience in the engineering, repair, and maintenance of offshore structures. MMS expects the CVA to independently assess the following factors:

1. Planning criteria
2. Operational requirements
3. Environmental loading data
4. Load determinations
5. Stress analyses
6. Material designations
7. Soil and foundation conditions
8. Safety factors
9. Other pertinent parameters of the proposed design[38]

If floating structures are installed, the CVA also must ensure that they meet U.S. Coast Guard standards for structural integrity and stability and that the mooring design is appropriate.

MMS regulations require the CVA to "verify, witness, survey, or check" certain key project milestones. Some of the CVA's observations must be performed in fabrication facilities to verify the quality of materials and welds, and others, such as inspections of foundations and component installations, will be at the construction site. Records demonstrating the project's conformity with the design documents must be produced and maintained.[39]

Before commercial operations may commence, the CVA must submit a report certifying that the facilities were fabricated and installed in accordance with accepted engineering practices, the approved SAP and COP, and the fabrication and installation report. (By implication, the CVA is also certifying that the project conforms to the design criteria set out in the facility design report.) MMS has 30 days to object to the CVA's report, after which acceptance is presumed, and the project can begin operating. Since a project cannot earn revenues without the CVA certification, developers must coordinate closely with the project CVA to ensure that costly errors and misunderstandings are avoided.

Environmental and Safety Rules Applicable to MMS Leases

In the interest of safe and environmentally responsible development, MMS has adopted best management practices drawn from Europe's marine renewable energy experience and the agency's own offshore oil and gas background (table 5–5). The best management practices may be incorporated as a binding stipulation in any lease agreement between MMS and a project developer. For the most part, the best management practices would apply even if not specified in the lease. This is because MMS regulations require all activities within the leased area to be conducted with technologies, precautions, and techniques that will minimize the likelihood of harm to human life and the marine environment.

Table 5–5. Summary of MMS best management practices

Topic	Recommended or required management practices
Preconstruction planning	Minimize area disturbed. Consolidate necessary infrastructure between projects where possible. Implement a monitoring program to identify and mitigate adverse impacts.
Seafloor habitats	Conduct seafloor surveys to: - Avoid/minimize impacts associated with seafloor instability and other hazards, and - Identify sensitive seafloor habitats and topographic features. Avoid locating facilities near, or anchoring on, sensitive habitats such as coral reefs and hard-bottom areas. Minimize seafloor disturbance and sediment dispersion. Use cable shielding to control electromagnetic fields. Control scouring around foundations to preserve structural integrity. Avoid use of explosives. Evaluate marine mammal use of project area and minimize the potential for mortality or disturbance. - Vessels should travel at reduced speeds (and maintain a reasonable distance) when cetaceans are observed. - Vessels should follow NMFS Regional Viewing Guidelines and operators should have training in the guidelines. Minimize disruption to marine life from sound emissions (e.g., pile driving) during construction. - Post a qualified observer approved by MMS and NMFS during construction activities.
Fish resources and essential fish habitat	Use existing data or conduct surveys to identify important, sensitive and unique marine habitats in the vicinity of the project and design to avoid, minimize or otherwise mitigate adverse impacts to these habitats. Minimize construction activities in areas with anadromous fish during migration periods. Minimize seafloor disturbance.
Sea turtles	Vessels should follow NMFS Regional Viewing Guidelines and operators should have training in the guidelines. Minimize disruption due to sound emissions during construction. Locate cable landfalls and onshore facilities to avoid impacts to nesting beaches.
Avian resources	Evaluate avian use of the project area and design project to minimize bird strikes and habitat loss. Reduce perching opportunities. Locate cable landfalls and onshore facilities to avoid impacts to nesting beaches. Comply with Federal Aviation Administration (FAA) and United States Coast Guard (USCG) requirements for lighting while using lighting technology that minimizes impacts to avian species.
Acoustic environment	Use the lowest sound levels necessary to obtain needed site information. Minimize disruption due to sound emissions during construction. To the extent practical, use state-of-the-art, low noise turbines to minimize operational noise.

Source: U.S. Department of the Interior, Minerals Management Service. 2007. Establishment of an OCS Alternative Energy and Alternate Use Program, Record of Decision: Attachment B, Decision Table for Best Management Practices.

Topic	Recommended or required management practices
Fisheries	Cooperate with commercial/recreational fishing interests to minimize potential conflicts associated with project construction and operation.
	Coordinate with fishing interests and port authorities to prevent unreasonable fishing gear conflicts (e.g., notify registered fishermen of the location and time frame of project construction activities).
	Use operating practices and procedures to reduce the likelihood of vessel accidents and spills.
	Mark offshore structures with USCG-approved measures to ensure safe vessel operation.
	Bury cables, where appropriate, to avoid conflict with fishing vessels and gear. Inspect buried cables periodically for adequate coverage.
Coastal habitats	Avoid hard-bottom habitats, including seagrass and kelp beds where practicable, and restore if damaged.
	Implement turbidity reduction measures to minimize effects on hard-bottom habitats.
	Minimize effects to seagrass and kelp beds by restricting vessel traffic to established routes.
	Maintain buffers around wetlands and use best practices for erosion and sediment control (including maintaining natural drainage patterns).
Electromagnetic fields	Use shielded submarine cables and bury cables in the seafloor where practicable.
Transportation and vessel traffic	Site projects to avoid unreasonable interference with major ports and USCG-designated traffic separation schemes.
	Follow FAA guidelines for siting and lighting facilities.
	Follow USCG requirements for lighting and signage on structures to aid navigation.
	Conduct necessary studies of potential interference of wind turbines with defense, air traffic control, and weather radar systems, and identify possible solutions.
Visual resources	Address key design elements, including visual uniformity, use of tubular towers, and proportion and color of turbines.
	Use appropriate viewshed mapping and simulation techniques to determine with reasonable accuracy the visibility of the proposed project. Simulations should illustrate sensitive and scenic viewpoints.
	Comply with FAA and USCG lighting requirements while minimizing impacts, for example, through the use of directional aviation lights.
	Seek public input on design of visual elements of project.
Cultural resources	Conduct magnetometer tows using 30m line spacing where there is a high potential for shipwrecks.
Operations	Prepare waste management plans, hazardous material plans, and oil spill prevention plans, as appropriate, for the facility.

Protected species

Protected species receive special attention, and MMS requires each lessee to describe the measures it will implement to avoid or minimize adverse effects on endangered and threatened species and marine mammals. MMS also requires measures to avoid damage to designated critical habitat for endangered and threatened species. Lessees must commit to MMS that they will monitor for incidental takes, including harassment, and abide by the terms and conditions of incidental take authorizations under the Endangered Species Act and the Marine Mammal Protection Act.[40] These laws are discussed in detail in chapter 7.

An early and thorough survey to identify the likely presence of endangered or threatened species, or designated critical habitat, in the project area can minimize future frustration, delay, and expense. Early identification helps the lessee to work with federal wildlife agencies to secure an incidental take authorization on the best terms. Such an authorization will typically impose conditions on site activities when the species is present or during certain times of the year, for example, when a migrating species uses certain affected critical habitat in the project area.

A lessee faces substantially greater likelihood of encountering an unexpected delay if it does not obtain an incidental take authorization prior to beginning construction and operation. MMS regulations provide that "if there is reason to believe" that a threatened or endangered species may be present during the conduct of MMS-approved activities, or that designated critical habitat may be adversely affected, the lessee must notify MMS. MMS will then consult with the appropriate wildlife agency and indicate "whether, and under what conditions, you may proceed."[41] MMS could suspend all project activities pending the results of a biological survey and the issuance of a takings authorization.

MMS also would consult with the National Marine Fisheries Services (NMFS) if activity on the lease reveals the presence of essential fish habitat or habitat areas of particular concern. The NMFS consultation may result in MMS incorporating into the lease certain conservation terms and conditions on fish habitat which the lessee must follow. For example, additional biological surveys may be required to define habitat boundaries and avoidance distances.[42]

Archaeological resources

Newly discovered archaeological resources may raise a similar issue. An archaeological survey would typically be conducted as part of a SAP, COP, or GAP. If an archaeological resource may be present, MMS will require that project activities maintain a minimum distance from the archaeological site. The lessee may avoid setback requirements by showing that an archaeological resource is unlikely to be present or demonstrating that, despite the presence of archaeological resources, seabed activities will be conducted in a manner that minimizes adverse effects on the resource. When an archaeological resource is discovered during project construction, however, the lessee must halt all seabed disturbances and avoid damaging the site. The lessee should report the finding to MMS and refrain from publicly disclosing the location of the archaeological site. MMS may require the lessee to investigate the site further to determine its historical significance and pay those reasonable costs incurred by MMS for its preservation.[43]

Safety management systems

Even under the best conditions, heavy objects, large equipment, working at heights, and other hazards add danger to construction projects. Open water construction adds weather and the pressure of meeting project deadlines given limited operational windows to the list of hazards. Workers may be injured in transfers to and from offshore structures, fall into frigid water and/or from heights, and may be stranded offshore in bad weather. Fortunately, the operational phase is much less risky than construction. During normal operation, offshore renewable energy facilities pose little risk to human life, since persons would not usually be on or about the facilities except during maintenance and repairs. The risk of catastrophic events on an offshore renewable energy facility also is low in comparison to oil and gas platforms, where flammable and explosive products are handled.

To address the risks of construction and operation, the COP or GAP must describe the lessee's safety management system. A SAP and GAP that includes the installation of complex or significant facilities also must have a safety management system. The system should include safety procedures, a protocol for testing the safety system, and a program for training facility personnel. Key safety equipment, such as remote monitoring and control equipment, which can shut down the machines during an emergency, and fire suppression equipment, if needed, should also be described.[44]

Safety management practices from Europe's offshore wind projects are instructive. Project planners of the U.K.'s North Hoyle wind farm, for example, performed detailed risk assessments and then sought ways to design out the risk. If a design solution was lacking, they explored ways to minimize the risk exposure and trained personnel in safe procedures. Depending on need, staff received from one to three days' training in sea survival, tower climbing, transfers between vessels and towers, and emergency procedures. During construction, only seamen and personnel who had received safety training certificates were permitted offshore. Extra safety gear, including full-body immersion suits and fall-arresting harnesses, supplemented the standard, more mundane equipment, such as safety goggles and ear protection.

The North Hoyle project also employed weather-driven rules to ensure that safety was not sacrificed to meet project schedules. Work in and over the water stopped when the weather turned rough. Transfers were restricted to conditions under 1.2 m significant wave height, diving was confined to currents below 1 m/s, and lifting was curtailed in high winds (8 m/s for blades and 12 m/s for towers and nacelles). Vessels also were subject to movement restrictions depending on wave heights.[45] The British Wind Energy Association's health and safety guidelines include an appendix of offshore wind facility safety procedures that may assist lessees in developing a safety management system.[46]

MMS regulations also address inspections conducted by MMS and the lessee. MMS inspections may be announced or unannounced and may cover all project facilities and leased areas and records of design, construction, operations, maintenance, repairs or investigations related to the project or leasehold area. Each lessee must also prepare an annual self-inspection plan covering all facilities above and below the water line. The plan should include at least annual inspections for corrosion and other factors that may affect the structural integrity of project facilities. Lessees must follow certain recommended practices of the American Petroleum Institute relating to offshore structures in conducting their facility assessments.[47] Annual reports to MMS detailing the facilities inspected and the method used are required.

In addition, MMS requires prompt reports of incidents involving fatalities, evacuation, fires and explosions, and collisions and other property damage above a $25,000 threshold. Incidents involving structural damage, crane accidents, damage to safety equipment, and other incidents involving environmental damage must also be reported promptly.[48] Lessees must have a spill response plan for accidents that may result in oil spills. Such a plan should state how the operator will mitigate or prevent a discharge of oil from the facility and respond in the event of a release. The plan should also describe the lessee's training program (including periodic drills) and equipment testing measures designed to enhance the ability to immediately respond to a spill.[49]

Decommissioning

Decommissioning offshore renewable energy facilities is the obligation of the lessee. Where a renewable energy facility makes alternate use of an existing structure on the OCS, the decommissioning liability is shared between the renewable energy lessee and the existing or previous "traditional" use lessee. The renewable energy lessee has responsibility for decommissioning the renewable energy–related structures placed on the preexisting platform, while the original leaseholder (e.g., an oil company) remains responsible for decommissioning the balance of the structure.[50]

The lessee must generally remove facilities within two years after the lease ends. Unless other arrangements for the facilities have been made, all facilities, including cables and pipelines, generally must be removed to a depth of 15 ft (4.57 m) below the mud line. MMS may authorize the facilities to remain in place to support other activities permitted on the OCS, or the agency may permit the conversion of a structure into an artificial reef.

Removal methods, schedule, transportation, and disposal details should be described in a decommissioning application filed with MMS. The application should include the results of recent biological surveys conducted in the project area and a description of the resources that could be affected by the decommissioning activities. Explosive foundation removal is a particularly sensitive decommissioning method due to the potentially lethal effects of explosive shock waves on fish and marine mammals. MMS will restrict the type and amount of explosives used and require that a report after decommissioning certify that only the authorized explosives and amounts were used.

Prior to approving decommissioning activities, MMS will compare the general conceptual description of decommissioning activities included in the SAP, COP, or GAP to the decommissioning application's detailed description to determine whether additional environmental review is required. If federal permits are required that were not previously identified in the plans, or the proposed decommissioning plan would result in a significant change in the impacts previously identified in the SAP, COP, or GAP, amendments to the plans that are subject to NEPA and other environmental reviews may be required. Lessees that do not comply with an MMS-approved decommissioning plan risk the forfeiture of the project bond or other financial assurance, as well as civil and criminal penalties under section 24 of the OCS Lands Act. A lessee that neglects its decommissioning obligations will continue to be liable for the costs of removing and disposing of the facilities. In addition, accidents and damage that result from the continued presence of the facilities on the OCS may create additional liability for the lessee.[51]

The FERC Hydrokinetic Licensing Process

Legal basis for regulating hydrokinetic energy projects

It is against the law to construct, operate, or maintain a hydropower facility in any navigable water of the United States except under the terms of a license granted by FERC under the Federal Power Act (FPA). Some have argued that FERC should not regulate marine energy technology because it is fundamentally different from the river-sited impoundments that FERC has historically regulated. As the law is written, FERC may "issue licenses…for the purpose of constructing, operating, and maintaining dams, water conduits, reservoirs, power houses, transmission lines, or other project works necessary or convenient…for the development, transmission, and utilization of power across, along, from, or in any of the streams or other bodies of water over which Congress has jurisdiction…, [subject to Army Corps of Engineers approval over] structures affecting navigation.[52] FERC interprets this authority very broadly. A wave energy converter is a *power house* in FERC's view. The agency has held that an ocean wave energy project located 3.17 nm offshore is in U.S. navigable waters and subject to FERC jurisdiction.[53]

As previously mentioned in the section on MMS authority, the Energy Policy Act of 2005 (EPAct) granted the secretary of the Department of the Interior authority to lease areas of the OCS for renewable energy production and corresponding regulatory authority. EPAct directs the secretary to conduct its offshore program in coordination with other federal agencies and state and local governments. MMS, acting on behalf of the secretary, and with the consent of the other agencies, functions as the lead agency with respect to renewable energy on the OCS. MMS must maintain a cooperative relationship with other federal agencies, because EPAct expressly preserves the existing authority of the agencies with concurrent jurisdiction under other federal law. EPAct's grant of authority concludes, "Nothing in this subsection displaces, supersedes, limits, or modifies the jurisdiction, responsibility, or authority of any Federal or State agency under any other Federal law."[54]

Since EPAct does not displace FERC's jurisdiction under the FPA, projects on the OCS that convert the energy of freely moving water into electricity (i.e., hydrokinetic energy projects) are subject to both FERC and MMS regulation. Those projects located in state waters, however, would not involve MMS, thus leaving FERC to take the undisputed role of lead agency. MMS and FERC staffs have tried to rationalize the overlapping jurisdiction. "I think we're certainly all committed in the federal family to working together," said FERC's John Katz.[55] On April

9, 2009, the meetings between agency staff yielded a compromise in the form of a MOU, the spirit of which is reflected in various provisions of the MMS Final Rules.

The MOU establishes a division of authority between MMS and FERC in the area of hydrokinetic facilities. Offshore wind energy projects, in contrast, are the exclusive bailiwick of MMS. Broadly stated, MMS has exclusive jurisdiction to issue leases, easements, and rights-of-way regarding OCS lands for hydrokinetic projects. FERC has exclusive jurisdiction to issue licenses and exemptions for hydrokinetic projects located on the OCS. MMS conveys the property right that hydrokinetic developers will need to build a project, and FERC gives authorization to build the project through its licensing process.

In practice, the division of responsibilities between the two agencies is more fluid and complex and a developer seeking to build a hydrokinetic project on the OCS will find itself substantially regulated by both agencies. The NEPA analysis, for example, will be done by both MMS and FERC, with each agency having the option to join as a cooperating agency in connection with the preparation of the other agency's environmental document.

In furtherance of the cooperative relationship, FERC will not issue preliminary permits (explained further below) for hydrokinetic projects located on the OCS. MMS will not conduct "off-the-record" communications relevant to the merits of FERC's licensing or exemption proceedings, if MMS has elected to be a cooperating agency for purposes of NEPA review. If MMS attaches terms and conditions to leases, easements, and rights-of-way issued for an OCS hydrokinetic project, FERC will include a requirement in any license or exemption that it issues for the project that the developer must comply with the MMS terms and conditions. Moreover, a FERC license or exemption may not be issued before MMS issues the related lease or grant. In addition, the FERC licensing process substitutes for the COP that MMS would otherwise require in connection with authorizing the construction of permanent commercial facilities on the OCS.[56] Lastly, both agencies will coordinate inspections of projects and jointly develop policies and regulations.

An interesting quirk arising from the jurisdictional split stems from the prohibition against MMS issuing leases on the OCS within the boundaries of a national park, national monument, national marine sanctuary, or national wildlife refuge. Section 4(e) of the FPA, however, authorizes FERC to issue licenses for projects that are located on reservations of the United States, which, under FPA § 3(2), includes U.S.-government-owned lands reserved for public purposes but does not include national monuments or national parks. FERC could, therefore, license hydrokinetic energy projects located in a national marine sanctuary or a national wildlife refuge. This circumstance was addressed by Finavera Renewables Ocean Energy Ltd., which sought authorization for a wave energy conversion project in part of the Olympic Coast National Marine Sanctuary, near Flattery Rocks National Wildlife Refuge.[57] It is unclear how the MOU would be applied to hydrokinetic projects that may be located in a national marine sanctuary or national wildlife refuge, or whether the MOU would apply at all.

The best counsel for developers seeking to build hydrokinetic energy projects on the OCS is to meet early with staff from both FERC and MMS to review the applicability of the rules, and develop acceptable protocols for coordination, consultation, and documentation, as well as project review timelines that minimize duplication and delay. Note that under FERC regulations, a license applicant is permitted to petition FERC for alternative license application processing procedures provided that the procedures meet standards intended to ensure effective consultation with all stakeholders.[58]

The remainder of this section discusses FERC regulation of hydrokinetic energy projects under the simplifying assumption that the project is located exclusively in state waters, where coordination with MMS would not be required.

Preliminary permits

Developers seeking to build hydrokinetic energy projects will typically want to identify a site suitable for their technology, acquire an interest in that site, conduct testing and surveys at the site, and then procure a license permitting project construction and operation. FERC provides a staged process to reach this goal (fig. 5–3). The developer's first step should be to obtain a preliminary permit. (As noted above, a preliminary permit is not available for hydrokinetic energy projects located on the OCS.) The permit does nothing more than preserve the permit holder's priority of right to apply for a license for the project being studied. A preliminary permit is not needed to conduct initial studies that will form the factual basis of a license application. However, having a permit protects the developer's investment in site characterization studies against competing claims by other parties that may seek to develop a hydroelectric project on the site.[59] In this respect, a FERC preliminary permit is more advantageous than an MMS limited lease, which does not afford the lessee any preference in the process of procuring the commercial lease necessary to build a project. MMS limited leases do, however, allow a lessee to place data measurement and other research facilities on site. In contrast, FERC's preliminary permit does not authorize the installation of measurement facilities or entry onto land for the purpose of conducting studies. Even after receiving a preliminary permit, the permittee would still need to obtain landowner consent and any federal, state, and local permits that may be required before proceeding with testing and surveys. Because it actually permits so little, FERC's preliminary permit is not a major federal action requiring NEPA analysis and may therefore be issued with relative speed.[60]

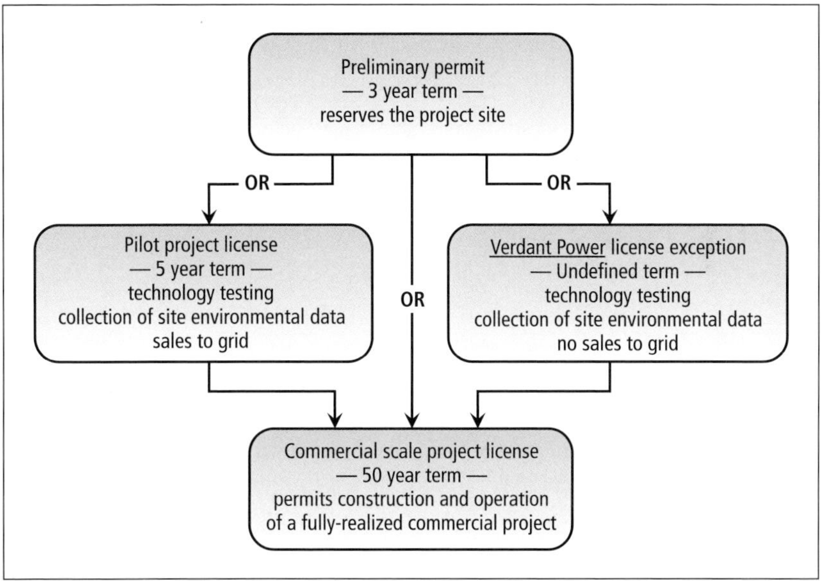

Fig. 5–3. Alternate paths to a FERC hydrokinetic energy project license

The place-holding function of the preliminary permit has been abused by some developers to claim priority rights to more sites and larger areas than they can reasonably hope to develop. FERC has attempted to control *site banking* by subjecting preliminary permit applications to strict scrutiny. FERC staff will review permit applications with a view toward limiting site boundaries. In addition, a standard condition in preliminary permit orders requires the submission of semiannual progress reports. A progress report must describe how the permit holder is meeting the prefiling requirements under FERC's hydroelectric licensing regulations, which require consultation with applicable resource agencies and Indian tribes, and planning for resource studies. If the permit holder needs permission to access land, it should describe in its report its progress towards obtaining such authorization. Under FERC's strict scrutiny policy, permit holders also must submit, within 45 days after the permit is issued, a schedule of activities leading to the filing of a development application. If the periodic reports fail to show sufficient progress, FERC may cancel the permit.

Preliminary permits may have a term of up to three years, and only the named permittee may claim the priority to file a license application afforded by the permit. Again, this differs from MMS leases, which generally are transferable. For this and other reasons, it is advisable to form a special-purpose project company to apply for the preliminary permit and conduct related project development work. Should the need or desire arise to transfer rights in the project site to another entity, it may be possible to transfer the entire special purpose project company (and its priority status) to a new owner, without losing the priority to file for a license.

The project area described in the preliminary permit application should not be overbroad such that it attracts staff objection but also not so narrow that it hampers the developer's ability to adjust the project's design. A developer may credibly claim a large project footprint in some cases. Resource studies and stakeholder consultation occurring after permit issuance will invariably lead to adjustments in a project's location. Verdant Power's experience is instructive:

> *We filed a preliminary permit application with a specific footprint. That was based on the knowledge we have of the velocity of the East River, the flows in the East River, and what we would need in order to generate ten megawatts, which we have described with our partner, [NYSERDA], who has provided us with funding on the project. As we got into the project and began the consultations, we had to move. The Coast Guard didn't want us in the middle of the river. We had to move closer to the shore. We had to deal with recreational impacts. We're now on the west side primarily of the East River channel. The end result is our license application is going to have a significantly expanded footprint over the preliminary permit area.*[61]

A seabed survey of the initial project site may reveal an ecologically sensitive hard bottom habitat. A prudent developer faced with such a fact might choose to move the project to a similarly energetic site over a nearby sandy bottom. A developer also may want to assess the performance of two or more different technologies in varying depths and wave conditions. If the project boundaries are too narrowly drawn, a new or amended permit may be required, resulting in a possible loss of priority. Permit applicants that claim large project areas, however, should be prepared to justify the size of their request.

Applicants also should consult with state resource agencies in advance of seeking a preliminary permit. Oregon, for example, has signed a MOU with FERC in which the state and the commission commit to work together to efficiently process hydrokinetic project applications. Under the MOU, Oregon would prepare a comprehensive plan for the siting of wave energy projects in its territorial sea, which FERC would then consider in connection with issuing a preliminary permit, pilot project license, or other license.[62] A developer interested in an Oregon project would want to first examine Oregon's comprehensive siting plan before beginning the FERC preliminary permit process.

Technology testing and site evaluation

A developer that has acquired a preliminary permit has several options for possible next steps. If the immediate objective is to test technology on a small scale, there are two choices. The developer may structure the project to qualify for an exception from the FPA or alternatively may seek a pilot project license. A developer planning to use a proven technology, however, may skip the testing phase and apply directly for a long-term license. Each of these options is discussed further below.

The *Verdant Power* license exception. In the *Verdant Power* case, FERC considered a developer's request to build a tidal current turbine array sited in New York's East River.[63] The developer sought an exception from the FPA hydropower licensing requirements to test its underwater turbines and gather operational performance data and information about the impact of the devices on marine life and navigation. In its initial order, FERC agreed to the exception based on three factors:

1. The technology in question is experimental.
2. The proposed facilities are to be utilized for a short period for the purpose of conducting studies necessary to prepare a license application.
3. Power generated from the test project will not be transmitted into, or displace power from, the national electric energy grid.

Verdant met the first two parts of the test but could not meet the third because its experimental turbines would be connected to two customers. Verdant explained that the induction generators in its turbines required excitation with reactive power that could only be provided with a grid connection. Verdant offered that it would not charge the two customers for the power and would make whole the utilities whose power would be displaced by Verdant's test power. FERC accepted these conditions as a modification to the third prong of its licensing exception test based on the theory that, under such conditions, there would be no net economic effect on the national grid.

The *Verdant Power* exception therefore allows developers to put test facilities in the water without the delays associated with preconstruction licensing. The downside, however, is that the company is barred from making even small power sales from the facility. Also, the time savings may be illusory, because the developer still needs other federal, state, and local authorizations before constructing its test facilities.

Pilot project licenses. The *Verdant Power* case helped FERC to better understand the needs of hydrokinetic energy developers. Verdant spent $4.5 million of its $20 million total project investment on studies needed to satisfy permitting authorities.[64] The substantial investment in time and money needed to obtain a full license for a small-scale experimental project, even one with expected minimal environmental effects, needlessly discouraged technology developers. Such was the by-product of using regulations designed for large hydroelectric dam projects for small-scale trials of hydrokinetic technologies.

In 2007, FERC acknowledged these issues with a new hydrokinetic pilot project licensing process. This process is intended for projects that are small (typically 5 MW or smaller), removable, or able to shut down on short notice, located in waters that have no sensitive designations, and for the purpose of testing new hydro technologies or determining appropriate sites for ocean, wave, and tidal energy projects. Projects licensed under the pilot project process may sell power into the grid, and the license term is typically five years.

The pilot project process adapts FERC's existing regulations under its integrated licensing process (ILP) to provide for expedited review. The path to a pilot license may be as short as six months, compared to five years or longer under the ILP. The pilot license process is premised on the belief that small and temporary test installations should have minimal environmental impacts. If an environmental assessment of the proposed project concludes in a *finding of no significant impact (FONSI)*, then the commission can grant the license quickly. But if the environmental assessment shows a potentially significant impact, the project would be disqualified from receiving a pilot project license as further prelicensing environmental review would be required.

In exchange for the abbreviated licensing process, FERC requires pilot projects to monitor their effect on the environment for the duration of the project. If the monitoring reveals an unanticipated hazard to public safety or harm to environmental resources, FERC may require the licensee to modify, shut down, or remove the device. FERC terms its approach *adaptive management*. Under this method of project administration, the relationship between developer and regulator must be carefully coordinated and extends for the full term of the pilot license. The developer studies the project's impact on the human and natural environment throughout the license term and reports its findings to the regulator. The regulator responds with suggestions or requirements regarding how the project should be built or operated. Developers still face the risk that adverse effects may be revealed that cannot be mitigated. As a result, the project may be forced to shut down, and the developer's investment may be lost.

To begin the pilot project licensing process, a developer first files a notice of intent to file an application and an accompanying draft application. These are known as the *preapplication documents*. The draft application should explain how the project's impact on the environment will be monitored and managed, as well as how public safety will be protected. In the event that a licensee decides not to seek a license for a commercial project, FERC will require the licensee to remove all devices and restore the project site before the license terminates. The application, therefore, must demonstrate that the licensee has made adequate provision for project decommissioning and site restoration.[65] Other key parts of the pilot project draft application include a description of the existing environmental baseline and details of the project, such as type, size, and location of facilities. It also includes an appraisal of the potential effects of the project that identifies possible sensitive areas and user conflicts, along with the consultation record. FERC regulations describe the general content requirements of a license application.[66] Applicants for a hydrokinetic pilot project license should also include in their applications the additional information required by appendix B of FERC's hydrokinetics white paper.[67]

Notably absent is an extensive environmental study plan that would describe the studies to be conducted prior to license approval and inform the environmental analysis. Instead, FERC and the stakeholders rely on the preliminary environmental information in the application and the applicant's proposed postlicense monitoring and safeguard plans developed in consultation with the stakeholders. The plans should establish measures and thresholds that would signal increased risk to public safety or the environment. This would in turn trigger actions to modify the project or even shut it down if it is not possible to mitigate the harm.

The developer must distribute its preapplication documents to interested parties and stakeholders, such as federal, state, and local resource agencies and affected Indian tribes. A public meeting and opportunity for stakeholder comments typically would follow, initiating the process of agreeing to appropriate safeguards and mitigation techniques. The developer should address concerns raised by the preapplication comments in its final application.

Stakeholder concerns are often addressed by agreeing to a laundry list of monitoring tasks, studies, and conditions designed to reduce potential adverse impacts. The five-year pilot project license issued in December 2007 to Finavera Renewables Ocean Energy, Ltd. is a case in point.[68] Finavera proposed to construct a 1.0-MW wave energy converter demonstration project in Makah Bay, 1.9 nm off of the state of Washington's coast. Most of the conditions in Finavera's license were expressed in terms of plans required to be developed and filed with regulators prior to commencement of construction, with actual implementation of the plan to follow during the lease term (table 5–6). (Appendix C of FERC's hydrokinetics white paper includes standard hydrokinetic pilot project license articles that echo many of the conditions imposed on Finavera.[69])

Table 5–6. Key conditions placed on Makah Bay wave energy five-year license

Safety	**Marine mammals**
Develop a project safety plan including procedures for public safety, emergency response and reporting, testing emergency equipment, and coordinating with response agencies. Prepare a navigation safety plan, which may include an exclusion zone, addressing vessel collision risks, fishing gear entanglement, anchor lines, dredging equipment and other devices that may become entangled with project equipment. Construct and maintain lights and other signals for the protection of navigation. Develop a project facility inspection plan, including plans for removing marine debris such as derelict fishing gear.	Develop a noise assessment plan that includes monitoring of facility sound emissions and responses by fish and marine mammals. Plan acoustic monitoring of marine mammal behavior and interaction with the wave energy converter buoy array. Develop a marine mammal entanglement mitigation and monitoring plan. Prepare a marine mammal and marbled murrelet [a diving seabird] observer plan with protocols to guard against vessel strikes and to minimize disturbance caused by construction activities. Monitor effectiveness of sea lion excluder devices used to prevent sea lions from resting on buoys.
Bottom habitats	**Water quality**
Conduct pre-construction eelgrass and epibenthic surveys to inform low-impact project design. Locate mooring and anchoring entirely over sandy substrate to avoid essential fish habitat. Plan an electromagnetic field analysis of the submarine transmission cable and, if there may be adverse impacts, develop a mitigation plan.	Develop an antifouling compound (paint) study plan which demonstrates that antifouling compounds/ methods comply with recognized standards and monitors for effectiveness and environmental effects. Develop a water quality monitoring plan to address the potential for leakage of drilling fluids during horizontal directional drilling for transmission cable installation. Prepare a fuel and oil spill control, prevention and countermeasures plan that addresses the potential for spills during construction, operation, and maintenance.
Lighting	**Other**
Prepare a night construction lighting plan to minimize disturbance to fish and wildlife.	Prepare a plan for revegetation and noxious weed control to restore land disturbed by construction activities with native vegetation. Prepare an underwater archaeological resources report (pre-construction) based on surveys a cultural resources management plan to protect archeological sites that may be unearthed by construction. Submit a project financing plan showing adequate funds to construct, operate and maintain the facility, purchase a bond for the cost of project removal, and submit a decommissioning plan which includes an environmental analysis of potential decommissioning impacts.

Assuming FERC finds the final application sufficiently complete, another opportunity for stakeholder comments follows. FERC staff then prepares an environmental assessment, and stakeholders have one last opportunity to comment on the proposed pilot project license. If the assessment concludes in a FONSI, the path is clear for FERC to grant the license. A finding of significant impact, on the other hand, would result in dismissal of the pilot project license application. In that event, a developer that wishes to proceed would switch to the standard license application review process discussed below.

Licensing a commercial project

When a pilot project license is not desired or cannot be obtained, a developer may instead apply for a long-term license that would permit the permanent installation of a large-scale project for a 30- to 50-year term. FERC has designed three licensing processes (traditional, alternative, and integrated [i.e., ILP]) that feature different degrees of collaboration with stakeholders and dispute resolution approaches. The *integrated process* features structured stakeholder consultation to promote early issue identification and to set deadlines for various stages of license review. The integrated process is perhaps less likely to become bogged down in litigation than are the other processes. The *alternative process*, in contrast, allows a degree of customization and flexibility in how stakeholders participate and studies are designed. The alternative process may work well for nonstandard technologies, such as hydrokinetics. However, the applicant must develop the process in collaboration with stakeholders and have the process approved by FERC—a potentially tall order for small companies building hydrokinetic projects. The *traditional long-term licensing process* limits FERC staff involvement in study plan development, dispute resolution, and deadline setting. Consequently, it may last longer than the other two process alternatives. The discussion that follows will focus on the ILP, which is FERC's default process (fig. 5–4).[70]

The preapplication phase. The preapplication phase of the ILP is concerned principally with developing a study plan. To gather all reasonably available information about the project site and the project's possible impacts, the developer starts with informal consultations with state and federal agencies, Indian tribes, interested environmental advocacy groups, and other stakeholders. In the next step, the developer organizes this information into a preapplication document (with specific contents as required by FERC regulations) that effectively communicates the context of the project proposal.[71] Ideally, this exercise establishes a baseline of shared knowledge so that subsequent stakeholder meetings can focus on narrowing the environmental issues and identifying the best ways to fill in data gaps. The developer need not conduct studies to prepare the preapplication document, but it must exercise due diligence when gathering all available information.

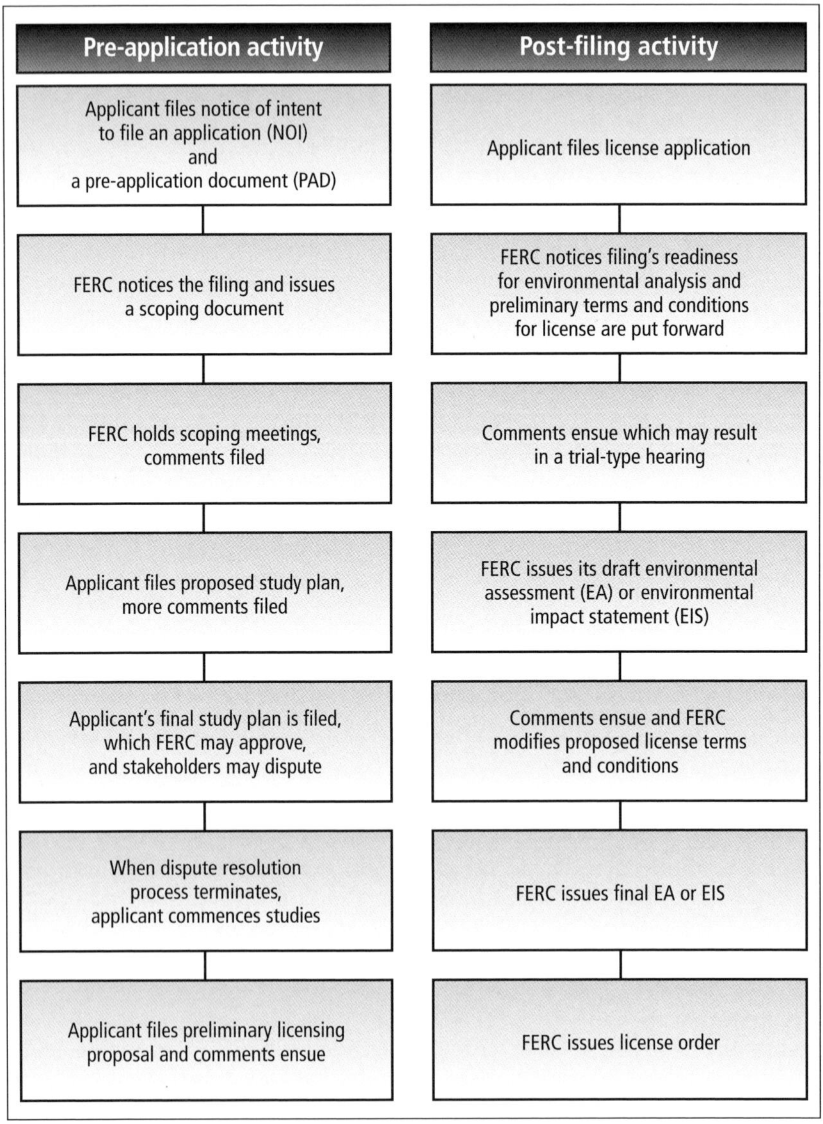

Fig. 5–4. FERC integrated licensing process summary

A developer's preapplication document should be thorough and understandable, as it is this document that serves as the foundation for discussions about shared knowledge and data gaps, which in turn give rise to study requests. Coordination and management of the stakeholder consultation process is the developer's responsibility; the FERC acts as a neutral facilitator to uncover issues and to resolve disputes. To promote collaboration and signal its willingness to listen to stakeholder concerns, the developer may establish an informal advisory board with representative stakeholder members. Bringing into the process resource working groups made up of stakeholders with common interest in a particular resource (e.g., marine mammals or fish) also can be an effective way to share information and design better study requests.

The developer should leverage the FERC staff's expertise to facilitate the process. For example, FERC may conduct an ILP training session, which would help stakeholders understand their role and the timing of the application process and could perhaps minimize delays from late study requests. FERC staff also may assist with developing a process schedule that integrates not only FERC's regulatory deadlines, but also other regulatory approvals that run on a parallel track. Even when working closely with FERC staff, a developer should nevertheless continue to consult with key stakeholders when developing the process schedule.

The process schedule will include dates for a site visit and scoping meetings, as well as methods of information dissemination. A site visit organized by the developer helps foster a better understanding of the project. Scoping meetings, on the other hand, are designed to focus the study plan, uncover previously unidentified issues, avoid duplication, and promote agreement on study methodologies. Developers should prepare visual materials such as site maps, posters illustrating the proposed hydrokinetic energy technology, and other aids for the meetings. A good document distribution plan serves not only to keep parties informed but also helps to manage costs. A project Web site is one way to provide access to public project documents, inform participants of upcoming meetings, and communicate progress under the project approval timeline. Regular e-mail distributions are another way to remind people of important meeting dates or the availability of new documents for review.[72]

Once the developer has collected the relevant background information and is confident of the project's feasibility, it should submit to FERC the preapplication document and a notice of intent to file a license application. The notice of intent includes an "unequivocal statement of the potential applicant's intention to file an application for an original license" and other basic project information.[73] The developer also should include in its submission a plan and procedural schedule that follows the protocol set out in FERC's regulations, including the location, date, and time for a site visit and scoping meetings. All documents provided to FERC also should be provided to the stakeholders.

Soon after receiving the developer's submission, FERC will hold scoping meetings, site visits, and study plan meetings, during which stakeholders and the developer identify and develop issues that need additional consideration. FERC will issue at least one scoping document identifying the issues that will require examination in a NEPA environmental assessment (EA) or environmental impact statement (EIS). The developer then issues a proposed study plan that, after stakeholders comment, is revised and approved by FERC.

The study plan describes the state of current knowledge on issues proposed for study and identifies the objectives and methodology of the studies proposed to fill in the information gaps.[74] Agencies with statutory authority to impose mandatory conditions on the project may appeal the authorized study plan using FERC's dispute resolution procedures. After any disputes have been resolved, the developer typically would commence the required studies. Interim reports on studies lasting more than one season may trigger additional stakeholder comments that lead to adjustments in the conduct of studies in the following seasons. The preapplication phase ends with the developer's submission of a *preliminary licensing proposal* (or a draft license application) and stakeholder comments on the document.

The postfiling phase. The second phase of the licensing process, known as the *postfiling phase*, commences with the developer's submission of a license application. This application, which must meet FERC requirements, should reflect the fully developed environmental picture and, it is hoped, offer a project design that is reasonably responsive to stakeholder concerns.[75] If the application is complete, FERC will issue a notice of acceptance and readiness for environmental analysis. Stakeholders will then have the opportunity to file comments and offer terms and conditions. Parties also may request that FERC hold a trial-type hearing. FERC staff also begins to prepare an EA or EIS depending on whether the proposed project would have a significant impact on the environment. The EA or EIS may be issued in draft form to permit further comments and modifications to the license terms and conditions. The findings of a trial-type hearing, if any, also would be incorporated into the terms and conditions. As a final step in this phase of the process, FERC would issue its final EA or EIS. With the record complete, FERC could then proceed to grant or deny the license application.

FERC's decision to grant the license depends on whether the project is "best adapted to serve the public interest," having regard for the following:

- The developer's abilities to comply with the terms of the proposed license
- The need for the electricity the project will generate, as compared to the cost and availability of alternative sources of power
- Other considerations, such as conservation and the effect on communities to be served by the project[76]

Early hydrokinetic energy projects are likely to produce power that is more costly than that produced by traditional generating facilities. Thus the developer should remember to explain the environmental and social benefits of promoting hydrokinetic energy in its application (see chapters 6 and 9).

License fees. FERC will assess licensees with plants that exceed 1.5 MW for both the administrative costs of its hydropower regulatory program and the use of federal lands. Administrative costs are apportioned annually among licensees based on the proportion that each licensee's combined installed capacity and energy output bears to the total of all licensees.[77] In setting fees for the use of government lands, FERC may take into consideration factors such as commercial value, the most profitable use for which the lands or other property may be suited, and the beneficial purpose for which the lands have been or may be used. FERC's current fee schedule does not list offshore areas, but the annual fee for areas otherwise not specifically designated is $5.74/acre.[78] This amount seems to be at the upper end of the fee range that FERC may impose on offshore license holders, given annual MMS offshore rental rates of $3/acre to $5/acre. When a FERC-licensed hydrokinetic energy project is located on an MMS-issued lease, MMS will set the operating fee on a case-by-case basis. This provides MMS flexibility to adjust the operating fee rate taking into consideration that hydrokinetic energy technology is in a nascent state of development and FERC may require payments from the project as well.[79]

This chapter considered two alternate agency schemes for granting a developer control over an offshore project site and authorization to build. The MMS and FERC regulations and policies will surely evolve as MMS and FERC acquire more experience with offshore wind and hydrokinetic energy technologies, and the agencies learn to work together when a project is subject to concurrent FERC and MMS jurisdiction. This chapter also introduced the environmental review process that the agencies employ. Because environmental review is so central to the project authorization process, the next chapter covers in detail how federal agencies promote sound decision making on environmental matters and build a record to support their actions that may affect the environment.

6 Action and Its Impacts: Project Review Under the National Environmental Policy Act

There is no such thing as absolute delay;

rather, one should think of "optimum delay."

—*Hans Lansberg, Energy: The Next Twenty Years*

To many environmentalists, the National Environmental Policy Act of 1969 (NEPA), the focus of this chapter, is the Magna Carta of U.S. environmental law. To many project developers, NEPA is synonymous with delay. In practice, a law intended to encourage thoughtful decision making has too often become an excuse for paralysis by analysis. The discussion below points out some pitfalls in the NEPA process to help project proponents navigate more successfully through this important law.

Signed into law in 1970 by President Richard Nixon, NEPA is unusual for environmental law in that it does not aim to protect a particular resource or species. The law's basic charge is as simple as the lesson parents try to teach their children: "Think before you act." NEPA challenges us "to create and maintain conditions under which man and nature can exist in productive harmony, and fulfill the social, economic, and other requirements of present and future generations of Americans."[1] The act in practice, however, is more mundane. Gather the facts about a proposed action, evaluate other reasonable alternatives, consider the impacts of the various alternatives on the natural and human environment, and then make an informed decision. In sum, NEPA creates a forum for the development of the factual record concerning potential environmental impacts of a proposed project. The facts thus determined influence decisions under a wide variety of other more specialized environmental laws.

Data collected in the NEPA assessment forms the basis for judging whether other federal and state environmental permits should be issued for a project and what mitigation measures may make a project environmentally acceptable. Also, some states have NEPA-like laws that drive the state permitting process. For these reasons, NEPA sets the tone for a developer's entire effort to secure project authorization. Clearly, some delay in permitting is prudent to gather and consider information from various sources and understand the complexities of a large project's possible impacts on the environment. The challenge for all involved in the NEPA process, however, is to move with purpose through the process toward the best possible decision under the circumstances, often with imperfect information. Paralysis deprives society and developers of the right to a fair hearing, signals a breakdown in the system for resolving disputes, and condemns the United States to the status quo, which may have its own severe environmental consequences.

Federal Environmental Requirements for Offshore Renewable Energy Projects

Given that renewable energy is, overall, the most environmentally sound source of energy, it is ironic that the various environmental laws are so significant a hurdle for developers of offshore renewable energy projects. Obtaining the necessary environmental authorization is costly, human resources intensive, time-consuming, and, especially for startup companies, daunting. In planning a project, developers must account for the federal, state, and local approvals necessary from the outset. They must also budget for the required assessments of the project's potential environmental impacts.

While the federal government is ultimately responsible for NEPA compliance, all agencies rely heavily upon the research, data, and assessments provided by the license or permit applicant to fulfill their statutory obligations. The laws listed in the table below will be described in detail in subsequent chapters (table 6–1). This chapter will demonstrate how the record used to make decisions under these laws is primarily developed under NEPA.

From the developer's perspective, ensuring that the environmental assessment is thorough pays dividends in the avoidance of unanticipated delay. Unanticipated delay may raise project costs substantially. This is especially true if the delay comes at a time when a significant investment in project construction has already been made and site work comes to a halt. In such circumstances, delay becomes an ongoing daily cash drain in the form of interest, wages, vessel hire rents, and other expenses that accrue on an unfinished project. Planning and budgeting for all the steps in the permitting and construction process allows the developer to order the project timeline so that major investments are postponed until all potentially "show-stopping" issues are resolved and critical permits have been obtained.

Table 6–1. Federal agencies and laws involved with offshore renewable energy projects

Agency	Law
Department of Interior, Minerals Management Service	Outer Continental Shelf Lands Act of 1953 Archaeological and Historic Preservation Act
U.S. Army Corps of Engineers (COE)	Rivers and Harbors Act of 1899 Clean Water Act Marine Protection Research and Sanctuaries Act of 1972
Federal Energy Regulatory Commission	Federal Power Act
U.S. Department of Commerce, National Oceanic and Atmospheric Administration	Coastal Zone Management Act National Marine Sanctuaries Act
U.S. Department of Commerce, National Marine Fisheries Service	Endangered Species Act Magnuson-Stevens Fishery Conservation and Management Act Marine Mammal Protection Act
Department of the Interior, U.S. Fish and Wildlife Service	Endangered Species Act Migratory Bird Treaty Act Fish and Wildlife Coordination Act Marine Mammal Protection Act
U.S. Environmental Protection Agency (EPA)	Clean Water Act Clean Air Act Act to Prevent Pollution from Ships
U.S. Department of Homeland Security, U.S. Coast Guard	Coast Guard and Maritime Transportation Act of 2006 Ports and Waterways Safety Act Maritime Transportation Security Act
Federal Aviation Administration	Federal Aviation Act of 1958
U.S. Department of Defense/Department of Homeland Security Long Range Radar Joint Program Office	National Defense Authorization Act of 2006
Advisory Council on Historic Preservation	National Historic Preservation Act Abandoned Shipwreck Act
U.S. Customs and Border Protection	Merchant Marine Act (Jones Act)

"Not-in-my-backyard" (NIMBY) resistance to a proposed project is a predictable expression of competing interests in an area. NEPA was designed to encourage broad consideration of all project impacts. The law refers to the "human environment," which regulations interpret to include "the natural and physical environment and the relationship of people with that environment."[2] It is to be expected that fishers would resist being excluded from fishing in an area where a renewable energy project is proposed to be built, or that some coastal residents might object to wind turbines as a visual intrusion on the seascape. Harmonizing opposing views is rarely a frictionless exercise. An understanding of the project evaluation process created by NEPA will not eliminate disagreements. However, it will make developers more effective advocates for their projects, reduce uncertainty and cost, and hopefully lead to more satisfaction with the outcome for all concerned.

The NEPA framework

NEPA requires that the federal government perform an environmental assessment for every major federal action that could significantly affect the quality of the human environment. The agency conducting an assessment must use "a systematic, interdisciplinary approach," incorporating the natural and social sciences and the environmental design arts in planning and decision making. "Presently unquantified environmental amenities" must be evaluated alongside traditional economic and technical considerations. Aesthetic values deserve weight in a balanced presentation of project impacts. To say that some weight should be accorded to aesthetics does not imply, however, that aesthetics should be weighted equally with other impacts, such as public health effects, to pick just one category. NEPA does not specify how much weight aesthetics should be accorded, nor does NEPA require a rigorous cost-benefit analysis. Perhaps most surprisingly, NEPA also does not require that decisions favor the course of action with the least harmful environmental impact. The permitting agency may place other values (e.g., national security or economic growth) higher than environmental values in making a decision. NEPA requires a process, not a substantive result. "NEPA itself does not impose substantive duties mandating particular results, but simply prescribes the necessary process for preventing uninformed—rather than unwise—agency action."[3]

NEPA applies when a federal agency is faced with a choice of one or more alternative means of accomplishing a goal that may significantly affect the environment.[4] Major federal actions subject to NEPA may include granting permits, leases, or other approvals for offshore energy projects. Proposals to construct renewable energy projects on, or adjacent to, the inland waters of the United States, and those in coastal waters extending to the seaward limits of the outer continental shelf (OCS), are subject to federal jurisdiction and environmental regulation under the Commerce Clause of the U.S. Constitution and various international treaties. (For definitions and a discussion of these water resource terms, see chapter 2.) These projects trigger NEPA review when a federal agency must decide whether to authorize a project or award a lease.

The federal agency with the principal responsibility for preparing an environmental assessment under NEPA is deemed the lead agency. Numerous other agencies will be designated as cooperating entities in the NEPA process. Development of a wind turbine renewable energy project located on the OCS would involve the Minerals Management Service (MMS) as the lead agency. MMS would perform the environmental assessment and then decide whether to issue a lease, easement, or right-of-way to the prospective developer. The U.S. Army Corps of Engineers (Corps) would take the lead federal role in the environmental assessment process for a wind farm located in navigable waters of the United States designated as state territorial waters. Under the Submerged Lands Act, the states are responsible for conveying the appropriate lease or other property rights necessary to build renewable energy projects in state territorial waters. (See chapter 8 for a discussion on areas where state and federal regulations converge in the project approval process.)

FERC and MMS have agreed to cooperate with respect to the environmental analysis of hydrokinetics on the OCS.[5] FERC has the lead agency role when a project is located in state waters. Accordingly, for projects on the OCS, the federal government (through either MMS or FERC) has primary responsibility for both the environmental assessment and the conveyance of any property or natural resource interest needed for a hydrokinetic project. In state waters, the federal government conducts licensing (and the associated environmental review through either FERC or the Corps), and the states generally convey the real property interest.

NEPA categorical exclusions, environmental assessments, and environmental impact statements

The White House Council on Environmental Quality (CEQ) has adopted regulations that provide general guidance in the application of NEPA.[6] CEQ guidance is often cited by the courts in deciding contested NEPA cases.[7] CEQ regulations are supplemented by NEPA-implementing regulations adopted by federal agencies to further their specific missions. NEPA, the CEQ regulations, and the appropriate agency's implementing regulations govern federal environmental assessments regardless of the proposed location of the project.

A project proposal subject to NEPA may follow one of three paths through the lead agency's analytical process (fig. 6–1). The agency's action could be any one of the following:

1. The project could qualify for a categorical exclusion.
2. The agency could require preliminary review in the form of an environmental assessment.
3. The agency could require in-depth analysis in the form of an environmental impact statement.

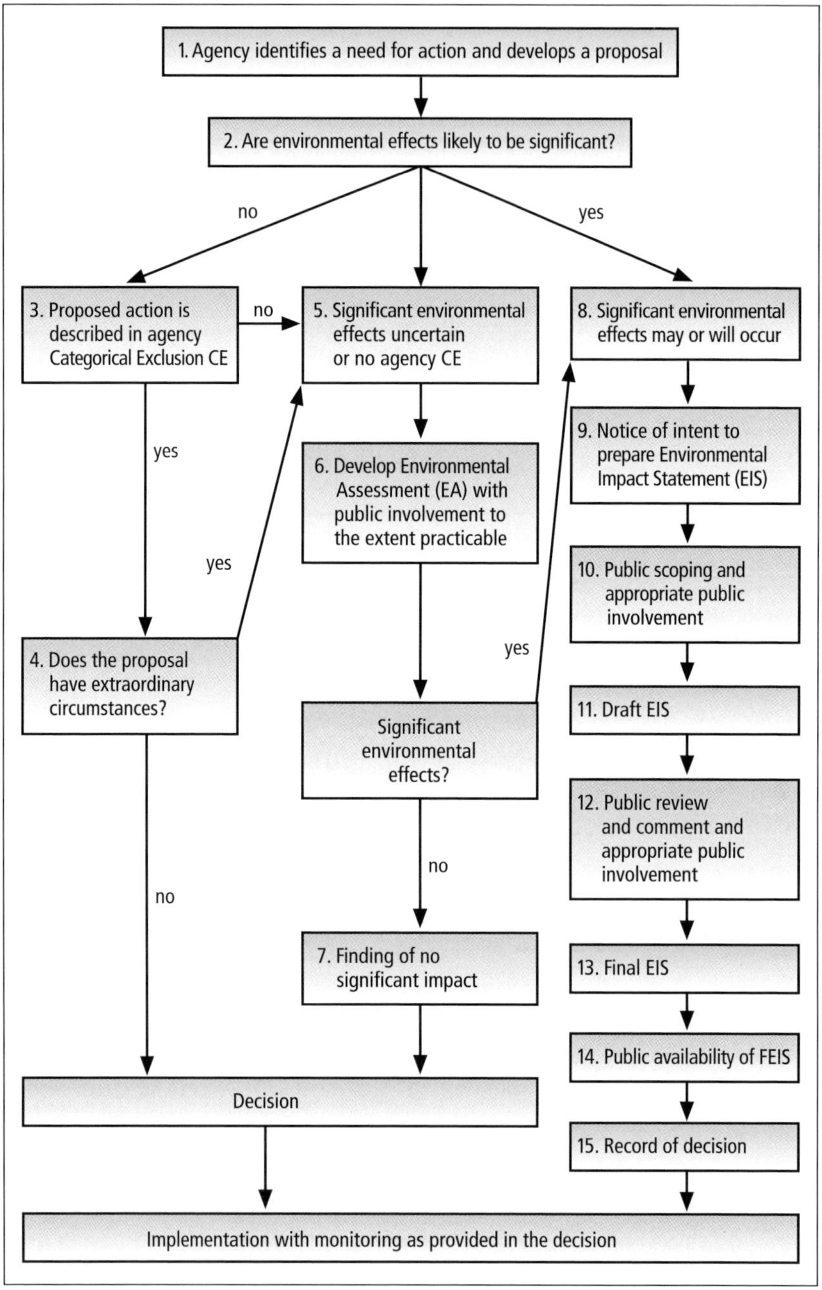

Fig. 6–1. The NEPA process

Categorical exclusions. A project that falls into a category of common activities that, based on past experience, are unlikely to have significant environmental effects may qualify for a *categorical exclusion (CE)* from NEPA. A CE may be provided in agency regulations for a category of actions that do not individually or cumulatively have a significant effect on the human environment. Therefore, they do not merit the more incisive analysis of an environmental assessment or an environmental impact statement. There are presently no CEs applicable to offshore renewable energy projects, though with time and experience, some CEs may be adopted by MMS and other agencies. Certain offshore site assessment and resource characterization activities may have small detrimental environmental effects and no lasting environmental impact. These activities would be good candidates for categorical exclusion.

Before applying a CE to a particular project element, the agency checks for extraordinary circumstances that might make the application of the exclusion environmentally detrimental in the particular circumstances. In the absence of extraordinary circumstances, the exclusion would apply, thus concluding the NEPA review.

Environmental assessments. When a CE is not available or does not apply in a particular case, the agency must review its action under the *environmental assessment (EA)* or *environmental impact statement (EIS)* process. EAs are intended to be concise. The purpose of an EA is to determine if the lease, license, permit, or other federal approval is a "major federal action significantly affecting the quality of the human environment." (Examples may be found in *Sierra Club v. Morton*, *Stewart Park & Preserve Coalition, Inc. ("SPARC") v. Slater*, and *Hudson River Sloop Clearwater, Inc. v. Dept. of the Navy*.[9]) The EA allows the federal agency to determine if a more detailed EIS must be prepared. An EA should:

1. Briefly provide sufficient evidence and analysis for determining whether to prepare an environmental impact statement or a finding of no significant impact (FONSI)
2. Aid an agency's compliance with NEPA when no EIS is necessary
3. Facilitate preparation of an EIS when one is necessary

The need for the proposed action, alternatives to the action, the environmental impacts of the proposed action and alternatives, and a list of the agencies and persons consulted should be included in an EA.[10]

If the agency's EA indicates that only "insignificant impacts" are likely to occur, the agency will issue a FONSI, and the developer may proceed to obtain the other required environmental permits and approvals. However, the agency EA may conclude that the approval is a major action with potential to significantly affect the environment.

In that case, a full examination of alternatives to the project, and the relative impacts of those alternatives, must be given a "hard look" by the agency in a formal EIS.

Developers often ask what makes a federal action "major" and how significant the impacts must be to "significantly affect the environment." Lawyers and courts have debated those issues since NEPA was first passed. Seemingly minor projects on the nation's waterways with few environmental impacts often trigger a full EIS. Large projects that are implemented in phases may require several EISs at the various agency permitting stages. MMS, for example, prepared a programmatic EIS to review the generic impacts of establishing an offshore renewable energy site leasing program.[11] The agency expects that specific projects will require two and perhaps three additional EIS reviews to determine the site and technology-specific impacts on the environment.

As a practical matter, without resolving the legal issues of what is a major federal action with significant environmental impacts, most, if not all, commercial renewable energy projects in U.S. waters will be considered major federal actions. This is based on MMS program statements, the Cape Wind experience, and NEPA case law, as well as the experiences of owners and operators of offshore oil and gas projects requiring federal permits. NEPA has been determined to apply to many discretionary actions subject to federal control and responsibility, including projects and programs financed, assisted, conducted, regulated, or approved by federal agencies, as well as new or revised agency rules, regulations, plans, policies, procedures, and legislative proposals. A proponent of an offshore renewable energy project should therefore anticipate an EIS and be prepared to assist in its development.

Environmental impact statements. In contrast to the abbreviated analysis conducted under a CE or EA, an EIS is based on a fully developed record. Public participation is one way to build the record. The process begins with a notice of intent (NOI) issued by the lead agency, which briefly states the proposed action and alternatives and describes the agency's proposed scoping process. In *scoping*, the agency enlists the aid of stakeholders to identify the range of actions, alternatives, and impacts that should be considered in the EIS.

After scoping, the lead agency prepares a draft EIS with the assistance of the applicant and cooperating agencies. The agency then publishes the draft EIS for comment. Public meetings or hearings also may be held to gather comments. In response to the comments, the agency may make changes to the draft. The agency may reissue the draft for additional comment if the changes are substantial. If the changes are minor, the agency proceeds to publish a final EIS in support of its record of decision on the underlying renewable energy project application.

A person who is dissatisfied with the final agency decision may file a challenge in federal court. In most jurisdictions, such a challenge is not ripe until a final agency decision is rendered on the permit or other authorization for the project. The most frequently cited grounds for legal challenge is an alleged inadequacy in the agency's consideration of alternatives to the proposed project or the negative environmental impacts that it will trigger. While the NEPA process is procedural, the courts will carefully scrutinize an EIS that is perfunctorily drafted without a good faith hard look at project options and environmental impacts, both direct and indirect. Disputes under NEPA are resolved by the federal courts using the "arbitrary, capricious, or otherwise contrary to law" standard established by the Administrative Procedure Act.[12] Agency decisions on issues of fact that require technical expertise are given a high degree of deference by the courts.[13] The remedy for inadequate consideration of project alternatives or their impacts is a court remand to the agency for additional NEPA analysis and documentation.

The majority of states also have NEPA-type statutes that to varying degrees track the federal requirements. The lead federal agency for a project ordinarily will designate states and other federal agencies with interests affected by the particular project as *cooperating* authorities or agencies. For offshore renewable energy projects, the federal agencies so designated, or at least consulted, are those with some level of jurisdiction over the nation's waters and related environmental issues: MMS, the Corps, other Department of Interior agencies (e.g., FWS), the Coast Guard, NOAA, and EPA. On the state level, the state equivalent of the EPA and various natural resources protection agencies will be consulted and often expected to participate as cooperating agencies. This joint NEPA process is designed to reveal all of the environmental concerns with the project from the outset and avoid the need for duplication or multiple EISs for a single project.

Crafting an EIS

The EIS should focus the attention of the federal government and the public on a proposed action so that the consequences of the action may be studied before the action is implemented.[14] NEPA's mandate to federal agencies is "essentially procedural…It is to insure a fully informed and well-considered decision."[15] While NEPA mandates procedures to be used by federal agencies in considering the environmental impacts of their actions, it does not dictate the substantive results. Agencies that have given the requisite hard look to the environmental consequences and various alternatives to a project are free to choose an option that fully achieves the project's purposes. This is the case even if that option has unavoidable negative impacts upon the environment.[16]

NEPA promotes sound agency decision making by requiring that the proposed and alternative courses of action are presented "in comparative form, thus sharply defining the issues and providing a clear basis for choice among options by the decisionmaker and the public."[17] To that end, the EIS must set out a detailed statement covering the following:

1. The environmental impact of the proposed action
2. Any adverse environmental effects that cannot be avoided should the proposal be implemented
3. Alternatives to the proposed action
4. The relationship between local short-term uses of the environment and the maintenance and enhancement of long-term productivity
5. Any irreversible and irretrievable commitments of resources that would be involved in the proposed action should it be implemented[18]

All reasonable alternatives should be rigorously and objectively evaluated, and the EIS should explain the reasons for eliminating any alternatives from detailed study. The agency's preferred alternative must be identified in the EIS. In addition, the EIS should identify mitigation measures that have not otherwise been identified in the proposed action or the alternatives.

One of the most important pieces of practical advice for a developer seeking federal approval of an offshore project is to make it easier for government reviewers to say "yes" to the project application. This includes providing sufficient documentation to the agency so that its decision will be sustained if challenged in court. The level of documentation provided should constitute the private sector equivalent of a comprehensive environmental assessment that demonstrates to the government that the benefits of the project outweigh any negative environmental consequences. Most agencies, including MMS, FERC, and the Corps, expect the project applicant to bear the burden of an initial environmental assessment. It is true that the agency cannot simply adopt the applicant's assessment as its EA or EIS and must independently conduct the impact analysis and weigh the alternatives. However, an applicant's provision of sound environmental information, data, and analysis to the responsible agency facilitates and expedites the process and establishes a positive context for the approval of the project.

Scoping

Developing a defensible environmental case for a renewable energy project starts with sound environmental science and an analysis of the site selected for the project. A well-planned environmental assessment by the project developer sets the stage for conducting the appropriate scientific/engineering analysis of alternatives and environmental impacts and facilitates the government's own scoping of its EA or EIS.

Developers have an interest in promoting effective scoping that helps identify issues early, thus minimizing the possibility for delay, the expense of unnecessarily broad studies, and litigation. When scoping fails to anticipate and address legitimate environmental issues, a project may require a supplemental EIS to correct gaps in the initial EIS. Stakeholder involvement in the scoping effort, where all participants are operating in good faith, has the advantage of engaging those parties that can act to address potential environmental issues before opinions and positions have hardened. Ideally, stakeholders who have participated in the scoping process will feel that their voices were heard and should be more satisfied with the eventual outcome of the process than stakeholders who join the NEPA process late.

Unfortunately, some parties will oppose almost any kind of development. Their motivation for participating is merely to cause delay, and bringing them into the process early may be counterproductive. Developers can counter obstructionist stakeholders by working with the lead agency to ensure that scoping meetings are run by a skilled meeting facilitator and follow a prepared agenda. Efforts at delay also are more easily recognized and dismissed if the developer has taken the initiative to form alliances and prepare a reasonable scoping plan with the participation and cooperation of other stakeholder representatives.

Although the lead agency is responsible for the EIS and manages the scoping process, developers should actively participate to accomplish the following:

- Identify persons and organizations interested in the proposed project
- Identify the significant issues to be analyzed in depth in the EIS and, in particular, help to focus the EIS so that all analyses and studies required to support federal, state, and local permitting requirements are conducted concurrently in an integrated manner
- Identify issues that should be eliminated from in-depth analysis because they are not significant or they have already been covered in prior environmental reviews, EAs, or EISs

- Assume part of the burden of preparing the EIS, although the lead agency will retain responsibility for the statement
- Contribute to the development and maintenance of a schedule for preparing the EIS and the agency's planning and decision-making process by encouraging page and time limits on environmental document preparation and structuring the public participation process[19]

To determine the scope of an EIS, it is helpful to consider three types of actions, three types of alternatives, and three types of impacts, as shown in the following:

Actions

- Connected (closely related or interdependent actions)
- Cumulative (actions that together may be cumulatively significant and should be discussed together in the EIS)
- Similar (actions with similarities that provide a basis for evaluating their environmental consequences together)

An EIS should define the actions with which it is concerned with sufficient breadth that interdependent and connected relationships are addressed.

Alternatives

- No-action alternative
- Other reasonable courses of action
- Mitigation measures that are not in the proposed action

Alternatives that are always evaluated include: the project as proposed; other alternatives that fully achieve the purposes of the project as originally proposed; and a so-called no-action alternative that assumes the project will not be built. To properly reflect the environmental benefits of a renewable energy project, the no-action alternative should either assume that the baseline environmental conditions continue, including the continued pollution from the existing generating mix, or that continued growth in load will require construction of additional polluting traditional generation. In so doing, the no-action alternative will better reflect the benefits to public health and the environment from the pollution that is avoided by the renewable energy project.

Impacts

- Direct
- Indirect
- Cumulative

A review of impacts should start by choosing the right geographic scope for the project evaluation. The direct impacts in the vicinity of all project structures—the generating equipment, the area surrounding all cables laid, and any onshore structures—must be evaluated. Since a renewable energy project will benefit an entire service area or region, however, a much broader area also may have to be evaluated for possible direct and indirect environmental impacts. This includes pollution reductions attributable to the displacement of traditional power generated in the region as a result of the renewable power to be generated by the project.

The EIS prepared by MMS for the Nantucket Sound offshore wind farm proposal considered several alternatives, including two alternate offshore wind farm sites and three variations on the proposed 130-turbine project. The variations considered included the following:

1. A smaller project, half the capacity of the proposed farm
2. A condensed array, with the full capacity of the proposed farm but occupying less area
3. A phased development alternative, in which the wind farm would be constructed in two phases of 65 turbines in each phase
4. A no-action alternative

According to MMS, "The No-action Alternative considers other strategies for addressing the demand for electricity in New England if the proposed action were not constructed, and the viability of those strategies and or impacts associated with those other strategies. This includes an assessment of energy efficiency, and the assessment of other energy options including fossil fuel technologies, and other alternative energy technologies."[22]

CEQ guidance provides that significant "direct," "indirect," and even "cumulative" environmental impacts from a project should be evaluated. By definition in 40 CFR, *direct effects* are those effects "which are caused by the action and occur at the same time and place."[23] *Indirect effects* are those "which are caused by the action and are later in time or farther removed in distance, but are still reasonably foreseeable."[24] *Cumulative effects* are defined as "the impact on the environment which results from the incremental impact of the action when added to other past, present, and reasonably foreseeable future actions regardless of what agency (Federal or non-Federal) or person undertakes such other actions."[25] Since theoretically one ripple from a single environmental impact reverberates throughout an ecosystem, it can be difficult to determine where to stop. Fortunately for proponents of offshore projects, MMS has examined a broad range of possible environmental impacts from typical offshore renewable energy projects and has identified generically which impacts are likely to be substantial in any given project area. The analysis in the agency's programmatic EIS for the offshore renewable energy program is therefore a useful gauge of relevant impacts.[26]

The most significant potential impacts of an offshore renewable energy project as identified by MMS include the following:

- Environmental and natural resources impacts resulting from the connection of the technology to the sea floor
- Impacts to currents, water flow, wave height, and water temperature characteristics
- Water and air quality impacts
- Physical, chemical, and acoustical impacts to marine and coastal biota, particularly marine mammals, fish, coral, and avian species, as well as their habitats
- Land use and existing infrastructure impacts
- Cultural resource and viewshed impacts
- Socioeconomic impacts to such systems as transportation, fishing, tourism, and recreation

The impact assessment must span the entire project life cycle, from construction and operation to eventual decommissioning.

The four impact levels used in the Nantucket Sound project EIS for biological and physical resources range from *negligible* to *major* (indicating limited options for effective mitigation and lasting, unavoidable disruption to the affected resource), as follows:

Negligible

- No measurable impacts.

Minor

- Most impacts to the affected resource could be avoided with proper mitigation, or
- If impacts occur, the resource would recover completely without any mitigation once the impacting agent is eliminated.

Moderate

- Impacts to the affected resource are unavoidable, and
- The viability of the affected resource is not threatened, although some impacts may be irreversible, or
- The affected resource would recover completely if proper mitigation is applied during the life of the proposed action or proper remedial action is taken once the impacting agent is eliminated.

Major

- Impacts to the affected resource are unavoidable, and
- The viability of the affected resource may be threatened, and
- The affected resource would not fully recover even if proper mitigation is applied during the life of the proposed action or remedial action is taken once the impacting agent is eliminated.[27]

(Note that MMS used the same four terms to characterize impacts to socioeconomic aspects of the affected environment, such as impacts to employment, fisheries, tourism, and recreation, but defined the terms slightly differently.) The Nantucket Sound project EIS considered these and other impacts, finding most of them negligible and minor (table 6–2).

Table 6–2. Summary of impacts of the Nantucket Sound wind energy project

Resource	Impacts	
	Construction Impacts	**Operation Impacts**
Regional Geologic Setting	minor	minor
Noise	Onshore: minor Offshore: minor Underwater: minor	Onshore: negligible Offshore: negligible Underwater: negligible
Oceanography	Currents: negligible Waves: negligible Salinity: negligible Temperature: negligible Sediment Transport: minor Water depth/bathymetry: minor	Currents: minor Waves: negligible Salinity: negligible Temperature: negligible Sediment Transport: minor Water depth/bathymetry: minor
Climate and Meteorology	minor	negligible
Air Quality	Public Health: negligible Visibility: negligible Emissions: minor	Public Health: negligible Visibility: negligible Emissions: minor (beneficial to climate change)
Water Quality	minor	negligible (with the exception of spills)
Electric and Magnetic Fields (EMF)	negligible	negligible
Terrestrial Vegetation	negligible to minor	negligible to minor
Coastal and Intertidal Vegetation	negligible to minor	negligible (negligible to minor for repairs, depending on location)
Terrestrial and Coastal Faunas other than Birds	negligible to minor	negligible (minor for migratory bats)
Avifauna	Terrestrial Birds: Raptors: negligible Passerines: minor Coastal Birds: negligible to minor Marine Birds: minor to moderate Pelagic Species: minor Waterfowl and Non-Pelagic Water Birds: minor	Terrestrial Birds: Raptors: negligible Passerines: minor to moderate Coastal Birds: negligible to moderate Marine Birds: negligible to major Pelagic Species: minor Waterfowl and Non-Pelagic Water Birds: minor
Subtidal Offshore Resources	Soft-Bottom Benthic Invertebrate Communities: minor Shellfish: minor Meiofauna: minor Plankton: negligible	Soft-Bottom Benthic Invertebrate communities: minor Shellfish: minor Meiofauna: minor Plankton: minor
Non-Endangered Species Act Marine Mammals	Acoustical Harassment: minor Vessel Strikes: minor Vessel Harassment: minor Temporary Reduced Habitat: minor Turbidity: negligible to moderate (due to pile driving) Pollution/ Potential Spills: minor	Acoustical Harassment: negligible EMF: negligible Pollution/Potential Spills: minor to moderate Vessel Strikes: minor Vessel Harassment: minor Fouling Communities: negligible to minor

Source: U.S. Department of the Interior. Minerals Management Service. 2009. Cape Wind energy project: Final Environmental Impact Statement, January 2009: E-11-E-12.

	Impacts	
Resource	**Construction Impacts**	**Operation Impacts**
Fisheries	Finfish: .... minor Finfish (juveniles): .... minor Demersal Eggs and Larvae: .. minor Commercial & Recreational Fishing/Gear: .... minor	Commercial & Recreational Fishing/Gear: ... negligible to minor Sound and Vibration: ..... negligible to minor Vessel Traffic: ....minor to moderate EMF: .... negligible Lighting: .... negligible/none Alterations to Waves, Currents, Circulation: .... negligible Habitat Change: .... minor Displacement of Prey: .... none
Essential Fish Habitat	Benthic/Demersal: ....minor Water Column: .. negligible to minor SAV/Eelgrass: .... negligible to minor	Benthic/Demersal: .... minor Water Column: .. negligible to minor SAV/Eelgrass: ... negligible to minor
Threatened & Endangered	Sea turtles: ...... negligible to minor Cetaceans: ..... negligible to minor Avifauna: .... negligible to minor Eastern Cottontail Rabbit: . . negligible	Sea turtles: ...... negligible to minor Cetaceans: ...... negligible to minor Avifauna: .... minor to moderate Eastern Cottontail Rabbit: . . negligible
Urban and Suburban Infrastructure	negligible to minor	negligible
Population and Economics	minor	minor
Environmental Justice	negligible (i.e., not a disproportionately high impact on minority or low income populations)	negligible (i.e., not a disproportionately high impact on minority or low income populations)
Visual Resources	minor	moderate impacts on shore (major impacts on water in close proximity to proposed action)
Cultural Resources	minor	Pending on the outcome of Section 106 process
Recreation and Tourism	minor	minor
Competing Uses of Waters and Seabed	minor	minor (except for impacts to Figawi Race which are moderate)
Overland Transportation Arteries	minor	negligible
Airport Facilities and Aviation Traffic	negligible to minor	minor
Port Facilities and Vessel Traffic	minor	Ship, Container and Bulk Handling Facilities: ...... negligible Cruise Ship Traffic: .... negligible Ferry Operations: .... minor Marinas and Recreational Boating: ... minor to moderate Commercial fishing: .... minor to moderate Search and Rescue: ...... negligible Ice: .... negligible
Communications: Radar, EMF, Signals, and Beacons	minor	minor (moderate for radar)

Each site has unique characteristics. Nonetheless, it makes good sense for a developer to structure its environmental assessment, at least in the first instance, around the impacts identified by MMS (or other lead agency such as the Corps or FERC) and then tailor the assessment to the unique characteristics of the project site and the technology employed. Agencies are allowed to tier various EISs prepared for different purposes. Therefore, a programmatic EIS from MMS will be a valuable initial analysis that may be tiered with more detailed site- and technology-specific analysis to streamline the overall impacts analysis.

Marine sanctuaries, parks, wildlife preserves, and areas populated or frequented by endangered or threatened species are particularly sensitive and treasured environments. Due to their special status, developers will generally encounter a much higher hurdle to justify locating in these special interest areas. In the case of MMS, a legal bar to issuing leases on lands within the boundaries of a national park, marine sanctuary, monument, or preserve exists. Therefore, while energy projects may not be absolutely prohibited in these areas, they should be avoided if possible because it will be difficult to obtain the necessary project authorizations. Even if authorizations can be obtained, they will likely include conditions that increase the difficulty and expense of operating on the site.

Environmental impacts expected from offshore renewable energy facilities

Positive environmental impacts. While most NEPA analyses focus upon negative environmental impacts, it is important not to neglect the positive impacts from an offshore renewable energy project. NEPA is designed to allow the federal agency to weigh all benefits, including environmental ones, against any negative environmental impacts. Each megawatt-hour of electricity produced from renewable sources is one less megawatt-hour that must be generated using traditional polluting energy sources. Typically, the traditional energy sources will have far greater negative environmental consequences than those stemming from a renewable energy source.

Traditional electric generation plants emit numerous air pollutants, including greenhouse gases, and the environmental impacts of these plants extend beyond merely the effects of combustion. Fuels must be mined or extracted, processed, and transported, and waste products (e.g., spent nuclear fuel and coal ash) must be stored or disposed of safely. Marine renewable energy, in contrast, uses a clean resource delivered by nature directly to the project site, avoiding the fuel cycle impacts associated with traditional electricity generation. The positive impacts of the renewable energy option should be evaluated in every environmental assessment performed by a proponent of an offshore renewable energy project.

Global warming impacts are becoming increasingly important. CEQ guidelines direct agencies to consider global warming in NEPA documentation.[28] The U.S. Supreme Court recently held that the "harms associated with climate change are serious and well recognized."[29] The EPA, building on this decision, has now proposed to find that greenhouse gas emissions from motor vehicles endanger public health and welfare and should be regulated under the Clean Air Act. Unless Congress acts with a comprehensive approach to greenhouse gas emissions, EPA regulation of stationary source greenhouse gas emitters, such as power plants, is likely to follow. The lower federal courts have always required agencies to consider direct air quality impacts of major federal actions. More recently, the courts have begun to require agencies to conduct climate change analyses pursuant to NEPA to determine if certain traditional energy projects contribute, even indirectly, to global warming.

Opponents of fossil fuel energy plants have framed the NEPA climate change issue in a variety of ways. Litigants may argue that an agency has inadequately evaluated less polluting energy alternatives to the proposed action, highlighting their reduced potential for negative effects on global climate change. Opponents also may claim that greenhouse gas emissions are a reasonably foreseeable indirect and cumulative impact that has been inadequately addressed, or not addressed at all, by the agency performing the NEPA analysis.

In *Border Power Plant Working Group v. Department of Energy*, DOE asserted that NEPA did not require the department to evaluate "questionable" or speculative effects of permitting cross-border transmission lines to power plants in Mexico.[30] The court for the Southern District of California, however, found nothing questionable about the emission of carbon dioxide from the power plants. Since it was determined that the power plants would in fact emit carbon dioxide, a greenhouse gas, the court held the failure to discuss and analyze those emissions before allowing construction of the transmission lines violated NEPA.[31]

Similarly, in *Mid States Coalition for Progress v. Surface Transportation Board*, the Eighth Circuit held that NEPA required the Surface Transportation Board (STB) to analyze "the effects that may occur as a result of the reasonably foreseeable increase in coal consumption" stemming from STB's approval of construction of hundreds of miles of new rail lines to reach coal mines. The plaintiffs argued that the EIS failed to consider the effects on air quality, including increased carbon dioxide emissions, which would result from greater use of low sulfur coal instead of other fuels such as natural gas. The court rejected the agency's argument that these effects were too speculative, concluding that degradation in air quality was a "reasonably foreseeable" indirect effect of the proposed rail lines, even if the extent of the impact could not be precisely quantified.[33]

The court also determined that STB failed to follow the CEQ's regulations establishing a process for evaluating reasonably foreseeable significant adverse impacts when the necessary information is incomplete or unavailable. If an agency drafting an EIS cannot obtain adequate information to resolve relevant environmental impact issues, the EIS must accomplish the following:

1. Explain the relevance of the missing information for assessing "reasonably foreseeable significant adverse impacts"
2. Summarize existing credible scientific evidence relevant to evaluating impacts
3. Evaluate the impacts based on theoretical approaches or research methods generally accepted in the scientific community

On remand, the agency prepared a supplemental EIS applying the principles articulated by the court and CEQ for circumstances involving uncertainty, and STB's action was subsequently upheld.[34]

Federal case law and CEQ's acknowledgement of climate change bode well for the emerging offshore renewable energy industry. MMS has recognized the potential environmental benefits of offshore renewable energy sources compared to coal, natural gas, and nuclear-fueled generation, as well as alternative energy sources such as conventional hydropower and onshore wind.[35]

The draft EIS prepared by MMS for the Nantucket Sound project discussed the use of traditional generating technologies to provide for regional electrical loads under the no-action alternative. MMS also considered energy efficiency and other renewable energy technologies as possible substitutes for the offshore wind project. However, MMS found that only the traditional fossil fuel generating technologies were feasible in New England on a scale comparable to the proposed offshore wind project. The broad effects of traditional generating options considered in the draft EIS are as follows:

1. **Land disturbance.** The area occupied by the plant and the surrounding area directly affected by plant air emissions.
2. **Air emissions.** Regulated pollutants (NO_x, SO_x and particulate matter [PM]), as well as mercury (Hg) and CO_2.

3. **Water use.** Consumptive water use for cooling and the effects on fish such as larvae entrainment.
4. **Solid waste generation and waste management.** Some fuels can produce significant quantities of solid waste, requiring onsite or offsite disposal.
5. **Water discharge.** Cooling water discharges may result in temperature pollution and contain chemical constituents.

Each of the traditional fossil fuel generating technologies (natural gas, coal, and oil), and the alternative of repowering existing facilities, was associated with significant adverse impacts. MMS concluded that "these facilities would result in substantial emissions of various air pollutants during their operational life span, affecting the air quality in the region and continuing to increase global CO_2 levels."[36] Notably, the substantial indirect public health impact attributable to pollution from fossil fuel generation was mentioned only briefly. Mercury was identified as a "persistent bioaccumulative toxin" without explaining the public health implications of environmental mercury pollution, such as developmental impacts on children and widespread fish consumption advisories. The draft EIS also stated that elevated particulate levels have been associated with "a variety of health effects such as respiratory ailments, especially in the young and the elderly." Public health studies have quantified the effects of particulate matter pollution in terms of work days missed, hospital admissions, and deaths. Placing the pollution from traditional fossil fuel generation sources more clearly in a public health context would have made an even stronger case for the proposed offshore wind project. Moreover, the draft EIS gave only brief mention of the homeland security implications of continued dependence on foreign oil suppliers.

These examples highlight the importance of assessing the positive local, regional, and global environmental impact of a project as part of a balanced NEPA analysis. The results also underscore an earlier point: NEPA is procedural in nature. So long as the agency takes a hard look at potential environmental impacts, a court will not substitute its judgment for the agency's substantive determination, even if a more environmentally sound alternative is available. Developers of marine renewable energy projects should encourage a balanced NEPA review by providing complete information to the authorizing agency to ensure that the EIS analyzes all positive and negative environmental impacts from the project. An agency that has met the hard look threshold and, despite having identified adverse environmental impacts, has authorized an offshore renewable energy project will not lightly be reversed by the courts.

Impacts to birds and their habitats. Avian impacts from both offshore and onshore wind projects worldwide have been shown in most studies to be minor. Bird mortalities average approximately 2.19 per turbine per year for land-based turbines and 1.2 per turbine per year for offshore wind turbines.[37] Most birds have extremely acute eyesight and are capable of avoiding fixed structures and slow moving rotors. Nevertheless, the unfortunate experience of raptor fatalities, particularly associated with wind power in California, has raised concern among wildlife advocates and regulators about the potential threat of bird strikes, especially in the case of endangered or threatened species. Wave and current technologies clearly present less of a problem but are not exempt from avian issues. The potential for diving bird collisions with the rotors of ocean current facilities has been raised.

European offshore wind energy experience indicates that, when properly sited, offshore wind turbines present only minor risks to birds, often much lower than other forms of energy generation onshore. Avoiding coastal bird flyways, roosting areas, popular feeding areas, and the designated habitats of endangered and threatened birds is the best way to limit adverse bird impacts. Bird migration pathways and locations where migrating birds gather also should be identified. These areas will need to be evaluated to determine whether prevailing flight altitudes and other factors might result in a significant collision risk. Even if the collision risk is low, the displacement of birds from their normal migration pathway or feeding grounds should be evaluated for adverse effects on sensitive species. These assessments should be considered in the project siting analysis.

Major impacts to birds from large spills of fuel, lubricating oils, or dielectric fluids are possible during construction or as a result of vessel allisions with operating wind turbines. These threats may be addressed with a spill prevention and response plan.

Bird advocacy groups should be expected to give heightened scrutiny to the avian impacts of a project and to be especially concerned about the cumulative impacts of new technologies on the little-understood marine environment. Developers should attempt to coordinate with these groups during the scoping process to agree on impact evaluation methods and appropriate studies that will address these concerns and support the preparation of a thorough EIS.

Noise, the marine acoustical environment, and marine mammals. Construction of wind, wave, and ocean current facilities could generate high-intensity noise from pile driving, drilling, and trenching. Removal activities during project decommissioning might include the use of explosives. Although the disruption caused by noise during construction and decommissioning is transitory, high-intensity noise may cause significant impacts on aquatic life, with the most pronounced effects in the immediate vicinity of the activity.

Though less intense, operational noise may be disruptive to wildlife feeding and mating patterns. Wind turbines will generate low levels of noise during operation as the vibrations of rotating machinery in the nacelle are transferred down the tower and into the water. Wave energy devices are expected to generate a minor amount of noise during operation emanating from the terminators, attenuators, and point absorbers that interact with the wave front. Ocean current technologies are expected to have the lowest levels of noise above background levels in comparison to the two other technologies.

Before expensive mitigation efforts are undertaken, it makes sense to measure the noise emissions from the proposed technology and compare the results to the known sensitivities of species likely to frequent the project site. Artificially induced noises may or may not fall within the auditory range of any species of concern. If mitigation is required, isolating vibrating machinery with shock absorbing mounting hardware is one possible noise reduction measure.

Mooring lines used in floating renewable energy facilities present other concerns, particularly as they may affect marine mammals. Moorings may themselves entangle a marine mammal or, more likely, they may become fouled with derelict fishing gear that then becomes a hazard to marine life. Mooring design should give attention to eliminating slack lines, which have a higher potential to loop around marine wildlife. Maintenance and inspection protocols also should be developed that include regular checking and clearing of fishing gear that becomes snagged on the marine energy device. Sonic pingers and other devices that may be installed to warn marine mammals of the location of a facility and keep them away are another possible mitigation method.

Impacts to the ocean floor. Wind turbine foundations may be of several designs, including concrete gravity base, monopile, and tripod structures. A monopile can be set into the seabed in several hours. Construction disturbances of the sea floor are transient, and repopulation of native species and regrowth of essential habitat begins as soon as the construction stops. Wave and current technologies may be attached securely to the sea floor by fixed foundations or flexible moorings, usually with embedded anchors. Mooring design should include a method of maintaining line tension, such as midline floats that suspend the line or chain off the seabed. In the absence of tension, slack lines and chains may drag on the seabed during operation and continually disturb bottom habitat. Foundations and anchors may displace rock, sediment, coral, and bottom dwelling (i.e., benthic) organisms occupying that habitat. Compared to the total surface area occupied by a typical renewable energy facility, only a very small fraction of the seabed would be disturbed by the attachment itself, even considering the aggregate disturbance of a large number of devices in a renewable energy installation.

Connecting the various generating devices to the grid with collection and transmission cables also has an impact on the sea floor, sediment, and biota. The impact can be reduced with modern pipe and cable laying methods commonplace in other offshore energy applications. Mitigation measures used by traditional oil and gas companies that may be used for renewable energy projects include the following:

1. Avoidance of sensitive areas and routes.
2. Aligning the offshore facilities in a way that reduces cable distances and connection routes.
3. Using the least disruptive transmission and cable laying technologies. Jet plowing and horizontal directional drilling techniques, for example, help minimize the disturbances during construction and promote resettling/regrowth after construction is complete.

Planning for the Nantucket Sound project's transmission cables illustrates this point. The developers proposed to interconnect the wind farm to the onshore substation with four, three-conductor insulated cables buried in two trenches, each holding two cables. The cable-laying vessel could install two cables simultaneously into one trench. The developer also considered using three, three-conductor cables with a larger cross section in place of the four cables, but the larger cables could not be installed in pairs. The three-cable alternative required three separate trenches, resulting in greater disturbance to the seabed. The four-cable alternative emerged as the preferred approach, since it was less costly and more environmentally sound.[38]

Air quality impacts. A clear advantage of renewable energy technology is the ability to virtually eliminate air pollution while generating electricity. There are nevertheless minor air quality impacts, particularly during the construction phase, when vessels and equipment are working on-site. Vessel and equipment emissions may contribute to nearshore nonattainment for certain pollutants. EPA and the states are required under the Clean Air Act (CAA) to designate areas that have not attained air quality standards for certain priority pollutants. (See chapter 8 for CAA permitting requirements for operational facilities, vessels (mobile sources), and other equipment.) *Nonattainment* areas seldom extend much beyond the shoreline area of the coast and do not reach the OCS. The nearshore impacts of vessel pollution will be greatest during the construction phase and should be more than outweighed by the pollution avoided by operating the renewable energy facility.

Water quality impacts. Marine renewable energy technologies do not generally produce a discharge of water pollutants as a result of normal operations. Bilge water, deck wash water, and other water discharges from vessels used in construction and maintenance may result in small emissions of oil. Small spills of fuel, lubricants, and solvents also may occur during construction. Reasonable measures to reduce these emissions should be implemented, but inadvertent emissions will typically be easily dispersed in open water settings. Plastic and other waste produced on vessels should be managed appropriately and in accordance with regulations. Sediment dispersal and turbidity in the water column will increase temporarily due to construction and cable laying. Directional drilling used to create a cable conduit without disturbing sensitive nearshore habitats also may cause water quality impacts if drilling mud seeps out of the drill bore. These impacts will generally be minor and transitory.

During operation, marine renewable energy machines may discharge lubricating oils or dielectric fluids in the event of a collision or fire, particularly if the electric service platform, which holds greater quantities of fluids, is involved. Offshore renewable energy facilities should have spill detection, prevention, and response plans in place to mitigate these impacts. Plans should incorporate regular training and drills to ensure rapid response to spills and effective coordination with agencies, such as the Coast Guard, that can provide emergency assistance. On balance, marine renewable energy technologies will eliminate many negative water quality impacts caused by the mining, extraction, transportation, and plant operations that accompany traditional energy production.

Impacts to terrestrial biota and coastal habitat. Only negligible to moderate impacts to terrestrial biota and coastal habitat related to the construction, operation, and decommissioning of onshore support facilities are expected for offshore projects. According to MMS, there are "potential minor to moderate impacts to bats and migratory birds from collisions with wind turbines activities—these impacts could be major for threatened and endangered species whose population levels are low."[39] Experience to date with foreign facilities located more than three miles offshore, however, suggests that heightened concern seems unjustified. Nevertheless, renewable energy developers should address impacts to land-based species that may come into the vicinity of an offshore project.

MMS expects negligible to moderate impacts to coastal habitat and wetlands from dredging, cable installation, onshore construction, and connection to electrical substations and the grid. Dredging and construction in wetlands adjacent to the nation's coastline requires a Corps permit (discussed in chapter 8), which will specify any necessary mitigation measures. If high-value wetlands cannot be avoided, mitigation such as directional drilling to avoid disruption to sensitive sea grass beds and corals may be an alternative. Revegetation with native plants of onshore areas disturbed by construction, combined with invasive species control, also may mitigate onshore impacts.

Fish resources and essential fish habitat (EFH). All else being equal, the farther from shore a renewable project is located, the less likely it is to have major impacts on fish and essential fish habitat. Given the vast expanses of the OCS, cabling and turbine attachments to the sea floor will have minor, mostly transient impacts described in the previous discussion of ocean floor impacts. Seafloor features that support life will be temporarily disrupted by foundation, mooring, and cable installation activities, but resedimentation and regrowth are expected at those sites.

MMS expects negligible to moderate impacts on fish from noise (e.g., pile driving or drilling) and the use of explosives during construction or decommissioning, but population-level effects are considered unlikely for most species and most offshore wind farm locations. The risk to fish from other marine renewable energy technologies is considered to be somewhat higher. Potential moderate impacts are expected if wave and current technologies entrain, impinge, or entrap fish or their prey base.

Impacts to sea turtles. The sea turtle habitat of greatest concern is, for the most part, near to the shoreline. Most offshore wind generating facilities should not impact sea turtles or their habitats with proper siting and construction techniques. Cabling for any renewable project that must transect nearshore waters and make landfall to connect to the grid is expected to have minor to moderate impacts during construction. Impacts from noise, onshore construction, onshore lighting, vessel and facility collisions, and entanglement in moorings could be major if turtle nests or numerous hatchlings are destroyed. These risks are further increased if the density of the wind, wave, or current energy devices is high in critical habitat locations. If it is not possible to avoid turtle habitat, other mitigation techniques may be employed. These include onshore and offshore lighting restrictions, particularly during times when hatchlings may become disoriented, and minimization of disturbance to seashore habitat during sensitive periods (e.g., when eggs are in nests).

Other mitigation techniques include vessel speed restrictions, inclusion of observers to minimize vessel-turtle collisions, and maintenance of offshore facilities to regularly remove derelict fishing gear.

Visual impacts. Existing onshore and proposed offshore wind energy projects have raised concerns regarding visual impacts in the project viewshed. The farther a wind project is located from shore, the less impact it has on the viewshed—a phenomenon attributable to the earth's curvature, which causes objects to eventually fall below the horizon as they recede into the distance (fig. 6–2). The exact distance at which a wind turbine will no longer be seen by an observer on shore depends on the height of the observer and turbine, and lighting, coloration, and atmospheric conditions. Visual impacts are much less for marine hydrokinetic facilities because of their low profile.

Fig. 6–2. Visual impact simulation at various distances. The simulated photograph has been rescaled to fit the space available, and therefore does not recreate the actual view from shore. The simulation does, however, convey the relative impact associated with distance from shore. *Photo courtesy of Deepwater Wind LLC.*

Site-specific impacts from wind facilities may be perceived as positive or negative by different viewers. Not only is beauty in the eye of the beholder, but perceptions of beauty may change over time. Denmark's offshore wind farm experience demonstrates that many individuals who initially opposed projects on aesthetic grounds changed their mind once the wind farm had been in operation. Closer to home, anecdotal evidence from Gulf Coast residents within view of oil and gas platforms demonstrates that many become habituated to the view and accept it as part of the economic fabric of the community.

Despite the highly personal nature of aesthetic sensibilities, there are measures for qualifying and ranking visual intrusions that may prove useful in the context of an EIS. An environment can be described in terms of the visual sensitivity of the resident public and in terms of the visual character, or dominant landscape type, of the environment.

Visual sensitivity refers to public attitudes about certain views (table 6–3). Impacts to views that fall into the high-sensitivity category are likely to engender a strong reaction at least from some members of the public. These views are unique or otherwise special to the region or locale. Views falling into the moderate or low categories are less special because changes to the views would affect fewer members of the public. A lower impact would be less likely to trigger opposition.

Table 6–3. Indicators of visual sensitivity

High Sensitivity
• Views of and from areas the aesthetic values of which are protected in laws, public regulations and policies, and public planning documents.
• Views of and from designated areas of aesthetic, recreational, cultural, or scientific interest, including national, state, county, and community parks, reserves, memorials, scenic roads, trails, interpretive sites of scientific value, scenic overlooks, recreation areas, and historic structures, sites, and districts.
• Views of and from areas or sites of cultural/religious importance to Native Americans.
• Views from national- or state-designated scenic highways or roads, or designated scenic highways or roads of regional importance.
• Views from resort areas.
• Views from urban residential subdivisions.
• Views from segments of travel routes, such as roads, rail lines, pedestrian and equestrian trails, and bicycle paths near designated areas of aesthetic, recreational, cultural, or scientific interest leading directly to them. Views seen while approaching an area of interest may be closely related to the appreciation of the aesthetic, cultural, scientific, or recreational significance of that destination.
Moderate Sensitivity
• Views from segments of travel routes near highly sensitive use areas of interest, serving as a secondary access route to those areas.
• Views from rural residential areas and segments of roads near them which serve as their primary access route.
• Views of and from undesignated but protected or popularly used or appreciated areas of aesthetic, recreational, cultural, or scientific significance at the local, county, or state level.
• Views from highways or roads locally designated as scenic routes and of importance only to the local population, or informally designated as such in literature, road maps, and road atlases.
• Views from travel routes, such as roads, trails, bicycle paths, and equestrian trails leading directly to protected or popularly used undesignated areas important for their aesthetic, recreational, cultural, or scientific interest.
• Views of and from religious facilities and cemeteries.
Low Sensitivity
• Views from travel routes serving as secondary access to moderately sensitive areas.
• Views from farmsteads, or groupings of fewer than four residences.
• Views from industrial research/development, commercial, and agricultural use areas.

Source: U.S. Forest Service 1995. Landscape Aesthetics Handbook. Agriculture Handbook No. 701.

Visual character is a measure of landscape harmony. The types of features prevalent in the area are one aspect of visual character. The pattern or distribution of features is another element, and the quality of the visual resource in terms of the presence or absence of incongruous features also plays a role in determining visual character. This framework would explain why an offshore wind farm located off the coast of Louisiana, among 4,000 offshore oil and gas platforms in the Gulf of Mexico, would more easily match the visual character of that environment than would a wind farm sited in Nantucket Sound. Visual modification class definitions may be used in an EIS to characterize how a proposed wind farm may be perceived in the context of the landscape (table 6–4).

Table 6–4. Visual modification class definitions

VMC	Definition
1	**Not noticeable** Changes in the landscape are within the field of view but generally would be overlooked by all but the most concerned and interested viewers; they generally would not be noticed unless pointed out (inconspicuous because of such factors as distance, screening, low contrast with context, or other features in view, including the adverse impacts of past activities).
2	**Noticeable, visually subordinate** Changes in the landscape would not be overlooked (noticeable to most without being pointed out); they may attract some attention but do not compete for it with other features in the field of view, including the adverse impacts of past activities. Such changes often are perceived as being in the background.
3	**Distracting, visually co-dominant** Changes in the landscape compete for attention with other features in view, including the adverse impacts of past activities (attention is drawn to the change about as frequently as to other features in the landscape).
4	**Visually dominant, demands attention** Changes in the landscape are the focus of attention and tend to become the subject of the view; such changes often cause a lasting impression on the affected landscape.

Source: Bureau of Land Management. Visual Resource Management Program. 1986. Bureau of Land Management Manual Series 8400, Handbooks 8410 and 8431-1, GPO No. 024-011-00116-6.

Minimizing lighting and painting wind turbines a light gray that blends into the horizon may help to manage dissonance with the prevailing visual character of a landscape. Note, however, that the wind turbines located in Copenhagen's harbor were painted white. The citizens of Copenhagen decided to celebrate the nearby turbines as kinetic sculpture rather than relegate them to the background.

MMS considers the visual impacts from wave and ocean current facilities to be negligible to minor. Wind energy project proponents, however, should expect opposition from coastal residents and recreational boaters, particularly for wind and other facilities located close to shore. Cape Wind's Nantucket Sound project provides an early object lesson of opposition based on viewshed impacts.

Other minor environmental and socioeconomic impacts. Depending on site-specific conditions, the following additional factors should be assessed in connection with a marine renewable energy project:

- **Seafloor habitat.** MMS expects negligible to minor impacts from construction noise and electromagnetic fields and potentially major impacts from installation of a facility on uncommon or sensitive seafloor habitat. Oil and gas developers have made it a practice generally to avoid such areas.
- **Areas of special concern.** Shipwreck sites and other sites of special interest should be avoided if possible.
- **Military use areas.** The U.S. Department of Defense will not clear military use areas for development if the private facility interferes with defense operations.
- **Transportation.** Established shipping lanes should be avoided. Major ports and harbors would generally be capable of accommodating additional traffic and should not be impacted by increased vessel activity associated with offshore project construction.
- **Socioeconomic resources.** Renewable energy developers should expect opposition from commercial fishers and related industry groups (e.g., fish processing), particularly when proposed projects would block access to high-volume or high-value fishing areas. Renewable energy facilities also may result in damage to or loss of equipment or vessels due to allisions. Site-specific socioeconomic impacts depend on the size of the population in the area onshore proximate to the offshore facility site. Positive educational and sociocultural impacts may be expected as regional residents become more accustomed to offshore renewable energy.
- **Cultural resources.** There may be negligible to moderate site-specific impacts from disturbance of intact culturally significant sites.

- **Tourism and recreation.** Minor impacts to beach recreation, sightseeing, diving, and recreational fishing are expected.
- **Nonroutine conditions.** Occupational injuries or fatalities also may be possible. Offshore renewable energy projects nevertheless present a low potential for human casualties because few personnel will be present at such facilities.

The *MMS Programmatic Environmental Impact Statement* provides a full discussion of these and other potential impacts.[40]

The environmental assessment process should guide developers to select sites that satisfy key project performance criteria and also present the smallest possible target for project opponents to attack during the NEPA review. Once the developer submits its assessment and application documents to MMS, FERC, or the Corps, the next step in the NEPA process lies with the agency.

Agency development and issuance of the final EIS

In preparing the NEPA analysis, the lead agency gathers the full spectrum of project design and environmental information from the developer, the agency's own staff and consultants, and cooperating federal, state, and local agencies. The lead agency must perform the NEPA analysis and document its efforts first in a draft EIS. Federal agencies use their professional staff and consultants (at times paid for by the applicant) to perform the analysis. The analysis may take months or years, depending upon the complexity of the project and the completeness of the application when received by the agency. The actual NEPA analysis calls for seasoned and expert technical judgment and requires the ability to weigh and balance costs and benefits of the various alternatives. In connection with the analysis, the agency selects a preferred alternative, which may be the project as proposed, some modified alternative, or the no-action option.

As noted previously, once an agency has developed a draft EIS, the agency publishes the draft for comment by interested parties and the public. In response to the comments, the agency may revise the draft. Substantial changes may cause the agency to issue a revised draft for additional comment. If the changes are minor, the agency proceeds to publish a final EIS in support of its record of decision on the underlying renewable energy application. Persons who are dissatisfied with the final agency decision may challenge it in federal court.

Assuming the applicant and the federal agency have done their jobs well, the record of decision on the final EIS clears the path for permitting agencies to make substantive determinations, issue leases, and authorize the permits necessary for a project to proceed. Leases, licenses, and permits issued by MMS, FERC, and the Corps are usually conditioned upon the applicant obtaining the necessary property rights and all applicable federal, state, and local permits and approvals. To avoid delays, applicants for a federally approved project usually seek related permits and clearances at the same time they file the application with the lead agency. The two chapters that follow address other federal permits, consultations, and approvals required for a typical offshore renewable energy project.

7 Nature and Machine Coexisting Offshore: Federal Protections for Natural and Cultural Resources

The nation behaves well if it treats the natural resources as assets which it must turn over to the next generation increased and not impaired in value.

—*Theodore Roosevelt*

It is possible for a terrestrial wind farm to be permitted and constructed within two years. To proceed as fast on the outer continental shelf would be highly unusual. The multiple consultations with local, state, and federal agencies under the many laws that apply to offshore projects are time-consuming. The approval process alone may stretch from three to five years or longer. A developer can shorten the timeline by anticipating the required approvals, arranging for the production of data and analysis of the proposed project's impact during the NEPA process, and coordinating the delivery of that information to all concerned agencies and stakeholders.

Chapter 6 described the process of preparing an environmental impact statement under NEPA, a law focused generally on federal agency decision-making procedures. This chapter examines some important federal laws and regulations that provide targeted protection for particular natural and cultural resources. Knowing how these laws work will help project developers and other stakeholders appreciate the perspectives of the various regulators who must authorize or comment on an offshore renewable energy project.

Fish, Marine Mammals, and Birds

Two federal agencies carry the principal responsibility for protecting fish and wildlife. The U.S. Fish and Wildlife Service (FWS), a branch of the Department of Interior, protects fish and wildlife located generally on land and in nearshore habitats. The National Marine Fisheries Service (NMFS), a branch of the National Oceanic and Atmospheric Administration (NOAA) within the Department of Commerce, regulates commercial fishing. The NMFS also manages and conserves living marine resources in the deeper offshore environs of the U.S. exclusive economic zone (EEZ). It is not unusual for both agencies to become involved in reviewing the effects of a proposed project on the natural environment, since fish and wildlife move freely across the artificial jurisdictional boundaries of the agencies.

Fish protection

The Fish and Wildlife Coordination Act requires consultation with the FWS and any affected state fish and game agency in connection with water resource development projects. Marine energy projects, including wind projects, fall under the purview of this statute, which applies when "the waters of any stream or other body of water are proposed or authorized to be...controlled or modified for any purpose whatever."[1] The law is permissive in that the recommendations that FWS makes with regard to measures that should be adopted to prevent or reduce damage to wildlife resources are not required to be followed by the agency proposing to take an action or issue a permit (the *action agency*). Nevertheless, as with the other consultation statutes, the consultation process creates a forum in which project developers, opponents, and other stakeholders may identify issues and work toward a reasonable mitigation strategy. Mitigation measures may include avoidance, minimization, and rectification of the harm to wildlife or compensation for harm that cannot be avoided. Finavera's attempt to install a WEC pilot project in Washington's Makah Bay provides a good example. As initially designed, Finavera's AquaBuOY would pump ocean water through a turbine to power a generator. After considering the impact on fish and fish larvae that could be entrained and killed in the mechanism, the developers subsequently modified their design into a closed-loop hydraulic pump system. The machine's mooring system also was modified with floats to keep the anchoring chains from dragging on and damaging seafloor habitat.

The impact of an offshore renewable energy project on fisheries is of concern to the NMFS. Under the Magnuson-Stevens Fishery Conservation and Management Act, the NMFS must be consulted by any agency that is proposing to issue a license or permit for a project that may adversely affect any "essential fish habitat" identified by the NMFS or one of the several regional fishery management councils.[2] If the NMFS (in coordination with the affected council) concludes that a project would adversely affect essential fish habitat, NMFS must recommend measures that the action agency should implement to conserve the habitat. The NMFS takes special interest in the possible effects of dredging or other sediment disruption, wetlands crossings, and oil spills that may be associated with project activities. Sediment disruption usually has a short-term impact on water quality. However, it may present a more significant longer term issue if disturbed sediment would cover sensitive seabed communities such as coral reefs or release toxins that previously were lying undisturbed below the surface. The action agency must respond to the recommendations in detail and describe proposed implementation measures to avoid, mitigate, or offset the activity's impact on fish habitat. If the action agency determines not to follow any of the NMFS recommendations, the action agency must explain its reasons.

The term *essential fish habitat* refers to parts of the water column and seabed that are important to the survival of fish species. These areas may include sea grass, salt marsh, coral reefs, kelp forests, and rocky intertidal areas. As fish progress through various life stages, they may move from one habitat to another. Habitats may serve breeding, spawning, nursery, feeding, and protection functions. Areas of particular importance, rarity, or vulnerability that are classified as *habitat areas of particular concern* may exist within areas designated as essential fish habitat. Siting offshore renewable energy projects to avoid particularly sensitive areas is advisable to limit NMFS objections to a project. A developer should consult with the fishery management council in the area of a proposed project early in the planning process. This will facilitate better understanding of the scope of any local fishery management plans (which describe the essential fish habitat and habitat areas of particular concern for each managed species) and identify critical habitats that are best avoided.

Marine mammal protection

The NMFS and the FWS also are involved in the protection of marine mammals under the Marine Mammal Protection Act (MMPA).[3] The MMPA prohibits taking marine mammals, except under certain narrowly defined circumstances. Note that this prohibition applies to all persons, even if they are building or operating an offshore renewable energy project that does not require a federal permit or license. A *taking* may occur in the form of harassment, hunting, capturing, or killing. (The Endangered Species Act and the Migratory Bird Treaty Act, also discussed in this chapter, use a similar taking concept.) Loud noises associated with offshore construction, for example, may injure the hearing of marine mammals or disrupt their mating or feeding behavior and thus constitute harassment. Harassment has been defined as "any act of pursuit, torment or annoyance which has the potential to injure a marine mammal or marine mammal stock in the wild or has the potential to disturb a marine mammal or marine mammal stock in the wild by causing disruption of behavioral patterns, including, but not limited to, migration, breathing, nursing, breeding, feeding, or sheltering."[4]

Penalties for unauthorized takings may be as high as $10,000 per violation ($20,000 for intentional violations) and may also include penalties of up to $25,000 levied against a vessel involved in an unlawful taking, as well as possible seizure and forfeiture of its cargo. Offshore project developers that anticipate disruptions to marine mammals during construction, or otherwise, should seek a permit authorizing the incidental taking. For disruptions of a limited duration and severity (not more than one year and not involving the potential for serious injury or mortality), as can typically be expected during construction, MMPA regulations provide for the issuance of an incidental harassment authorization. To qualify for the authorization, the harassment must be incidental to a specified activity other than commercial fishing, only small numbers of marine mammals may be affected, and the harassment must have only a negligible impact on the affected species or stock.[5] Where the affected marine mammals also are used by subsistence hunters, the impact on the animal community should not be so great that it cannot be mitigated.[6] An incidental harassment authorization will prescribe how the permit holder should minimize its impact on the species of concern and how the taking by harassment should be monitored and reported. NMFS may place an observer aboard the permit holder's vessels to monitor the impact of activities on marine mammals. It is possible to obtain an incidental harassment authorization within about four months (public notice of the proposed issuance is required).

For takings that may extend beyond one year or involve more than mere harassment (i.e., serious injury or mortality could result from the proposed activities even after mitigation), a developer may instead seek a *letter of authorization (LOA)* from NMFS. The procedure for a LOA is somewhat more complicated than the incidental harassment authorization process and takes about 6 to 12 months. In granting an LOA, NMFS must issue specific regulations applicable to the proposed taking.[7]

A project developer can expect NMFS to impose noise limits on acoustic sounding (used to characterize the seabed for foundation engineering) and pile driving. NMFS has required the Cape Wind project, for example, to refrain from emitting sounds in excess of 180 dB to protect sea turtles from stress or injury. A safety zone (typically a 500-m radius around the construction activity) also may be required to keep sensitive species at a distance from intense sound sources so as to lessen the chance of injury. Loud sounds may disrupt the behavior of and cause temporary or permanent hearing loss in whales, dolphins, and seals. Military testing of high-energy sonar, for example, has been associated with whale strandings. Certain fish species such as cod, whiting, and eel also may be affected by percussive sound forces, as the rapid compression caused by high-pressure sound waves may tear their vulnerable swim bladders, causing death.

Disruptions caused by foundation installation and other construction activities may be lessened by various technical approaches. In shallow water it may be possible to use precast concrete gravity base foundations (placed on a level stone pad) instead of monopiles. In relatively softer soils, vibrating pile drivers set piles with much less noise than do impact pile drivers. A bubble curtain around the pile also can lessen the transmission of sound into the surrounding waters. Vessel speed limits, lookouts, and traffic plans that restrict vessels to designated transportation corridors may reduce the likelihood of collisions with marine mammals. Burying cables at least 1 m below the seafloor and ensuring that they are effectively grounded also may reduce electromagnetic field emissions that may disorient certain marine mammals.

Bird protection

One of the nation's oldest conservation laws, the Migratory Bird Treaty Act (MBTA), owes its interesting structure to a constitutional dispute.[8] The act's predecessor, the Migratory Bird Act of 1913, protected all "migratory game and insectivorous birds that...do not remain permanently the entire year within the borders of any State or Territory." The law was immediately challenged as an unjustifiable intrusion into state's rights, and at a time when the federal government's power under the Constitution's Commerce Clause was not sufficiently established, it seemed that this argument would prevail. The law's supporters determined to recharacterize their effort under the federal government's considerably stronger power to make treaties with foreign nations. As a result, at the urging of Senator Elihu Root in 1913, the Senate passed a resolution asking the president to negotiate treaties for the protection of migratory birds. The first treaty with Great Britain, on behalf of Canada, was ratified in 1916. Treaties with Mexico, Japan, and, lastly in 1976, Russia followed. The result is a peculiar piece of environmental law that protects only those birds included under the individual treaties.

Under the MBTA it is generally unlawful "by any means or in any manner, to pursue, hunt, take, capture, kill," or attempt to do the foregoing, to any migratory birds listed under the treaties between the United States and Canada, Mexico, Japan, and Russia.[9] There are currently more than 800 species of birds covered under these treaties.[10] Regulations adopted under the MBTA provide exemptions and allow persons to take listed species if they have applied for and received a permit from the FWS. In the regulations, take means to pursue, hunt, shoot, wound, kill, trap, capture, or collect, or attempt to do any of the same.[11] Further interpretation of this term makes it clear that *take* applies to both intentional and unintentional activities that wound or kill birds.[12]

A knowing violation of the MBTA involving the sale of a bird is a felony punishable by fines, up to two years of imprisonment, or both. Other intentional and unintentional violations of the MBTA would be misdemeanors that could be penalized with a fine of up to $15,000, imprisonment of up to six months, or both.[13] Most violations of the MBTA involve hunters who have intentionally shot at birds. FWS regulations under the MBTA provide for various incidental take permits for activities such as scientific collection or the conduct of military readiness activities. Unfortunately, they do not provide a permitting process for incidental, unintentional takings of migratory birds that may occur, for example, as a result of collisions with wind turbine generators. A corporation may be held liable under the MBTA even if it takes steps to avoid harm and had no intention of causing bird deaths. The conviction of the pesticide manufacturer FMC Corporation for killing migratory birds at its plant in Middleport, New York is a case in point.[14]

FMC produced carbofuran, an insecticide used on corn. The manufacturing process required washing plant equipment to protect employee health. The wash water was collected in a sump where carbofuran residues were supposed to decompose before being released into an open 10-acre containment pond. Unknown to FMC, the sump treatment system did not work as designed, with the result that levels of untreated carbofuran in the pond reached 200 times greater than the level that could cause a significant probability of death to birds. Numerous dead birds were found at the pond, arousing the concern of FMC and wildlife agencies, and birds continued to die over a period of several weeks as the cause of the deaths was investigated.

To mitigate the situation, FMC posted guards and tried foam floats and *zon guns* (loud cannons) to frighten the birds. Rather than frightening the birds, the floats seemed to attract them. The zon guns disturbed the plant's neighbors at night, and the guards were found dozing on the job. An alarm that simulated a bird's distress cry was more effective. Eventually, FMC identified the cause of the problem and added chemicals to the water to break down the carbofuran.

The court in *United States v. FMC Corp.* found that the corporation had failed to act to prevent a toxic hazard that caused the death of protected migratory birds. The court held that the MBTA is a *strict liability* statute, under which criminal liability may be established without the need for proof that the actor knew or should have known that he or she was engaged in an illegal act. Yet the *FMC Corp.* opinion also states that an interpretation of the law "that would bring every killing within the statute, such as deaths caused by automobiles, airplanes, plate glass modern office buildings or picture windows in residential dwellings into which birds fly, would offend reason and common sense."[15] How then should the potential liability of an offshore project developer be estimated for unintentional bird deaths?[16]

Strict liability often applies when people engage in inherently hazardous activities such as using explosives or keeping dangerous wild animals. In *FMC Corp.*, the manufacture of toxic chemicals was seen as an inherently dangerous activity. The production of offshore renewable energy is not an inherently dangerous activity. This distinction, and the exercise of reasonable care in the design and operation of offshore facilities to minimize bird deaths and injuries, should provide refuge from prosecution under the MBTA to companies engaged in offshore renewable energy generation.[17]

Another case involving birds electrocuted on utility-owned transmission lines brings issues under the MBTA closer to home. The government argued in *U.S. v. Moon Lake Electric Ass'n, Inc.* that Moon Lake Electric Association, a rural electric cooperative, failed to install inexpensive equipment on its distribution system poles, resulting in numerous electrocution deaths of protected birds. The court found Moon Lake guilty of violations of the MBTA (and the Bald and Golden Eagle Protection Act) and disagreed with the utility's assertion that the MBTA prohibited only intentional conduct of the kind normally associated with hunting or poaching. *Moon Lake* reinforces the holding in *FMC Corp.* that a company may be held strictly liable for unintentional bird deaths.[18]

The holding in *Moon Lake*, however, distinguishes when an unintentional killing under the MBTA should be considered a violation. The court held that the government must prove that the actor was the *proximate* cause of the harm. In other words, it must be reasonably foreseeable that a wrongful act would produce the injury in the normal course of events without interruption by any intervening cause, and the wrongful act must actually cause the harm. Congressional hearings on the Bald and Golden Eagle Protection Act, which is similar to the MBTA, considered the liability of utilities under the legislation. In response to congressional questioning, the Department of Interior opined:

> *The proposed legislation, in accordance with the protection provided by the third clause of Section 9 of Article I of the United States Constitution, could not be interpreted as operating ex post facto. This means that the power companies would not be liable for acts committed prior to the date of enactment. However, since power lines have a tendency to destroy eagles, such lines erected after the date of enactment should provide such safeguards as are available in order for the power*

> *companies to avoid the charge of acting with "negligent disregard for the consequences" of their acts. This obligation would be no more of a burden upon power companies than upon any other person or organization performing operations which had a tendency to destroy wildlife. In every case, reasonable precautions would have to be taken to prevent the killing of eagles.*[19]

Moon Lake, therefore, is a caution to renewable energy project developers to act with reasonable care. To be in the best position under the law, project developers should take appropriate actions to minimize reasonably foreseeable harm to protected migratory birds. Such actions will help to establish a good faith defense to possible future enforcement efforts. Efforts to protect birds should include consulting with FWS early in the project planning phase to evaluate which birds may be common in the project area and implement good project design and lighting (for example, placing the array so that it minimizes interference with known bird flight paths). Consultation with the FWS also can help determine how the impact on birds during project operations will be monitored, recorded, and reported. If the project developer/owner follows FWS recommendations, FWS would typically be expected to exercise its discretion not to pursue enforcement action in the event of unintentional bird deaths.

Understanding the likely or reasonably foreseeable impact of a project typically requires baseline information about the existing conditions at a proposed project site. Due to the popularity of duck hunting, the FWS has deep knowledge about ducks and their habitat. The Duck Stamp Act in particular can be credited with generating funds that have been used to better understand duck migration patterns and to acquire and protect important habitat. Information on other bird species is generally less available, particularly if a species' flight pattern takes it far from shore. To address the information gaps, MMS has proposed to conduct several studies to learn more about the offshore habitats of birds, fish, and mammals.

> *The evaluation of the impact of a wind project on avian resources requires information about avian resources pre- and post-construction. Potential impacts to birds from offshore wind development involve two major factors: direct collision with wind turbine generators (WTG) or behavioral changes due to the presence of WTGs. Effects of WTGs may include: avoidance of the specific area, barriers to movement (migration, feeding flights), increased energy expenditure, and attraction (feeding and resting sites).*

> *To assess those effects, there is currently a lack of information in key areas. First is knowledge of avian resource (endangered and non-endangered) distribution, movements, and behavior in the specific regions of likely offshore wind energy development. There are currently large gaps in data on the general distribution and abundance, flight patterns (during good weather, nocturnal and inclement weather), and behavior during migration, wintering, foraging, and staging for most avian species that utilize the offshore environment. The second is an understanding of the mechanisms underlying the attraction or avoidance of avian resources to offshore wind facilities.*[20]

The proposed MMS studies include the following:

- A field survey to determine the abundance, distribution, and flight patterns of waterbirds, seabirds, and seaducks in the nearshore Atlantic
- An examination of the potential for bat interactions with offshore wind facilities
- Inquiry into potential artificial reef effects of offshore wind facilities
- An evaluation of the soundscape and potential impacts of noise on marine mammals and turtles from offshore wind facilities

Surprisingly, European studies of bird behavior in the presence of offshore wind turbines show both avoidance and attraction behaviors. Sea ducks have been observed to fly around wind farms. This behavior lessens the opportunities for collisions, but it also may be a problem if the birds must expend significant additional energy to make the detour or if the birds are now avoiding a place where they once fed. Attraction behavior is triggered by the presence of additional feeding opportunities when new populations of fish and invertebrates colonize the hard surfaces provided by wind turbine foundations. Aviation and navigation lights placed on turbines also may attract certain birds. Boat landing platforms at the base of wind turbine foundations and transmission platforms may provide perching opportunities that attract birds to a wind farm, where they may be hit by spinning blades. Further research into these and other areas is necessary, particularly to provide a better understanding of the possible cumulative impacts of offshore renewable energy development (table 7–1).

Table 7–1. Hazards to birds from offshore wind farms

Category	Impact	Population Effect
Collision	Birds may collide with rotor, tower or other structural element, or be injured by air turbulence created by rotor wake.	Population decline.
Avoidance	Birds may avoid wind farm, expending additional energy to fly around the obstacle.	Significant detours may possibly weaken bird population energy levels and cause a population decline.
	Birds may not feed in areas occupied by wind farms and suffer an energy deficit if other adequate feeding areas are not available.	Possible population decline.
Habitat loss	Wind farm foundations may physically destroy habitat used by feeding birds, possibly causing an energy deficit if other adequate feeding areas are not available.	Possible population decline.
	Wind farm foundations may provide hard substrate leading to habitat creation through "reef effect" which may increase available biomass and food.	Possible population increase.

Source: Dong Energy, Vattenfall, Danish Energy Authority, and Danish Forest and Nature Agency. 2006. Danish offshore wind: Key environmental issue: 95, fig. 7.1. http://www.ens.dk/graphics/Publikationer/Havvindmoeller/havvindmoellebog_nov_2006_skrm.pdf

Project developers will fund a good portion of the bird research effort to address stakeholder concerns and thereby minimize delay in the environmental review process. Early consultation with regulators and stakeholders is helpful in designing relevant studies that produce reliable data. The effect of wind farms in the Gulf of Mexico, for example, has been of concern to bird advocacy groups due to the large numbers of birds that migrate over the Gulf to winter in Central America, South America, and the Caribbean. Wind Energy Systems Technology (WEST) faced this issue with its proposed Galveston Offshore Wind (GOW) project.

> *Houston Audubon supports alternative energy sources. This project, however, has the potential to negatively impact millions of birds that nest throughout the eastern half of North America and deserves close attention to details...We are pleased that GOW/WEST has been willing to meet with stakeholders and stated their willingness to share the monitoring data. We understand that the project developers will monitor wind resources and avian activity at two locations on the lease site during this research phase. Houston Audubon would like to see the plan and specifications for the monitoring equipment as we share GOW/WEST's stated goal of obtaining the hard facts regarding the interaction of birds and offshore wind farms.*[21]

Bird advocates want more than access to avian study data. They clearly desire a role in structuring the research plan.

Few would call the investment bankers at Texas Pacific Group (TPG) and Kohlberg Kravis Roberts & Co. (KKR) naïve, yet before announcing their plans to purchase the giant Texas utility TXU, the bankers met with representatives of Environmental Defense Fund and the Natural Resources Defense Council. The environmental groups had long opposed TXU's plans to build 11 pulverized coal plants across Texas. TPG and KKR sought to reach terms with these two potentially major opponents to the transaction and find a compromise position that the environmentalists could support. As a result of the meeting, TXU dropped plans for 8 of the proposed coal plants and agreed to the following:

1. Reduce the utility's carbon emissions
2. Become a leader in the fight against global warming
3. Tie executive compensation to climate stewardship
4. Increase spending on energy efficiency
5. Establish a "Sustainable Energy Advisory Committee" to help guide the utility toward a more environmentally sustainable business model

The environmental groups agreed to serve on the advisory committee and to publicly voice their support for TXU's efforts to address global warming.[22] The TXU and GOW/WEST examples illustrate that, when properly managed, the participation of representatives from wildlife and environmental advocacy groups can support the developer's credibility and lead to higher public satisfaction with the project permitting process.

Endangered Species

When research on the fish, marine mammals, birds, and other wildlife that inhabit the site of a proposed offshore renewable energy project indicates the presence of one or more species with a tenuous hold on existence, an additional level of wildlife protection known as the Endangered Species Act (ESA) may apply.[23] The objective of the ESA is to conserve endangered and threatened species and the ecosystems upon which they depend. *Conserve* is defined under the act as all methods and procedures necessary to bring an endangered or threatened species to the point where the protections provided under the act are no longer necessary. Rather than simply directing federal agencies to do no harm, the ESA sets out an affirmative objective of species recovery. Agencies can promote recovery through "all activities associated with scientific resources management such as research, census, law enforcement, habitat acquisition and maintenance, propagation, live trapping, and transplantation, and, in the extraordinary case where population pressures within a given ecosystem cannot be otherwise relieved, may include regulated taking."[24]

An *endangered species* is a species that is in danger of extinction throughout all or a significant portion of its range, other than certain insect pests. A *threatened species* is a species that is likely to become endangered within the foreseeable future throughout all or a significant portion of its range. The act covers plants as well as birds, fish, mammals, and other life (vertebrates and invertebrates). The secretary of the Department of the Interior and the secretary of the Department of Commerce determine which species are threatened or endangered and publish a list in the *Federal Register.* Interested persons also may petition for a species listing. The list currently includes more than 1,900 threatened or endangered species. *Critical habitat* refers to the specific geographic area occupied by a threatened or endangered species at the time of its listing, plus areas outside of the geographical area occupied by the species at the time of its listing that are essential to its conservation. A critical habitat determination focuses on identifying the physical and biological features that a species needs to survive and recover, and may require special management considerations or protection. The secretary of the Department of the Interior should designate a species' critical habitat at the time of its listing, but in fact fewer than one-half of the listed species have designated critical habitats.

An agency that is proposing to authorize or fund a project must consult with the FWS and NMFS to ensure that the agency action is not likely to jeopardize the continued existence of a threatened or endangered species or result in the destruction or adverse modification of the species' critical habitat. It is likely that both agencies would be involved in the consultation for an offshore project, as the FWS and NMFS have split jurisdiction over marine mammals and joint jurisdiction over endangered salmon. NMFS has jurisdiction over endangered corals as well. For simplicity, the discussion that follows refers only to the FWS, although parallel procedures also would apply to an NMFS consultation. (A useful resource on the FWS/NMFS consultation process is the *Endangered Species Consultation Handbook.*[25])

The consultation process begins when, in connection with proposed *major construction activities*, the action agency asks the FWS whether any species that is listed (or proposed for listing) is present in the project area.[26] Major construction activity is a construction project (or other undertaking having similar physical impacts) that would be considered a major federal action significantly affecting the quality of the human environment under NEPA.[27] With some projects, the consultation ends there if no protected species or critical habitat is present in the project area (fig. 7–1). If the FWS advises that a protected species may be present, the action agency must then prepare a *biological assessment*. This study is for the purpose of identifying any endangered or threatened species likely to be affected by the proposed action and the extent of the impact on the species that may be caused by the proposed project.

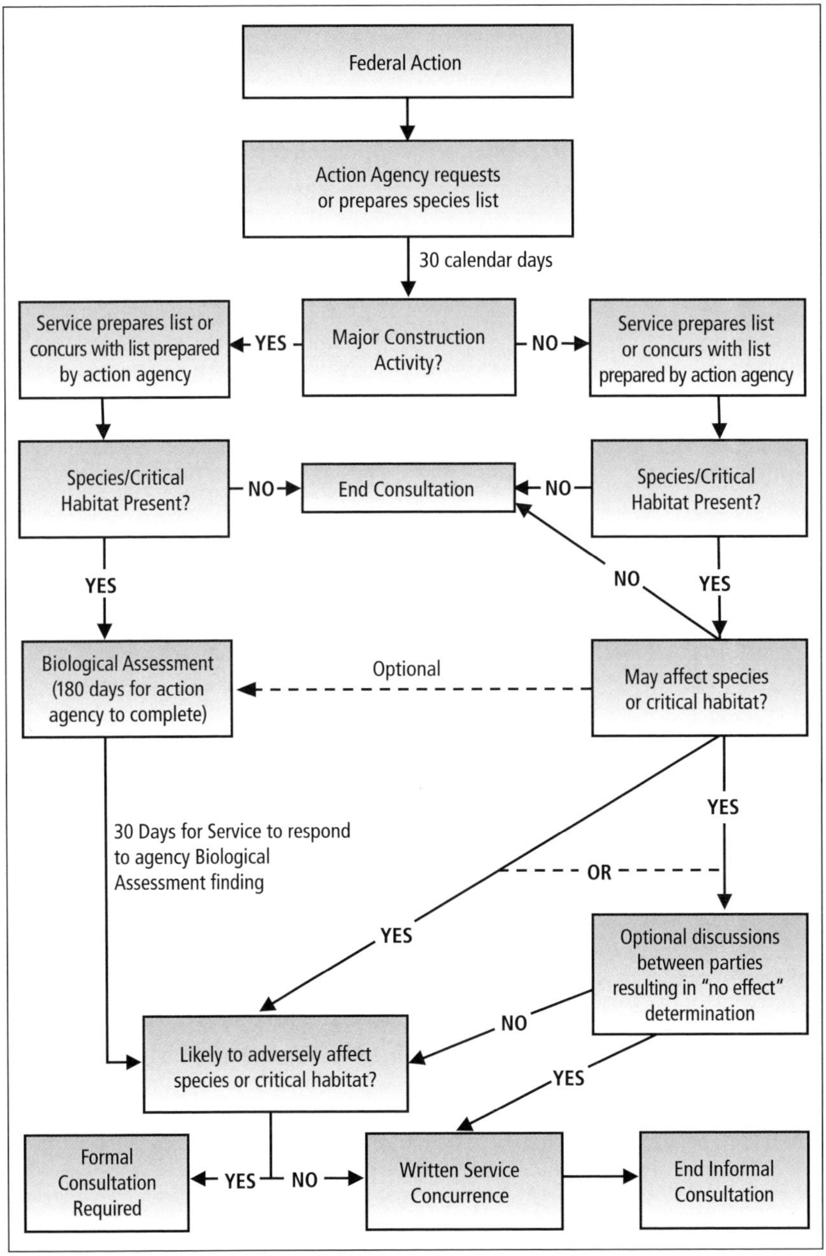

Fig. 7–1. Informal consultation process
Source: U.S. Fish and Wildlife Service

Because a listed species may exist currently in only a fraction of its historical range, it is possible that even if a thorough biological survey does not find threatened or endangered species at a proposed project site, the site may nevertheless be part of one or more species' critical habitat. The destruction or adverse modification of the habitat would then present a significant obstacle to project approval. As noted, fewer than one-half of the listed species have corresponding critical habitat designations, so it is helpful, but not sufficient, to avoid siting in designated critical habitat areas. A project that would encroach on undesignated critical habitat may have such deleterious effects on the recovery prospects of a species that a *jeopardy finding* would result, barring the project for that reason.

Decisions regarding the impact of a project on a threatened or endangered species are required to be made based on the best scientific and commercial data available. Thus it is reasonable to ask how regulators deal with cases where there is little existing research on a species or the habitat in which a project is proposed to be sited. In these cases, the FWS will ask the action agency to agree to an extension in the timelines prescribed under the ESA to allow for further research and a more complete analysis. If the action agency resists, the FWS will proceed to make its determinations based on the limited information available but will resolve any doubts in favor of a cautious approach that provides the most protection for the species at issue. The result of the second option is more likely to be a jeopardy finding that blocks the project.

Thorough species studies may take a long time. Species may be present in an area during only part of the year, and habitat usage may change from year to year depending on weather, food availability, and other factors. The Massachusetts Audubon Society conducted a preliminary survey of the endangered roseate tern in connection with the Nantucket Sound project. The society used observers in boats and planes to record the time, number, and location of terns in an attempt to better understand their distribution throughout the Nantucket Sound region. It was particularly important to learn where the birds stage in advance of their migration, and also where they feed, to determine whether the proposed wind farm would disrupt this behavior or present a significant collision risk. While the absence of tern concentrations close to the shoal where the project would be sited was a promising initial finding, the society concluded that a longer study would be required.

> *We believe that adequate assessment of the use of the Sound by terns will require at least three years of surveys, during spring migration, the breeding season, and through the full staging period. For example, tern distribution may shift annually as a function of shifts in the local*

> *distribution of fish, and the fish distribution may in turn be influenced by factors such as annual variations in water temperatures. A minimum of a three-year study period may enable us to detect these annual variations and to distinguish between alternative explanations for the patterns that we have observed.*[28]

The findings of the biological assessment determine whether formal consultation under ESA section 7 is necessary. The biological assessment may demonstrate that the proposed action is *not likely to adversely affect* listed species or critical habitat. Thereafter the FWS may concur in a written statement, and the consultation would end.

A taking under the ESA can occur merely through harassment of certain protected animals, such as when construction activities change feeding, mating, or other behavior patterns. When an incidental taking is anticipated to occur as a result of the proposed action, the biological assessment will likely conclude that the proposed action is *likely to adversely affect* the subject species.

When the action agency finds a likelihood of an adverse effect, a formal consultation with the FWS begins (referred to as a *section 7 consultation* after the applicable ESA section). The action agency, the project developer, and the FWS attempt to agree on reasonable measures to minimize the impact of any incidental taking on endangered or threatened species (fig. 7–2). The mitigation plan also will include terms and conditions related to implementing the protective measures and reporting requirements. Once the FWS concludes that, subject to the plan's protections and conditions, the proposed action would not be likely to jeopardize the continued existence of the species or adversely modify its critical habitat, it will issue a written statement that describes the affected species, its population dynamics, and the effects of the proposed action. This part of the written statement is referred to as a *biological opinion*. The written statement also will describe, in what is known as the *incidental take statement*, the anticipated incidental taking of the species and its effects, and memorialize conditions imposed on the action agency and project developer to mitigate the adverse effects. Both documents are often loosely referred to as the *biological opinion* or *BiOp*. A BiOp may conclude that, in light of the mitigation agreed pursuant to the incidental take statement, the proposed action will not jeopardize the continued existence of threatened or endangered species (and will not adversely modify critical habitat). This opinion is a green light for the action agency to proceed with permitting the project.

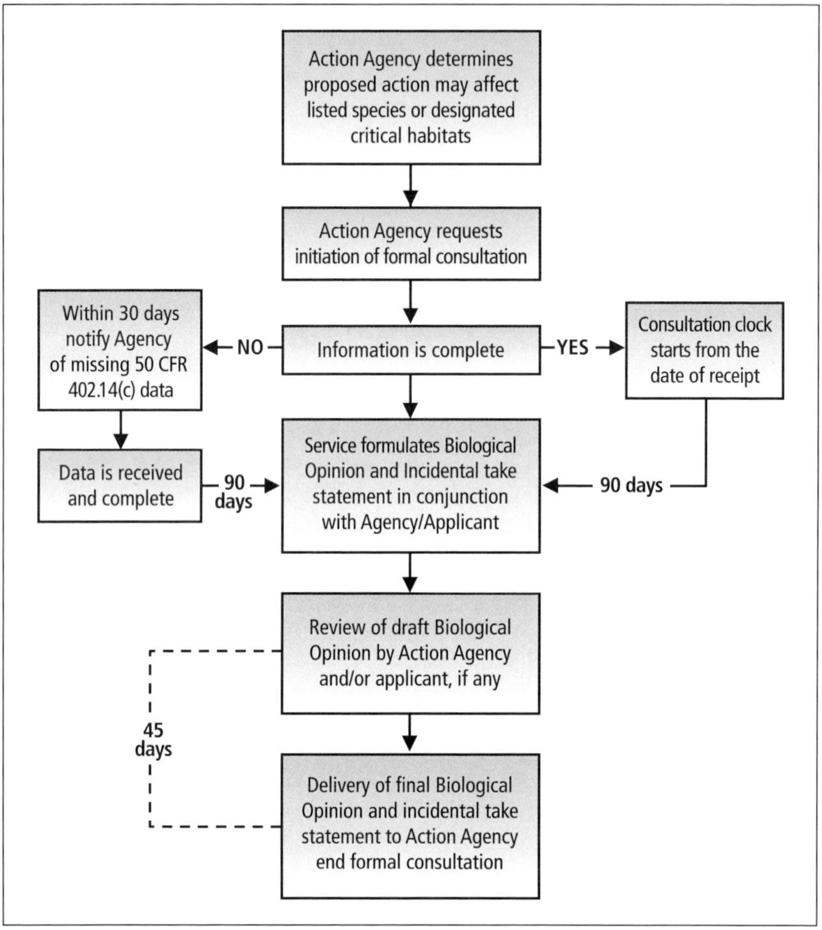

Fig. 7–2. Formal consultation process
Source: U.S. Fish and Wildlife Service

The project developer has an important role in the consultation process. The developer should coordinate with the action agency to submit information to the FWS staff as required, obtain a draft of the BiOp, and provide comments on the BiOp. Guidelines for FWS/NMFS biologists suggest:

> *Section 7 consultation is a cooperative process. The Services do not have all the answers. Actively seek the views of the action agency and its designated representatives, and involve them in your opinion preparation, especially in the development of reasonable and prudent alternatives, reasonable and prudent measures, terms and conditions to minimize the impacts of incidental take, and conservation recommendations.*

> *Use all aspects of section 7, especially opportunities for informal consultation where solutions can be worked out prior to the structured process mandated by formal consultation. Be creative, and make the process work to the species' advantage.*[29]

When the action agency and project developer act within the scope of an incidental take statement, they are protected from liability under section 9 of the ESA, which prohibits any person from taking a listed species. (The ESA provides that "any taking that is in compliance with the terms and conditions specified in a written statement provided under subsection [b][4][iv] of this section shall not be considered to be a prohibited taking of the species concerned."[30]) Contrast this with a developer that obtains an FWS concurrence that the project is not likely to jeopardize a threatened or endangered species or adversely modify its critical habitat. If the project subsequently results in a taking in fact, the developer could be held in violation of the ESA and possibly also subject to a project-halting injunction.

Of course, if the impact of a proposed project on an endangered or threatened species cannot be mitigated, the FWS will issue a BiOp concluding that jeopardy to the survival of the species would result. An appeal to the so-called God Squad would be a possible, though rarely used, alternative in such cases. Formally known as the *Endangered Species Committee*, the God Squad is composed of various cabinet secretaries and members from states affected by the proposed action. The God Squad nickname stems from the committee's ability to sign the death warrant for a species—it can reject an FWS jeopardy decision and allow the action agency to proceed with a project that may in fact permanently eliminate a species. An action agency that intends to seek a God Squad ruling would file an application for exemption with the Secretary of the Interior, who would then hold a hearing and file a report with the committee. The committee may grant an exemption if it finds that

> *(A)(i) there are no reasonable and prudent alternatives to the agency action;*
>
> *(ii) the benefits of such action clearly outweigh the benefits of alternative courses of action consistent with conserving the species or its critical habitat, and such action is in the public interest;*
>
> *(iii) the action is of regional or national significance; and*
>
> *(iv) neither the Federal agency concerned nor the exemption applicant made any irreversible or irretrievable commitment of resources prohibited by subsection (d) of this section; and*

> (B) *it establishes such reasonable mitigation and enhancement measures, including, but not limited to, live propagation, transplantation, and habitat acquisition and improvement, as are necessary and appropriate to minimize the adverse effects of the agency action upon the endangered species, threatened species, or critical habitat concerned.*[31]

Although an action agency may be cleared to proceed with permitting after engaging in a section 7 consultation, additional steps are still required for full ESA compliance. Persons, corporations, and other entities also may find themselves regulated under ESA section 9. As a general matter, unless a permit from the FWS has been obtained under section 10, it is unlawful to import or export, possess, sell, transport, or take a listed species. *Take* is broadly defined to include "harass, harm, pursue, hunt, shoot, wound, kill, trap, capture, or collect, or to attempt to engage in any such conduct."[32] *Harass* as used in the ESA's definition of take means an intentional or negligent act or omission that creates the likelihood of injury to wildlife by annoying it to such an extent as to significantly disrupt normal behavioral patterns, including, but not limited to, breeding, feeding, or sheltering. *Harm* is similar to harassment but more serious in that it may result in injury or death. *Harm* has been defined as an act that actually kills or injures wildlife, including significant habitat modification or degradation where it actually kills or injures wildlife by significantly impairing essential behavioral patterns, including breeding, feeding, or sheltering.[33] Pile driving when endangered marine mammals are present is an example of prohibited harassment. A collision between an endangered whale and a vessel, even if the operator was proceeding with caution and had no reason to believe endangered species were present, causes harm and constitutes a prohibited taking.

Intentional violations of the ESA may carry civil fines of up to $25,000 per violation, while an unintentional violation exposes the actor to a fine not to exceed $500 per violation. Intentional violations also carry a risk of criminal prosecution and, upon conviction, a fine of up to $50,000, one year in prison, or both. Vessels and other equipment used to aid an illegal taking also may be forfeited upon conviction of a crime.

Private parties may take part in enforcing the ESA. Citizens may petition for the listing of species and seek injunctions against persons alleged to be in violation of the act.[34] The act requires the secretary to consider petitions for the listing of species and authorizes civil suits to enjoin persons from violating the ESA. It also compels the secretary to perform nondiscretionary duties under the ESA. This *private right of action* provides an additional avenue for project opponents to prevent, delay, and interfere with an offshore renewable energy project. Lastly, the U.S. attorney also may seek an injunction against any person alleged to be in violation of the ESA.

Persons who aid in the enforcement of the ESA may earn a reward, so work site violations of the ESA are unlikely to remain secret for long. Consider this hypothetical example: A company manager in charge of offshore foundation installation is substantially behind schedule due to prolonged bad weather. At last the weather clears and workers and equipment are permitted under safety rules to return to the project site. Work is about to commence when a marine biologist positioned high above the deck of one vessel spots a group of endangered northern right whales that appear to be engaged in feeding behavior. The biologist advises the project manager of the whales' presence and the need to postpone pile driving until the whales have cleared the area. The project manager nevertheless orders his workers to proceed and the whales leave the area, presumably repelled by loud noises. This series of plausible events could expose a company and its executives to criminal prosecution, fines, an injunction, and vessel seizure. It is important, therefore, that project development companies establish reporting relationships that separate environmental compliance responsibility from project construction and operations responsibility. Furthermore, firms must place a priority on appropriately training their employees and communicating to them the importance of environmental compliance.

There is an interesting quirk in the taking prohibitions of section 9 of the ESA and the incidental take permitting authority in section 10. Section 9(a)(1) provides in relevant part that it is unlawful, except as provided in section 10, for any person subject to the jurisdiction of the United States to "(B) take any such species within the United States or the territorial sea of the United States; [or] (C) take any such species upon the high seas."[35] Section 10(a)(1)(B) allows the FWS to issue a permit, which may include limiting terms and conditions, for "any taking otherwise prohibited by section 9(a)(1)(B) if such taking is incidental to, and not the purpose of, the carrying out of an otherwise lawful activity."[36] Section 10 does not include authority for FWS to permit an incidental taking on the high seas that would otherwise be unlawful under section 9(a)(1)(C) of the ESA.

The U.S. territorial sea extends from the baseline (which is the coastal lower low waterline shown on official U.S. nautical charts) seaward 12 nm. The furthest extent of U.S. jurisdiction is the EEZ, which encompasses an area between the most seaward extent of the territorial sea and 200 nm seaward of the baseline. The term *high seas* generally refers to areas beyond a nation's jurisdiction, which means areas of the ocean beyond the EEZ. This parsing of the statute presents at least two questions. First, does the United States assert jurisdiction over takings between 12 and 200 nm from the baseline? And second, if jurisdiction over takings extends to this intermediate area and beyond to the high seas, can a project developer obtain a permit for incidental takes that occur throughout the whole area?

The practical answer is, of course, that the United States asserts jurisdiction over all takings, even in the intermediate zone between 12 and 200 nm. The FWS considers the high seas for purposes of the ESA to start where the territorial sea ends (i.e., 12 nm from shore). By interpreting section 9 of the ESA in this manner, there is no gap in the prohibition on takings. The answer to the second question is that, as a practical matter, even if section 10 does not provide authorization for incidental takings on the high seas, the FWS can provide such authorization in the biological opinion issued pursuant to a section 7 consultation. It would be difficult to imagine an offshore renewable energy project in U.S. waters beyond 12 nm that would not involve a federal permit and at least a preliminary section 7 consultation. Incidental takings associated with projects on the high seas, therefore, should be authorized in a biological opinion issued under the section 7 consultation process.

National Parks, Wildlife Refuges, Marine Sanctuaries, and Monuments

The MMS does not have authority to issue leases on the outer continental shelf for areas within the boundaries of any unit of the National Park System, National Wildlife Refuge System, National Marine Sanctuary System, or any national monument.[37] Given the absence of leasing authority and the likely public and regulatory opposition that a proposal to site a renewable energy facility in an area of natural and cultural significance would encounter, this section focuses on issues that may arise when projects are sited close to, but not inside of, a protected area.

The United States has more than 700 national parks, wildlife refuges, marine sanctuaries, and monuments, so there is a good chance that one will be proximate to an offshore project. Almost anything sited off the California coast, for example, could impact the California Coastal National Monument, which is comprised of more than 20,000 small islands, rocks, exposed reefs, and pinnacles along 1,100 miles of California's coast, out to a distance of 12 nm from the shoreline (fig. 7–3). The monument includes only those public lands that are exposed above mean high tide.

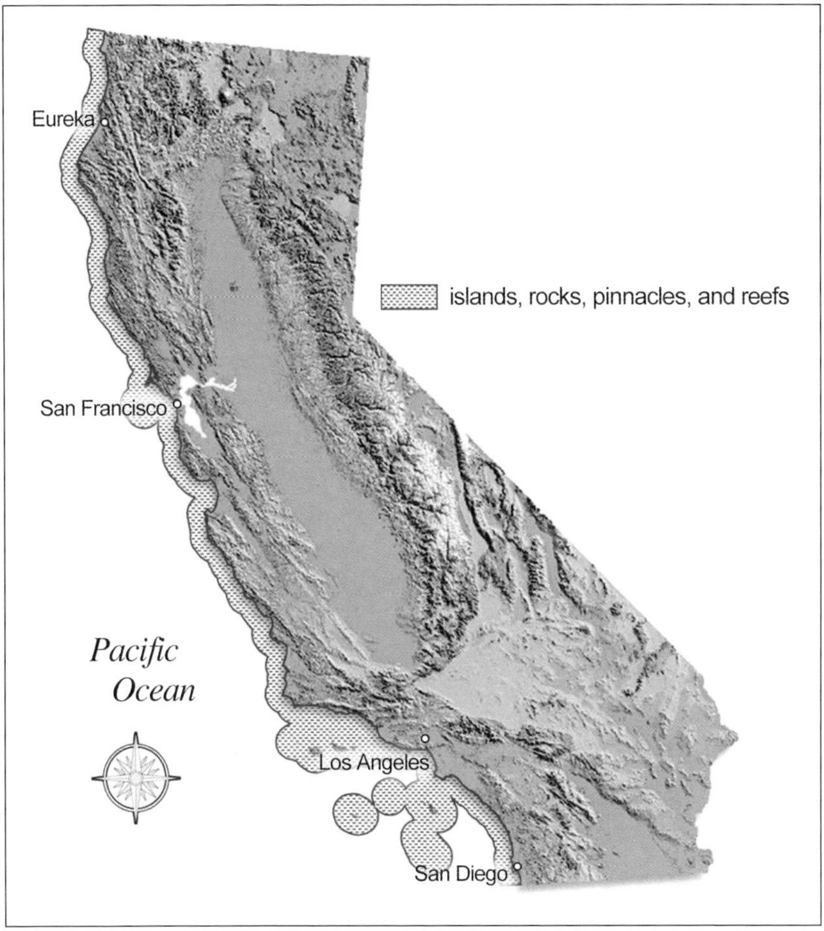

Fig. 7–3. The California Coastal National Monument
Source: U.S. Department of the Interior, Bureau of Land Management

It would seem at first glance that, as a technical matter, MMS could lease areas of the seabed and water column that lie in the midst of the islands that form the California monument. The Outer Continental Shelf Lands Act, however, prohibits MMS from leasing sites on any part of the outer continental shelf *within the exterior boundaries* of any unit of the national parks, wildlife refuges, marine sanctuaries, or monuments.[38] In the case of the California Coastal National Monument (and presumably other places with complex boundaries), MMS leasing would be permitted only in areas that are *exterior* to the entire assemblage of protected islands taken as a whole. This reading of the law may effectively limit renewable energy development off the California coast to sites further than 12 nm from shore. Whether the Bureau of Land

Management, which administers the national monument, or another agency such as FERC or the Corps could authorize a lease on the outer continental shelf in waters between the islands comprising the monument is a separate and interesting question.

There are 13 national marine sanctuaries. Among them, a large part of the northern coast of Washington state extending 25 to 50 miles from shore is under the protection of the Olympic Coast National Marine Sanctuary. On the East Coast, the Stellwagen Bank National Marine Sanctuary sits at the mouth of Massachusetts Bay between Cape Cod and Cape Ann. A significant part of the waters surrounding the Hawaiian Islands is protected as a humpback whale sanctuary, and a sanctuary surrounding the Florida Keys was established to protect unique coral, mangrove swamp, and other habitats.

Offshore renewable energy projects that propose to locate near a marine sanctuary must comply with the National Marine Sanctuaries Act (NMSA).[39] NMSA provides that "federal agency actions internal or external to a national marine sanctuary, including private activities authorized by licenses, leases, or permits, that are likely to destroy, cause the loss of, or injure any sanctuary resource are subject to consultation with the Secretary" of the Department of Commerce.[40] Further, no person may destroy, cause the loss of, or injure any sanctuary resource that is managed under the law or regulations for that sanctuary. Penalties for violations of the NMSA may be quite significant, including imprisonment and civil penalties up to $100,000 per violation. It should draw the attention of both fishers and builders of offshore energy projects that fines may be assessed as a maritime lien against vessels involved in the commission of an offense.[41]

The use of mooring chains on floating wave power devices raises the issue whether whales and turtles, for instance, might become entangled. Potential entanglement could be a threat to a sanctuary resource that would require a consultation between the permitting agency and NOAA, the agency under the Department of Commerce that manages the National Marine Sanctuary Program. Vessel collisions also might injure or kill marine mammals. Construction and operations traffic around an offshore renewable energy facility creates a risk of collision and therefore should be considered in the consultation.

NOAA is required to administer national marine sanctuaries in a manner that accommodates public and private uses of the sanctuary that are compatible with the sanctuary's resource conservation mission. NOAA prepares sanctuary management plans to accommodate these uses. A review of the relevant management plan may help project developers avoid or mitigate conflicts between a project and a neighboring marine sanctuary. Sometimes activities in the sanctuary are necessary in connection with a project (e.g., a transmission cable crossing). In these cases, the NMSA provides that NOAA may issue special use permits, provided, among other things, that the use is compatible with the protection of sanctuary resources.

Under the consultation procedure, NOAA reviews the agency's explanation of the proposed project and the potential effects on sanctuary resources. NOAA then responds with *reasonable and prudent alternatives* to protect the sanctuary, including possibly recommending that the project be located elsewhere. The agency must consult with NOAA regarding the alternatives and explain its reasons in writing if it chooses not to follow NOAA's recommendations.

In most cases the agency will follow NOAA's recommendations, because if it does not do so, and sanctuary resources are subsequently lost or destroyed, the agency must "promptly prevent and mitigate further damage and restore or replace the sanctuary resource" in a manner approved by NOAA.[42] It is unlikely that the permitting agency would override NOAA's objections to the siting of an offshore renewable energy facility. If the facility did indeed result in the loss or destruction of, or injury to, sanctuary resources, it may be impossible to mitigate further damage short of requiring the facility to cease operation and be dismantled.

Historic Objects

Objects of archeological or historical significance are protected under several federal laws. A shipwreck is the most common archeological object that is likely to be found in connection with an offshore renewable energy project. The Abandoned Shipwreck Act transfers title to shipwrecks located on state submerged lands to the states and encourages them to develop underwater parks for the protection and enjoyment of historic shipwrecks. Abandoned shipwrecks on federal submerged lands remain the property of the U.S. government. Historic shipwrecks are generally catalogued in the National Register of Historic Places, though consultation with a state's historic preservation office is advised to learn about wrecks that may be known but not yet fully evaluated for historical significance. The National Register Information System provides a listing of historic shipwrecks in a searchable database.[44] The National Park Service also has a helpful listing of historic shipwrecks by state.[45]

Maine's rocky, 3,500-mile coastline, with more than 2,000 islands, is fertile ground for wrecks. The Maine Historic Preservation

Commission's shipwreck inventory has more than 1,300 entries. This list includes ships from the time of the nation's founding to more modern times. Among the historically significant falls the *Defence*, a wooden brigantine from the Revolutionary War era that sank while privateering in the shallow waters of Stockton Springs Harbor in 1779.

Even when historically unimportant, the position of all wrecks near a project site should be known. The freighter *Empire Knight*, for example, struck a ledge near Boon Island, Maine in a February 1944 storm. The ship broke in half and sank, scattering its cargo of war materials in 260 ft of water. Due to the high mercury content of some of the cargo, the U.S. Coast Guard has established a permanent safety zone around the wreck in which fishing, diving, cable laying, dredging, and other activity is prohibited.[46]

A marine renewable energy project may influence the views from historic coastal properties or, in an unusual case, change sediment transport and erosion patterns along a shoreline threatening a historic structure such as a lighthouse. It is more likely, however, that onshore construction, for instance, building an electrical substation, would be the source of the most significant impacts on historic properties.

The National Historic Preservation Act (NHPA) requires a federal agency to take into account the effect of its activities, including permit issuances, on properties included in, or eligible for, the National Register of Historic Places.[47] The agency is required to make a good faith effort to identify historic properties through appropriate efforts such as background research, consultation, oral history interviews, and field surveys. The burden of much of this work is likely to fall on the applicant for the agency permit. Once identified, reasonable efforts must be made to avoid negatively impacting the protected property. An applicant for a federal permit should not intentionally damage historical properties in an effort to circumvent the NHPA process. In such cases, the federal permitting agency would be barred from granting the applicant's permit, and the applicant's objective would be frustrated.[48]

Like NEPA, NHPA requires an agency to understand the ramifications of its actions and to consider options to mitigate detrimental effects. However, provided that the agency's evaluation is thorough, the agency is not barred from proceeding with an action. If a historic place is affected by a project, the federal agency must provide the Advisory Council on Historic Preservation with an opportunity to comment. The Advisory Council is an independent federal agency charged with promoting the preservation of national historic resources and advising the president and Congress on this topic. An affected state's historic preservation office, local and Indian tribal officials, and members of the public also should be given the opportunity to comment.

An agency action may have an adverse effect on a historic property if the action diminishes "the integrity of the property's location, design, setting, materials, workmanship, feeling, or association."[49] Adverse effects may include the "introduction of visual, atmospheric or audible elements that diminish the integrity of the property's significant historic features."[50]

An agency's consultation with the Advisory Council must occur prior to the issuance of a project permit affecting a historic property.[51] Ideally, the agency should review the impact of the proposed action on historic properties in connection with the NEPA determination whether the proposed action is a major one with significant effects on the quality of the human environment. It is important to note, however, that a finding of an adverse effect on a historic property does not by itself amount to a significant impact on the human environment that would necessitate the preparation of an EIS.

The Archaeological and Historic Preservation Act also may apply when a federally licensed activity would cause "irreparable loss or destruction of significant scientific, prehistorical, historical, or archeological data."[52] The action agency involved in such a project should notify the Secretary of the Interior, and either the action agency or the Secretary should recover or preserve the cultural data if the Secretary determines that the data is significant and may be irrevocably lost or destroyed. The original title of the law, the Reservoir Salvage Act, hints at its principal purpose—to save parts of the U.S. cultural heritage that would be destroyed in connection with dam building. Notably, the act provides that persons who suffer delays in construction as a consequence of data recovery efforts may receive compensation. However, it also provides that federal licensees and permittees may be charged reasonable costs for data recovery activities within project areas as a condition of the issuance of a license or permit.[53]

Lessons about the Effects of Offshore Renewable Energy on Wildlife

A developer preparing a biological assessment or preparing for a consultation with a wildlife agency should gather relevant information and identify important knowledge gaps. The first stop for this information should be the MMS work *Worldwide Synthesis and Analysis of Existing Information Regarding Environmental Effects of Alternative Energy Uses on the Outer Continental Shelf.*[54] This review of biological literature provides a perspective on what scientists know and still need to learn about the possible impact of renewable energy on the marine environment. The *Worldwide Synthesis* summarizes many North American wildlife studies and references European studies that document practical experience with existing offshore wind farms.

Wind turbines have operated in European waters for more than 15 years. The most comprehensive of the European studies comes from Denmark, where wind power supplies 20% of Danish electricity consumption. Because Denmark's government wants to significantly increase the share of renewable energy in the future, Danish researchers have given particular attention to identifying and managing the cumulative effects of offshore wind energy development. Denmark's experience demonstrates that properly sited offshore wind farms can be operated without substantial damage to the marine environment.[55]

Researchers studied Denmark's two largest offshore projects, Nysted (165.5 MW, 72 turbines) and Horns Rev (160 MW, 80 turbines), to evaluate their effect on birds, fish, and mammals. Between 2000 and 2006, researchers documented the ecological baseline conditions, observed the impact of wind farm construction, and then followed up with further observations of the ecological conditions surrounding the wind farms in operation. The research was reviewed by the International Advisory Panel of Experts on Marine Ecology, who found it "very much at the forefront of research into the environmental effects of offshore wind farms."[56]

U.S. developers with an interest in fostering an environmental review process that is perceived as fair by all stakeholders should note the manner in which the Danish research proceeded. A green advisory coalition representing the World Wide Fund for Nature, the Danish Society for Conservation of Nature, the Danish Outdoor Council, Greenpeace, the Danish Ornithological Society, and the Danish Organisation for Renewable Energy provided input on how the studies should be structured.

The most dramatic changes observed by the researchers occurred on the seabed. The turbine foundations and their surrounding rock rubble scour protection changed the seabed in the immediate vicinity of the foundations from the typical sandy sediments to a hard bottom. The hard surfaces provide habitat for various species, and a more heterogeneous mix of sea life developed within the wind farm boundary. Biomass increased in the area of the turbine foundations by 50 to 150 times compared to the native undisturbed communities. More biomass, of course, means a greater supply of food for birds and fish.

Surprisingly, the turbine foundations did not appear to attract more fish as has been observed with the artificial reef effect around oil and gas structures. The researchers guessed that either the mussels that colonized the foundations were not attractive to the fish, or the reef simply needed more time to develop. Electromagnetic fields associated with collection cables within the farms and transmission cables stretching to shore had little pronounced effect. Some fish species seemed attracted to, while others avoided, the cable paths. The study found that migration of some fish species may be impaired, but migration would not be blocked by cable electromagnetic fields.

The efforts to study marine mammals employed sophisticated techniques. Acoustic monitoring, remote-controlled video, and tagging with satellite transmitters were used to supplement traditional visual observations. Noise during construction operations discouraged seals and porpoises from using the area in the vicinity of the wind farms. Indeed, porpoises as far as 9.3 miles (15 km) from the wind farm were observed to react from the noise of pile driving. "During the construction phase, the number of porpoises at the farms decreased immediately when noisy activities commenced, alleviating fears that marine mammals would remain in the area and so might be hurt by the intense pressures generated by pile driving."[57]

The number and behavior of seals appears to have returned to normal postconstruction conditions. Porpoises, however, are taking longer to reach their preconstruction numbers in the vicinity of the Nysted wind farm, while at Horns Rev, porpoise abundance has recovered.

Researchers from Denmark's Aarhus University used infrared video monitoring, radar, and visual observations to track bird behavior. In the vicinity of the Nysted and Horns Rev, wind farms behavior was species specific, although most of the more numerous bird species tended to avoid the facilities, resulting in an overall low expected frequency of collision with the turbines. Based on radar observations, the Danish researchers were able to build a model to predict bird collision risk. The model ran 1,000 iterations and predicted with 95% certainty that out of 235,000 passing birds during a single autumn migration, 0.018%–0.020% would collide with the turbines in the farm, resulting in the death of 41–48 individual birds (0.57 to 0.67 birds per turbine per season).[58]

A separate study of offshore wind farms in Sweden's Kalmar Sound observed 1.5 million migrating waterfowl and detected no wind turbine collisions. The study estimated the bird/turbine collision risk at 1.2 birds per turbine per year.[59] Birds tended to fly around the periphery of, rather than through, the wind farms, and this extra flight distance was judged too small to be a meaningful impediment to migration (fig. 7–4). "Birds flying through the wind farm tend to alter altitude to avoid the risk of collision. Under adverse weather conditions, which were thought to be likely to increase collision risk, results show that waterbirds tend to avoid flying. The strong avoidance behaviour results in very low estimates of collision risk but of course increases habitat loss and increases costs of travel."[60]

The Danish study found that wind farm lighting did not attract large nocturnal migrating waterbirds, which had been a concern. Habitat loss as a result of the avoidance behavior was also not significant in the context of the large areas of habitat remaining outside the wind farm boundaries. Notably, neither wind farm was located close to nesting areas, thus avoiding effects on breeding and reproduction.

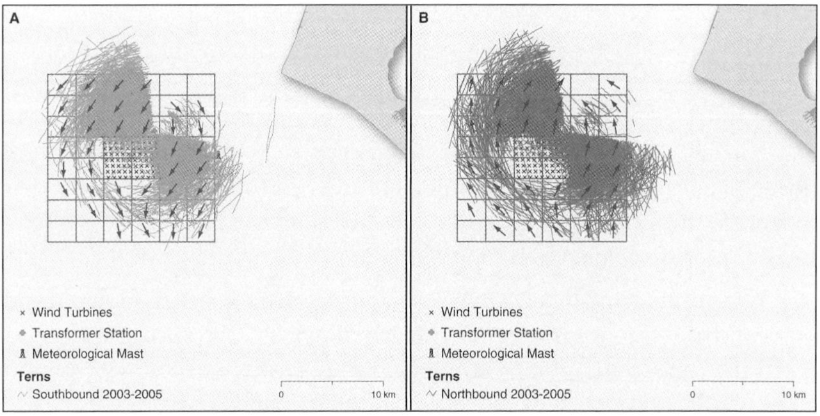

Fig. 7–4. Bird tracks during migration from radar study. *Courtesy of Thomas Kjaer Christensen, National Environmental Research Institute, Aarhus University, Denmark*

Environmental management systems, which included procedures and instructions on how workers should address environmental issues, were used by both projects. Waste handling, contingency plans for accidents such as oil spills, noise measurement, and instructions on scaring off marine mammals prior to the initiation of noisy activities were among the topics addressed by the environmental management systems. Marine mammals were scared off with acoustic deterrent devices commonly called *pingers* and *seal scarers*, and pile driving commenced with a *soft start*, in which full impact forces were gradually reached to allow sensitive sea life time to flee the area. In addition, a transportation corridor was established to minimize the disruptive effect of construction vessel traffic.

It would be hazardous to rely too much on a study of two wind farms in Denmark in predicting how wind farms may affect wildlife in the United States, particularly on a cumulative basis. Making generalizations about the environmental impact of hydrokinetic energy projects based on offshore wind project information also presents difficulties because of the large variety of wave and current energy technologies. Nevertheless, many of the concerns with respect to noise, habitat displacement, electromagnetic fields, and collision risks are generally similar. Clearly, there is still much to learn about how offshore renewable energy facilities may affect the complex web of marine life over time. When construction disturbs sediments, for example, bottom dwelling organisms may be affected. This may in turn lessen food available to certain fish species that are prey for waterbirds and marine mammals. Careful before-and-after studies, such as those conducted in Denmark, are most likely the best way to discover whether marine renewable energy development, which seems benign in the short-term, has lasting ecological effects.

An absence of comprehensive information should not block well-designed projects pending the completion of exhaustive studies. Development should proceed with the best available information, conducting reasonable before-and-after studies to advance understanding of the impact, both positive and negative, of offshore renewable energy on species and ecosystems over time. The preliminary information is encouraging. As Denmark's experience shows, properly sited offshore wind farms can provide sustainable energy without causing significant environmental harm.

The laws described in this chapter for the protection of wildlife and cultural resources are not the only requirements applicable to offshore projects. In the chapter that follows, other regulatory schemes are considered that offshore project developers will likely confront. The next chapter also looks at how state regulators may play an important role in the project approval process, even when the project is located in federal waters.

8 Still More Permits: The Roles of Other Federal and State Agencies Reviewing Proposed Offshore Projects

The free benefit of the wind ought not to be denied to any man.

—Herbert of Bury, Suffolk, c. 1180

The federal and state governments own or hold the rights to use the marine environment, including submerged lands, off the U.S. coastline. Renewable energy project developers must obtain leases, licenses, and permits from the appropriate "landlord"—either the federal government or the state, or both. The process of obtaining development rights from MMS and FERC is described in chapter 5; this chapter delves into the roles of other permitting agencies and statutory schemes that may affect project authorization.

The discovery of oil offshore following World War I led to a series of disputes between the federal government and coastal states. California's attempt in the 1930s to issue leases for offshore exploration and drilling for oil, gas, and other natural resources triggered a lawsuit by the federal government. In that case the Supreme Court affirmed the United States' property rights in the seabed and natural resources from the tidewaters and shore to the inner boundary of the OCS (the three-mile belt adjacent to the coastline). This was based upon the federal government's paramount rights and responsibilities for national defense and international affairs. (See chapter 2 for a more precise definition of the limits of state and federal control. State-controlled submerged land within the three-mile band of ocean adjacent to coast is sometimes referred to as *state territorial seas* or the *marginal seas*.) The Court held that those rights included "full dominion over the resources of the soil under that water, including oil."[1]

Needless to say, the Court's decision regarding federal primacy over offshore rights was not received well by the states, and the issue continued to fester. In passing the Submerged Lands Act of 1953 (SLA), Congress sought to avoid continuing disputes with the coastal states by partially undoing the earlier Supreme Court decision.[2] The SLA grants to the states title to the adjacent "lands beneath navigable waters" and gives the states control over their associated natural resources, including (but not limited to) oil, gas, other minerals, and marine animal and plant life.[3] (The grant is generally for land under water up to 3 miles offshore.) In general, the states now exercise property rights over submerged land within 3 nm of the coastline. The federal government reserved its jurisdiction over these nearshore submerged lands to regulate navigation, flood control, and hydroelectric facilities.[4] Moreover, the Supreme Court has also noted that the grant of submerged land is still subject to congressional authority to regulate interstate commerce.[5] The SLA retains federal control over the natural resources of the continental shelf's subsoil and seabed that extend seaward from the outer boundary of state submerged lands. This is established by the SLA itself and confirmed by the Supreme Court in the *Maine case*, which held that the federal government has exclusive sovereign rights over the seabed and subsoil of the continental shelf underlying the Atlantic Ocean.[6]

The Public Trust Doctrine and Water-Dependent Uses

In general, states regulate the use of nearshore submerged lands by the following means:

1. With instruments such as licenses, leases, or easements to private parties
2. By regulating private and commercial activities that do not require conveyance of property interests with various permit systems
3. By using some combination of the two

Instances of these regulations can be found in *N.Y. Environmental Conservation Law 15-0503(1)(b)*, requiring either a property interest or permit before placing a structure on lands underwater, and *Massachusetts General Laws Annotated*, ch. 91, authorizing licenses for construction of structures.[7] Massachusetts, for example, has reserved submerged lands primarily for *water-dependent uses* and requires structures on state-owned submerged lands to serve a public purpose that provides a greater public benefit than detriment.[8] Other states have similar requirements. Commercial fishing and fish processing activities, marinas, harbor and dock facilities, and power plants that require access to cooling water are examples of water-dependent uses.

Generally, an activity that requires direct access to the water to accomplish its primary function is a water-dependent use. A seaside restaurant or hotel, for example, would not qualify, since each business is only enhanced by, but not dependent on, proximity to the water.

Ocean current, tidal, and wave energy facilities clearly appear to qualify as water-dependent uses. Wind power facilities are less obviously water dependent—the technology has fared quite well in arid West Texas and on the Great Plains.

An argument for the water dependency of offshore wind turbines should make three points. First, ocean winds are generally stronger than terrestrial winds, and wind turbine generators must be located offshore where the resource is available. Just as fishermen must go offshore to fish, because that is where the resource is, wind energy project developers must pursue strong winds offshore, where they may be found. That traditional electric generation requiring access to cooling water may be considered water-dependent tends to support the case for similarly classifying offshore wind power. Thermal electric generation can be designed with cooling towers to achieve much lower water use, though at a greater cost. Onshore wind may be feasible inland in certain coastal regions but typically not at the scale possible offshore. Furthermore, weak onshore winds will cause turbines to operate at reduced efficiency and may make them uneconomical. Second, offshore wind turbines are dependent on water transport. These machines are too large to transport on land and therefore require direct water access for construction. Third, the purpose of a water-dependent use restriction is to promote the preservation of coastal areas, public access to the water, and the economic vitality associated with a working waterfront. Offshore wind energy projects support these objectives by accomplishing the following:

- Moving electric generation off the coast, thus freeing the coastal zone for preservation
- Providing jobs for coastal communities and sustainable energy to power local homes and businesses

Reserving submerged lands for activities that are inextricably connected to the water is a natural extension of the public trust doctrine. This doctrine holds that a state's interest in submerged lands and waters above them is held in trust for the public. State management of trust lands should benefit the public, and transfers of state submerged lands that inhibit public uses of the waters may be invalidated. The case of *Illinois Central Railroad Company v. Illinois*, decided by the Supreme Court in 1892, illustrates the public trust doctrine and casts light on present-day state actions in the coastal zone. In that case, the state of Illinois had granted the Illinois Central Railroad title to a portion of the bed of Lake Michigan, and the city of Chicago granted the railroad rights to develop track and associated facilities along Chicago's Lake Michigan shore. Acting under the grants, the railroad filled in a large section of the lake and built tracks, piers, warehouses, and terminals, thus converting the lakebed to private use. The Supreme Court found that Illinois had acted beyond its authority in divesting itself of these public lands, and reversed the state's grant.

> *The ownership of the navigable waters of the harbor and of the lands under them is a subject of public concern to the whole people of the state. The trust with which they are held therefore is governmental, and cannot be alienated, except in those instances mentioned of parcels used in the improvement of the interest thus held, or when parcels can be disposed of without detriment to the public interest in the lands and waters remaining. This follows necessarily from the public character of the property, being held by the whole people for purposes in which the whole people are interested.*[9]

The import of the public trust doctrine for today's renewable energy projects is that developers should not expect to obtain title to state submerged lands to build their projects. Only leases, licenses, easements, and similar interests that do not unreasonably interfere with or constrain the public interest in navigation, fishing, and other important water-dependent uses of the ocean environment would likely be granted. A state may invoke the public trust doctrine in an effort to block critical elements of a renewable energy project from state waters. Thus developers should be prepared to explain how their projects impose only minimal restrictions on public uses of the marine environment while advancing the public interest. Such benefits to the public include environmental improvements, fisheries enhancement (e.g., the reef effect on turbine foundations), and navigation improvements (such as turbines marking a dangerous shoal). (In general, federal regulation of an activity preempts state regulation that relates to the same activity or field. As discussed further in this chapter, however, federal statutory schemes may also be administered by the states. The state of Connecticut's refusal to issue a certification under the Clean Water Act in *Islander East Pipeline Company* is a case in point. In such cases, state opposition to a project may be difficult or impossible to overcome.)

Coastal Zone Management

Renewable energy projects may be sited in state waters, or, if the project's principal facilities are located in federal waters, related facilities such as transmission cables, dock facilities, and electrical substations still may be located in state waters or the coastal zone. Under the Coastal Zone Management Act (CZMA), coastal states may develop a comprehensive management program for the development and preservation of their coastal resources.[10] Actions by federal agencies affecting the state coastal zone must generally be consistent with the enforceable policies of a state-adopted coastal management program. Regulations of the National Oceanic and Atmospheric Administration (NOAA) implement the CZMA and establish a process for reviewing federal licensing and permit activities that have potential coastal effects. This is generally addressed in 15 C.F.R. pt. 930. According to 15 C.F.R. §§ 930.11(g); 930.151, "Effects are not just environmental effects, but include effects on coastal uses. Effects include both direct effects which result from the activity and occur at the same time and place as the activity, and indirect (cumulative and secondary) effects which result from the activity and are later in time or farther removed in distance, but are still reasonably foreseeable."[11]

An applicant for a federal permit, or a federal agency that is itself taking action that affects the coastal zone, establishes a project's consistency with the state coastal management program through a certification or consistency determination submitted to the state coastal management agency. The state coastal management agency may either concur with or object to the consistency certification or determination. (If the state disagrees with the applicant's consistency certification, the regulations also require the state to object within six months of the day that the state receives a complete coastal consistency certification from the applicant.[12]) In the case of an objection to a federal agency project, the agency may proceed with the project provided that it has complied with the state coastal management program policies to the maximum extent practicable. If the state objects to a project or activity proposed by an applicant for a federal license (e.g., a renewable energy project developer), the applicant's ability to proceed will be blocked unless, upon appeal to the secretary of the Department of Commerce, the secretary finds that the activity or project is consistent with the objectives of the CZMA or otherwise necessary in the interest of national security. (Resolution of any state objection on coastal consistency grounds would follow the procedures established in the rules, which include the opportunity for voluntary mediation before the secretary of Commerce or NOAA. Appeals from any state objection initially would be made to the secretary of Commerce. To initiate the appeal, the federal agency or applicant would need to file a notice of appeal with the secretary of Commerce within 30 days of receipt of a state agency objection.[13]) If the secretary of Commerce makes such a finding with respect to a proposed facility, the relevant federal agencies could then approve the facility despite a state's objections.[14]

The case of *Islander East Pipeline Company, LLC*, involving an offshore energy pipeline project, illustrates the CZMA appeal process. Connecticut has long opposed traditional energy projects proposed to be located in Long Island Sound and challenged Islander East's proposed pipeline on both CZMA and Clean Water Act water quality grounds. The state has asserted, among other allegations, that the project was inconsistent with its coastal zone management plan's emphasis on preservation of natural resources in the Sound.[15] On appeal of Connecticut's opposition based on the CZMA, the secretary of Commerce overruled the state's objections, finding the project to be consistent with the objectives of the CZMA since all of the following applied:

1. The proposed activity furthered the national interest in a significant and substantial manner.
2. The national interest furthered by the proposed activity outweighed the activity's adverse coastal effects.
3. No reasonable alternative was available that would permit the proposed activity to be conducted in a manner consistent with the enforceable policies of the CZMA.[16]

That decision was subsequently appealed to the District Court of Connecticut, and later the Second Circuit. (Further discussion of *Islander East*, including the disposition on appeal in the context of state CWA compliance certifications, is given below.)

A state's coastal zone typically extends from the most seaward extent of state waters (generally the 3-nm limit) to a certain distance inland depending on the state. (Individual state coastal zone boundaries are available on the Internet from NOAA.[17]) Renewable energy facilities sited in the coastal zone must be consistent with a state coastal management program, but more distant projects also may need to be reviewed for consistency if they have reasonably foreseeable effects on any land or water use or natural resource of the coastal zone. Whether a specific federal agency approval, license, or activity may affect any natural resource, land use, or water use in the coastal zone is determined by the responsible federal agency (MMS, FERC, or the Corps). Congress intended the CZMA consistency determination to include all effects in the coastal zone that the federal agency may reasonably anticipate as a result of its action, including cumulative and secondary effects. Therefore, the term *effects* is construed broadly to include direct effects caused by the activity and occurring at the same time and place, and indirect effects that may be caused by the activity and are later in time or farther removed in distance, but still reasonably foreseeable.[18] Generally speaking, these coastal effects, both direct and indirect, will have been analyzed during the NEPA process.

A single project also may be subject to the coastal management program of more than one state. In order for a state to review federal license or permit activities that occur outside of the state's jurisdiction for coastal consistency, NOAA's regulations first require the state to prepare, for NOAA's review and approval, a list of federal license or permit activities (including the identification of geographic locations for each type of listed activity) that the state intends to review for *interstate coastal effects*.[19] The Coastal Zone Act Reauthorization Amendments of 1990 clarified that a consistency determination is required for a project that would have interstate coastal effects, regardless of its location.[20] Interstate coastal effects include "any reasonably foreseeable effect resulting from a federal action occurring in one State of the United States on any coastal use or resource of another State that has a federally approved management program."[21]

Coastal management program policies are said to be *enforceable* if they are embodied in legally binding state law, including regulations, court decisions, and zoning ordinances, and are expressed in a state coastal management program that has been federally approved. The enforceability of a zoning ordinance under a state coastal management program was at issue when AES Corporation sought to build an LNG offloading terminal in the Chesapeake Bay near Sparrows Point. The project faced strong resistance, and Baltimore County passed a zoning ordinance prohibiting LNG terminals in the portion of the bay at issue. The state, however, neglected to have the new ordinance approved by NOAA as an amendment to Maryland's coastal management program. In litigation, the appeals court held that FERC authority to site LNG plants preempted the county zoning ordinance because the ordinance had not properly been incorporated into Maryland's coastal management program. The Court found that "the County has no authority under the CZMA to enact a ban on LNG terminals unless, at a minimum, that ban is enacted pursuant to the procedures established by the CZMA."[22] The AES case is a reminder of the distinction between state law and the coastal management program, which reflects only a subset of state law. The case also illustrates the value of monitoring local and state laws that may enter a state's coastal management program (through an amendment approved by NOAA) for developments that may restrict marine renewable energy projects. States can use consistency review to constrain federal action only after NOAA has approved the state's costal management program.[23]

Energy projects should be expressly covered in a state's coastal program. In particular, the CZMA requires that coastal management programs should include a planning process for energy facilities, including renewable energy facilities, likely to be located in, or which may significantly affect, the coastal zone. The term *energy facilities* explicitly includes "electric generating plants" and "any equipment or facility used primarily in the…transportation of any energy resource."[24] State management programs also must balance and provide for "adequate consideration of the national interest involved in planning for, and managing the coastal zone, including the siting of facilities such as energy facilities which are of greater than local significance."[25] Coastal management programs should, according to the CZMA, reflect "ecological, cultural, historic, and esthetic values as well as the needs for compatible economic development." New commercial and industrial developments should generally be encouraged to be sited in or adjacent to areas that are already similarly developed.[26] NOAA's regulations implementing the CZMA are in 15 C.F.R., and these requirements will influence the siting of renewable energy projects.[27]

As a practical matter, complying with the CZMA in the case of a complex offshore renewable energy project requires the developer to meet with the coastal management program office staff in every state affected by the project.[28] These meetings should uncover all the enforceable policies under each state's coastal management program applicable to the project. For example, cable laying may require wetlands permits and erosion and sediment control plans. The developer also should confirm the information needed by the state to determine consistency with its policies. The project plan and environmental assessment filed with the lead federal agency should be the starting point for the information submitted to the state. It may be supplemented by the following:

- Maps, diagrams, and technical data
- Information specifically identified by the state's coastal management program as required for an applicant's consistency certification, which may include completed state or local government permit applications that are required for the proposed activity (though actual issued permits are not required)
- An evaluation relating the coastal effects of the proposed project to the relevant enforceable policies of the state's coastal management program

A full description of the information that a developer may be required to provide to the state coastal management program office in connection with a consistency determination is provided in 15 C.F.R.[29] This process culminates with the developer's delivery of a fully developed consistency certification that concludes with the following standard language: "The proposed activity complies with the enforceable policies of (name of State) approved management program and will be conducted in a manner consistent with such program."[30]

Obstructions to Navigation and Dredged Materials Disposal

Very often, the federal action that triggers a review for consistency with a state coastal management program is the issuance of a permit by the U.S. Army Corps of Engineers under the Rivers and Harbors Act of 1899.[31] Section 10 of the Rivers and Harbors Act requires Corps authorization for the placement of certain structures (e.g., piers, wharfs, breakwaters, bulkheads, jetties, weirs, and transmission lines) and for activities such as dredging or disposal of dredged material, or excavation, filling, or other modifications affecting the navigable waters of the United States. The placement of renewable energy facilities offshore will require a permit from the Corps to ensure that the facilities do not constitute a hazard to navigation.

The Corps administers section 404 of the Clean Water Act (CWA), which prohibits the discharge of dredged or fill material into the inland waters of the United States unless a Corps permit has been obtained. The Corps also implements provisions of the Marine Protection, Research and Sanctuaries Act of 1972 (MPRSA), which imposes similar restrictions on ocean dumping.[32] The excavation of foundations for renewable energy facilities, and the placement of rubble scour protection around foundations, could require a materials disposal permit.

Construction activities such as cable laying within wetlands adjacent to or in the navigable waters of the United States will likely require a CWA section 404 permit from the Corps. (Section 502[7] of the Clean Water Act defines *navigable waters* as "waters of the United States including the territorial seas." Navigable waters of the United States extend to the 3-mile limit of the territorial seas.[33]) Modern cabling techniques help minimize impacts to wetlands and sensitive coastal habitats, as well as the size of the area actually requiring a permit. Horizontal directional drilling, for example, may be used to prepare a small conduit, starting on land and continuing under coastal wetlands and the nearshore seabed, for the placement of offshore transmission cables. If dredged material must be hauled by vessel and disposed of at sea, ocean disposal restrictions also come into play. Corps permits are required for the transportation of dredged material for the purpose of dumping it in ocean waters at *dumping sites* designated under the MPRSA. By interagency agreement between the Corps and EPA, the discharge of dredged material in the territorial seas and the ocean is regulated under section 103 of the MPRSA, rather than the criteria developed under CWA section 404.[34]

Permit applications are processed by the Corps district engineer with authority over the region where the project is proposed to be sited. Generally, a developer would meet with the Corps district staff for an informal preapplication consultation to review the merits and demerits of the proposal before preparing detailed designs or committing significant resources to the project. A meeting is a valuable opportunity to discuss project siting alternatives and impact mitigation measures, as well as to learn about the criteria the Corps will weigh most heavily in evaluating the proposal. The Corps evaluation will consider project benefits and detriments to determine if the project is in the public interest. Decision criteria generally include conservation, economics, aesthetics, wetlands, cultural values, navigation, fish and wildlife values, water supply, and water quality.[35]

If an applicant seeks a dredged material dumping permit, certain regulations would apply depending on whether dumping would occur in the inland waters or the ocean.[36] These regulations were developed jointly by EPA and the Corps to minimize the potential effects of disposal on marine life, marine sanctuaries, and water quality. The excerpt below from the Corps guidelines under CWA section 404 illustrates ecosystem impacts that may accompany dredged materials disposal.

> *The discharge of dredged or fill material can result in greatly elevated levels of suspended particulates in the water column for varying lengths of time. These new levels may reduce light penetration and lower the rate of photosynthesis and the primary productivity of an aquatic area if they last long enough. Sight-dependent species may suffer reduced feeding ability leading to limited growth and lowered resistance to disease if high levels of suspended particulates persist. The biological and the chemical content of the suspended material may react with the dissolved oxygen in the water, which can result in oxygen depletion. Toxic metals and organics, pathogens, and viruses absorbed or adsorbed to fine-grained particulates in the material may become biologically available to organisms either in the water column or on the substrate. Significant increases in suspended particulate levels create turbid plumes which are highly visible and aesthetically displeasing. The extent and persistence of these adverse impacts caused by discharges depend upon the relative increase in suspended particulates above the amount occurring naturally, the duration of the higher levels, the current patterns, water level, and fluctuations present when such discharges occur, the volume, rate, and duration of the discharge, particulate deposition, and the seasonal timing of the discharge.*[37]

Permits for dredged material dumping that violate state water quality standards will not be issued.

Once the formal application is filed by the developer, the Corps will issue a public notice, evaluate the project's impacts and the public comments, and request modifications to the project if necessary. The Corps may consider alternative locations and methods to accomplish the project's goals if resource use conflicts would otherwise block the project as proposed. If the Corps is the lead federal agency on the project, it also will manage the NEPA environmental review process, including preparing an EIS if one is required. A public hearing may be held if comments raise substantial issues of fact that such a hearing could help resolve. The Corps will issue a permit for a project that it finds to be in the public interest. The Corps' standard permit form states that the federal permit is contingent on the applicant's receipt of required state or local permits (including the CZMA consistency determination). The Corps permit would thus be deemed denied if any state and local permits are denied.

Prior to enactment of the Energy Policy Act of 2005, the Corps, and not MMS, was the only federal entity authorized to permit construction of renewable energy projects on the OCS. In 2001, Cape Wind Associates, LLC submitted an application to the Corps for a permit under section 10 of the Rivers and Harbors Act to construct and operate an offshore data tower in an area of federal submerged land in Nantucket Sound known as Horseshoe Shoal. (Section 10 delegates authority to the Corps to issue permits for projects that impact the navigability of U.S. waters.[38]) The proposed structure was to consist of a platform and a tower approximately 170 ft high, supported by three steel piles driven into the ocean floor. Various instruments on the tower would gather data for use in determining the feasibility of locating a wind energy plant on the shoal.

Environmental groups challenged the Corps' authority to permit the project.[39] Both the Federal District Court of Massachusetts and the First Circuit Court of Appeals rejected the groups' arguments and held the Corps did have jurisdiction under the OCS Lands Act, which in 1953 extended the Corps section 10 regulatory authority "to prevent obstruction to navigation in the navigable waters of the United States …to artificial islands and fixed structures located on the [OCS]."[40] In 1978, this grant of authority was again amended to apply to "the artificial islands, installations, and other devices referred to in subsection (a) of this section."[41] Subsection (a), in turn, extends federal jurisdiction to "all artificial islands, and all installations and other devices permanently or temporarily attached to the seabed, which may be erected thereon for the purpose of exploring for, developing, or producing resources therefrom, or any such installation or other device (other than a ship or vessel) for the purpose of transporting such resources."[42] The court disagreed with the environmental groups' argument that this language precluded Corps jurisdiction over structures such as wind energy data collection towers because wind energy data is unrelated to *mineral* resources.[43] (The environmental groups' argument was based on the various definitions contained in the OCSLA. While the term *resources* is not defined in OCSLA, *exploration*, *development*, and *production* are all defined in terms of *mineral*, which is in turn defined as "including oil, gas, sulphur, geopressured-geothermal and associated resources, and all other minerals which are authorized by an Act of Congress to be produced from 'public lands.'" No mention is made of wind or wind energy sources.[44]) The court found that Congress clearly intended the Corps to protect navigability in general, and to continue to exercise its authority over all structures on the OCS, regardless of whether they relate to mineral resources.[45]

The role of the Corps in granting access to sites on the OCS has been displaced at least partially by the authority granted to MMS under the Energy Policy Act of 2005. Though MMS now has clear authority to lease sites on the OCS, Corps authority to authorize structures in U.S. navigable waters and to address water quality issues continues. The Corps also retains primary federal jurisdiction over wind energy facilities to be built in state waters. However, the 2005 amendments to the OCS Lands Act specifically did not grant MMS authority to lease sites within the boundaries of a national park, marine sanctuary, wildlife refuge, or monument. Thus, it is possible that a developer bold enough to attempt to place a project in one of these areas may seek authorization from the Corps.

Compliance with the Clean Water and Clean Air Acts

Renewable energy projects are expected to result in a net reduction in water and air pollution when compared with other energy production facilities. Nevertheless, transient water and air impacts may result. For example, offshore renewable energy transformer platforms may house diesel generators for backup power, and project construction and maintenance will cause air and water emissions. Even the discharge of pure, clean water into navigable waters by a marine energy device may trigger a requirement for a certification of compliance. The following discussion considers the system of federal and state programs and rules under the Clean Air Act (CAA) and the Clean Water Act (CWA) as applied to marine renewable energy projects.

Clean Water Act

The restoration of the nation's surface waters, including its coastal waters, to fishable and swimmable quality, and the elimination of pollutant discharges, are the objectives of the Federal Water Pollution Control Act, commonly referred to as the Clean Water Act.[46] The CWA regulates direct discharges from *point sources* such as sewage outfall pipes under a permitting regime known as the *National Pollutant Discharge Elimination System (NPDES)*. Indirect, diffuse discharges such as storm water runoff are managed through state programs established under section 319 of the CWA.

Regulation under the CWA is divided between the states and the federal government. Before a federal agency may issue a license or permit that may result in a discharge to U.S. waters, the agency must obtain a certification from the state where the project is proposed to be located that the discharge is consistent with specified sections of the CWA and other appropriate requirements of state law. The state certification process has been used to require FERC to include conditions in hydroelectric dam licenses to prevent impairment to state waters, such as the kind discussed below in *National Wildlife Federation v. Gorsuch*.[47] Each state is authorized by the CWA to impose comprehensive water quality standards and establish water quality goals for all intrastate waters.[48] The goal for some waters is remediation to improve impaired waters. Water quality objectives may be expressed in terms of desirable uses for water bodies and in the form of concentrations of pollutants, temperature, pH, turbidity units, toxicity units, or other quantitative measures. The CWA also includes an *antidegradation policy* that requires state standards to be sufficient to maintain existing beneficial uses of navigable waters, preventing their further degradation.[49]

State water quality standards are established taking into consideration each waterway's use and value. Uses may include public water supplies, propagation of fish and wildlife, recreational purposes, agricultural and industrial needs, and navigation.[50] A state must submit its proposed water quality standards to EPA for federal approval, and upon approval the state standard becomes "the water quality standard for the applicable waters of that State."[51] EPA guidelines for the protection of aquatic life and human health form the basis for most state standards, although "states may develop water quality standards more stringent than required by this regulation."[52] *Total maximum daily load (TMDL) limits* for relevant pollutants are typically established to manage pollution levels. Portions of the TMDL are allocated among sources of the relevant pollutants, including categories of nonpoint sources.

In addition to the water quality–based limits intended to promote a water body's designated use, regulators also may impose specific technology-based limits for priority pollutants. These technology-based limits are tailored to particular industries. Generally, a point source discharger must obtain a NPDES permit from the state environmental agency where the project is located before initiating any discharges into U.S. waters. The permit will be issued subject to discharge limits and conditions determined by the applicant's industry (and available control technologies) and the use classification of the receiving body of water. Pollution limits are ordinarily expressed in terms of aggregate quantities delivered into a water body and concentration levels. The limits are designed so that compliance cannot be achieved merely by diluting pollution with clean water.

A wide variety of facilities qualify as point sources. NPDES permits cover industrial and municipal discharges, discharges from storm sewer systems, storm water associated with numerous kinds of industrial activity, mining operations, and certain agricultural activities such as animal feedlots and aquaculture facilities that exceed a size threshold. Of interest to developers of offshore renewable energy projects is that runoff from construction sites disturbing more than one acre also is considered a point source. Construction of the terrestrial components of offshore projects, such as transmission lines and substations, will usually require a NPDES permit. The permit will incorporate a *storm water pollution prevention plan (SWPPP)* for the site, as well as monitoring and recordkeeping requirements. An SWPPP addresses erosion and sediment control and must include the following:

- A site description
- A map that identifies sources of storm water discharges on the site
- Anticipated drainage patterns after major grading
- Areas where major structural and nonstructural measures for sediment control will be employed
- Surface waters, including wetlands
- Locations of discharge points to surface waters

An SWPPP should specifically describe the best management practices that will be employed to control erosion and water pollution. These practices include maintenance of buffer zones around water bodies, efforts to minimize disruption to existing vegetation wherever possible, and plans to stabilize disturbed areas of the site as quickly as practicable after construction activity has ceased. NPDES permits are valid for up to five years. Related permits for construction in tidal or nontidal wetlands also may be required under state law. Issues of erosion control, damage to native species, and avoidance of invasive species will arise in connection with these permits.

Water passing through a hydrokinetic energy facility is generally not altered, and the operation of the facility (as distinguished from its construction) should not require a NPDES permit. This view follows from *National Wildlife Federation v. Gorsuch*, a 1982 case concerning hydroelectric dams, and *South Florida Water Management District v. Miccosukee Tribe of Indians*, a 2004 Supreme Court case involving an Everglades water pumping station.[53] In Gorsuch, the National Wildlife Federation (NWF) sought a court order directing the EPA to require dam operators to apply for NPDES pollutant discharge permits under section 402(a) of the CWA. NWF argued that dams cause several water quality changes that should be considered tantamount to pollution and regulated as such. In particular, NWF identified five problems caused by dams that occur in the reservoir and the downstream river water:

1. **Low dissolved oxygen.** Water stratifies by temperature in reservoirs, and deep layers may become oxygen depleted. If the deep water layer is released from the dam, low oxygen levels may occur downstream and adversely affect fish survival.
2. **Dissolved minerals and nutrients.** Minerals such as iron, manganese, and phosphates may leach from mud on the bottom of a reservoir under anaerobic conditions, causing harm to fish, unpalatable drinking water, and undesirable plant growth.
3. **Water temperature changes.** Reservoir water tends to be stratified, with warm water at the surface and colder water at the bottom. Particular fish species may be harmed by the discharge of either warm or cold water.
4. **Sediment release.** Sediment settles behind a dam, and the low-sediment water that is released from the dam scours the riverbed downstream to reacquire its equilibrium sediment load.
5. **Supersaturation.** Turbulent water in the river below a dam outlet can become aerated in excess of normal conditions. Supersaturation can be fatal to fish.

Gorsuch illustrates why many of the current generation of hydrokinetic energy technologies are dramatically different from traditional impoundment hydroelectric facilities. A free-flow hydrokinetic energy machine does not impound water, so water stratification and the related temperature, oxygenation, mineral leaching, and sediment issues do not arise. Free-flow turbines also produce minimal turbulence, which is the cause of supersaturation at traditional hydroelectric facilities.

Gorsuch also is notable for its interpretation of the CWA prohibition on the "discharge of a pollutant" from a point source without a permit. According to the court, for a point source to require a NPDES permit, "five elements must be present: (1) a pollutant must be (2) added (3) to navigable waters (4) from (5) a point source."[54] A dam may fall within the statutory definition of *point source*, which includes "any discernible, confined and discrete conveyance, including but not limited to any pipe, ditch, channel, tunnel, conduit, well, discrete fissure, container, rolling stock, concentrated animal feeding operation, or vessel or other floating craft, from which pollutants are or may be discharged. This term does not include agricultural stormwater discharges and return flows from irrigated agriculture."[55]

A dam often has tunnels and spillways through which water flows and is confined. Many free-flow hydrokinetic energy devices do not have such features and thus may not be considered point sources. *Navigable waters* are defined under the CWA as the waters of the United States, including the territorial seas. Courts have held that actual navigability is not required.[56] The court in *Gorsuch* also considered the CWA definition of *pollutant*, which means "dredged spoil, solid waste, incinerator residue, sewage, garbage, sewage sludge, munitions, chemical wastes, biological materials, radioactive materials, heat, wrecked or discarded equipment, rock, sand, cellar dirt and industrial, municipal, and agricultural waste discharged into water."[57]

The court declined to expand the restrictive phrasing of the definition of *pollutant* to include the five problems caused by dams and cited by NWF, finding these to be more in the nature of conditions, rather than substances added to the water. The court took this view even though the CWA's separate definition of *pollution*—"the man-made or man-induced alteration of the chemical, physical, biological, and radiological integrity of water"—clearly is much broader than the list of substances set out under the definition of *pollutant*.[58] Finally, the court interpreted the term *addition*, agreeing with EPA's view that addition from a point source occurs only if the point source itself physically introduces a pollutant into water from the outside world. The court's views in *Gorsuch* suggest that most hydrokinetic energy devices should not be considered point sources. In normal operation, hydrokinetic energy devices should not emit pollutants into U.S. waters and should not therefore require an NPDES operating permit.

In *Miccosukee Tribe of Indians*, the Supreme Court considered a pumping station that moved already polluted water from a canal into less-polluted receiving waters of the Florida Everglades. The tribe argued that the local water management agency was required to obtain a NPDES permit to operate the pumping station. It was undisputed that the pumping station was a point source used to transfer phosphorous-laden surface runoff from an area that had been reclaimed from the Everglades over a levee and into a nearby remnant of the giant swamp. In contrast to the holding in *Gorsuch*, the Court in *Miccosukee* held that a point source could be regulated under the NPDES permitting system if it serves to convey pollution to the water, even if the pollution does not originate at the point source. The Court declined to resolve the matter, however, and instead remanded the case to the district court for further fact finding on whether the canal and the Everglades actually are parts of the same water body. If so, the Court reasoned that the mere transfer of water between two parts of the same water body cannot constitute the *addition* of pollutants. "If one takes a ladle of soup from a pot, lifts it above the pot, and pours it back into the pot, one has not 'added' soup or anything else to the pot."[59] In 40 C.F.R., discharges from a water transfer are exempted from the requirement to obtain an NPDES permit. "Water transfer means an activity that conveys or connects waters of the United States without subjecting the transferred water to intervening industrial, municipal, or commercial use. This exclusion does not apply to pollutants introduced by the water transfer activity itself to the water being transferred."[60]

Miccosukee, like *Gorsuch*, supports the view that the mere passage of water between the input and output ends of a hydrokinetic energy device, such as an in-stream turbine, would occur in the same body of water and should not be considered a pollution discharge, absent the addition by the device itself of an outside pollutant. Most hydrokinetic energy devices in operation, therefore, should not require an NPDES permit.

A caveat should be offered, however, with respect to biocide coatings that may be used on hydrokinetic energy machines to discourage fouling by marine organisms. Fouling can be controlled during periodic maintenance by physically removing the device from the water and scraping off the growth. Maintenance schedules can be lengthened and the performance of the device between maintenance events can be improved with the use of biocide coatings. Coatings have typically incorporated toxic compounds such as cuprous oxide, metallic copper, and tri-butyl-tin oxide (TBT), which are slowly released and unfortunately accumulate in the environment. For this reason their use has raised concern. The Anti-Fouling Treaty initiated by the International Maritime Organization (IMO), for example, calls for a complete ban of TBT coatings used on ocean-going vessels by January 1, 2008.

Silicone coatings are a nontoxic alternative that provide a very slippery surface on which marine growth has difficulty adhering. The durability of the easily damaged silicone compounds, however, needs improvement. Other less toxic compounds such as aluminate polymers are in development and have demonstrated good fouling resistance in marine exposure tests.

It is not entirely clear how EPA would regulate the use of biocide coatings on hydrokinetic energy devices. The example of vessel discharges is helpful but not conclusive. Under EPA regulations, the discharge of sewage from vessels, effluent from properly functioning marine engines, laundry, shower, and galley sink wastes, and any other discharge incidental to the normal operation of a vessel does not require a NPDES permit.[61] This exemption also would seem to apply to biocide coatings used on vessels, except, however, when vessels are "operating in a capacity other than as a means of transportation such as when used as an energy or mining facility."[62] Biocide coatings are regulated by EPA under the Federal Insecticide, Fungicide, and Rodenticide Act (FIFRA). FIFRA requires an applicant registering a pesticide for sale to demonstrate that, when used according to specifications, the pesticide "will not generally cause unreasonable adverse effects on the environment."[63] Wildlife regulators also are sensitive to the biocide coating pollution issue and, as would be expected, prefer a solution that minimizes toxic emissions to the environment. Developers of hydrokinetic energy devices should become familiar with EPA and IMO regulation of biocide coatings and be prepared to demonstrate why their solution to marine fouling is consistent with sound environmental practices.

State CWA compliance certifications

Applicants for federal licenses or permits to conduct an activity that would result in any discharge into U.S. navigable waters must provide to the licensing or permitting agency a certification from the state in which the discharge will originate that the discharge will comply with applicable provisions of the CWA.[64] The developer of a project built under the jurisdiction of the Corps, for example, including any renewable energy structure built in U.S. waters, is required to provide the Corps with a water quality certification from the states where water quality may be impacted. MMS and FERC require the same. The federal agency may not grant a license or permit if the state certification request is denied. If a state fails to act on a request within a reasonable period of time (not to exceed one year), the certification requirement is deemed waived. If a certification has been obtained, or if the requirement has been waived, the federal agency may issue the approval, but the federal license or permit must include any conditions that the state imposes in its certification.[65] In *PUD No. 1 of Jefferson County v. Washington Department of Ecology*, the Supreme Court found that the certification requirement may apply in the context of a broad variety of federal licenses and permits. Such licenses and permits include those issued by FERC under the Federal Power Act, the Corps under the Rivers and Harbors Act, and the secretaries of Interior and Agriculture relating to projects built on federal land.[66]

A discharge that triggers the state certification requirement may be nothing more than sediment discharges consequent to construction, and it may even be water that has passed unaltered through a hydrokinetic energy facility. In *PUD No. 1 of Jefferson County*, which involved the construction of a hydroelectric dam in Washington State, the Supreme Court held that discharge in the context of the CWA section 401 certification requirement is synonymous with activities that may affect water quality:

> *Petitioners concede that, at a minimum, the project will result in two possible discharges—the release of dredged and fill material during the construction of the project, and the discharge of water at the end of the tailrace after the water has been used to generate electricity.*
>
> *Our view of the statute is consistent with EPA's regulations implementing § 401. The regulations expressly interpret § 401 as requiring the State to find that "there is a reasonable assurance that the* activity *will be conducted in a manner which will not violate applicable water quality standards.*[67]

More recently, in *S.D. Warren Co.*, the Supreme Court considered the relicensing of hydroelectric projects in Maine and found, consistent with *PUD No. 1*, that water flowing out of the facilities also was a discharge that required a state certification under CWA section 401.[68] "When it applies to water, 'discharge' commonly means a 'flowing or issuing out,'...and this ordinary sense has consistently been the meaning intended when this Court has used the term in prior water cases."[69] Given these cases, it may be difficult for the developers of marine renewable energy technologies that capture water or flow water through a device located in state territorial waters to avoid the state certification requirement, since there arguably would be a discharge.

The protracted *Islander East Pipeline* litigation, discussed briefly in the previous section on the CZMA approval process, illustrates how a project may be defeated due to the water quality certification requirement.[70] Islander East sought to construct an interstate natural gas pipeline from New Haven, Connecticut, across Long Island Sound to Brookhaven, New York, but was unable to obtain the necessary water quality certification from Connecticut as required by FERC regulations under the Natural Gas Act. The company, however, did obtain both a CZMA and a CWA water quality certification from the state of New York, and the secretary of Commerce overruled Connecticut's objections to the project on CZMA grounds.

Islander East petitioned the Second Circuit for review of the order of the Connecticut Department of Environmental Protection (CTDEP) denying the company's application for a water quality certificate for discharge into the waters of Long Island Sound. (Islander East benefited from a provision inserted into section 19 of the Natural Gas Act, 15 U.S.C. § 717r[d][1], by the Energy Policy Act of 2005, that provides natural gas companies with a direct path to the U.S. Court of Appeals if a state agency denies a permit required under federal law other than the CZMA. Developers of offshore renewable energy projects would potentially be faced with a much longer sequence of appeals that could well kill a project through delay.) CTDEP's principal objections to the project rested on the potential harm to valuable shellfish beds that the state alleged could be caused by sedimentation from the pipeline trenching operation. This concern persisted even though Islander East had planned to use directional drilling to avoid sensitive nearshore habitat and implement other best practices to reduce construction impacts.

The court applied an *arbitrary and capricious* standard of review to the state agency's decision. Under this standard, the reviewing court must be satisfied that the agency's decision reveals a "rational connection between the facts found and the choice made," and that based on the record, the agency appears to have examined the relevant data and articulated a satisfactory explanation for its action.[71] Islander East succeeded with its first appeal of the CTDEP's very brief denial, when the court remanded the matter to CTDEP with instructions that it conduct a "complete and reasoned review."[72] Islander East was less fortunate in challenging the CTDEP's second, 80-page denial. On the second go-around, the CTDEP had succeeded in assembling sufficient evidence in the record to support a rational conclusion to deny the water quality certification, and the court refused to overturn the state's ruling.

Islander East's application was filed with FERC in 2001 and had been determined to be environmentally acceptable under NEPA in 2002 with additional mitigation steps required. After the appeals described above, in September 2008, Islander East asked the Supreme Court to review the case. The Court denied the company's appeal in December 2008.[73] Yet, the Islander East case is more than another lesson in the power of delay. When the legal standard of review is whether the agency determination was arbitrary and capricious, a developer has almost assuredly lost as soon as a project enters litigation. A determined state agency usually can find sufficient evidence in a complex record to cobble together a reasoned decision that will be extremely difficult to overturn on appeal.

Projects succeed when they have engendered broad-based political and community support that can be effectively impressed upon state agencies holding the authority to issue the required permits and certificates. A developer that fails to muster such support may expend substantial time and resources on a project only to find success elusive.

Oil spill prevention and response

The CWA expresses U.S. policy that there should be no discharges of oil into the navigable waters of the United States or in connection with activities under the OCS Lands Act.[74] Persons who have knowledge of spills are required to immediately notify the appropriate agency, and failure to do so may subject the owner and operator of the offshore facility or vessel to fines or imprisonment of up to five years. (EPA regulations at 40 C.F.R. 110.6 require immediate notification of a spill to the National Response Center [NRC] in Washington, D.C., at 800–424–8802. If direct reporting to the NRC is not practicable, reports may be made to the Coast Guard or EPA predesignated on-scene coordinator (OSC) for the geographic area where the discharge occurs.) Facility owners and operators also may be liable for actual costs incurred by the government to clean up a discharge.

The CWA requires certain facilities to prepare and maintain "a plan for responding, to the maximum extent practicable, to a worst case discharge" of oil or other hazardous substances.[75] Among other things, a response plan must include the following:

1. Identify the person with full authority to implement cleanup actions.
2. Require that person to communicate immediately with the appropriate federal officials, discharge response personnel, and equipment providers in the event of a spill.
3. Identify, and ensure by contract or other means, the availability of private personnel and equipment necessary to limit any discharges and to remove, if necessary, a worst-case discharge.
4. Describe the training, equipment testing, periodic unannounced drills, and response actions to be carried out under the plan to ensure safety and mitigate or prevent a spill.

The EPA's rules for oil spill response plans apply to activities on offshore facilities that are not regulated by MMS. Under a memorandum of understanding between EPA, MMS, and the Department of Transportation (DOT), MMS regulates any offshore facilities seaward of the coastline (e.g., wind farms located in state waters beyond the baseline and on the OCS). EPA regulates nontransportation-related offshore facilities located landward of the coastline (e.g., a wind farm located on the Great Lakes). DOT regulates both transportation-related facilities (including pipelines) located landward of the coastline, and deepwater ports (i.e., offshore facilities for the importation of oil and LNG).[76] MMS regulations also require offshore renewable energy projects to have an oil spill response plan; the MMS and EPA regulations are similar. (Compare 30 C.F.R. pt. 254 [MMS oil spill response regulations] to 40 C.F.R. pt. 112 [EPA regulations].)

Offshore wind turbines would not ordinarily have large quantities of oil onboard. Gearbox oil is the principal fluid that might be spilled in the case of a significant turbine failure. Together with hydraulic fluids and greases that also would be present inside the nacelle, the aggregate quantity of fluids would generally be less than 300 gallons per turbine. In addition, most offshore turbines are sealed to limit salt air intrusion and maintain a controlled environment that minimizes condensation and corrosion. Most spills from the turbine machinery therefore would likely be contained within the turbine's nacelle. An offshore wind turbine park's transformer station is another possible source of a spill, but technology and countermeasures also will likely substantially reduce the severity of most events. Like the nacelle, the transformer substation is sealed to resist corrosion. Unlike transformers on land, which are usually oil-cooled, offshore transformers are likely to be cooled with sulfur hexafluoride gas, which is inert, nonflammable, and nontoxic. The cabling used for offshore collection and transmission systems will typically incorporate solid polyethylene insulation which, unlike oil-cooled cables, cannot leak if the cable is damaged. A backup diesel generator on the transmission substation will require the storage of a relatively small quantity of fuel on the platform, typically in the range of 4,000 to 5,000 gallons. The fuel tank would most likely feature a double-wall design and an automatic leak detection alarm system. The generator would power aircraft and navigation marker lights and facilitate wind turbine start-up. Equipment such as sorbents and spill containment booms would be stored on the platform and readily available in the event of a spill.

Clean Air Act

The CAA directly addresses emissions from facilities located on the OCS. The act's requirements may affect offshore renewable energy projects in two ways. First, emissions from vessels traveling to and from offshore facilities are aggregated with the emissions from the facility itself. If the aggregate emissions are substantial, the facility may require a permit. Second, the air pollution controls of the state adjacent to the proposed project generally apply as though the facility was proposed to be located onshore. The imposition of state clean air standards may have a significant impact, particularly if the project is proposed to be located offshore of a regional air quality nonattainment area, where lower emission thresholds apply. To better appreciate the regulation of emissions from offshore renewable energy facilities, a brief background on CAA regulation is in order.

Under the CAA, the federal government establishes national ambient air quality standards for key pollutants with adverse health effects known as *criteria pollutants*. On the regional level, states report whether they are in attainment of the standards and adopt state implementation plans that seek to bring nonattainment areas into compliance and prevent deterioration of air quality in areas that currently have relatively clean air. Under this federal/state regulatory scheme, the states typically set emissions standards for stationary sources, except for certain *new source performance standards*. The standards for new emissions sources are targeted to particular industries and are based on available emissions control technologies. Though the new source performance standards are set by the EPA, the states still play a major role in their enforcement through the state implementation plans.

Under the CAA's operating permits program, major sources of emissions should obtain a state operating permit. A major source is one that has the potential to emit one or more hazardous air pollutants in excess of certain amounts specified by EPA regulations.[77] The operating permit is designed to cover all state and federal air emissions requirements in connection with the facility. An operating permit typically would include information on the particular pollutants released by the facility, emissions limits, required emissions control measures, and requirements for measuring and reporting emissions.

The diesel-powered emergency generators that would typically be found on offshore transformer platforms are subject to EPA's emissions standards for mobile nonroad engines.[78] The manufacturer of the engine is responsible for certifying compliance with these standards. Marine vessels that may travel to and from an offshore renewable energy facility also would be subject to emissions restrictions under EPA regulations. (Marine diesel engines under 37 kW output are regulated at 40 C.F.R. pt. 89; other marine diesel engines are regulated at 40 C.F.R. pt. 94.) Pollution from ocean-going ships has over time become increasingly subject to worldwide regulation through the International Maritime Organization, a U.N. organization. The International Convention on the Prevention of Pollution from Ships (known as MARPOL), which has been adopted by the United States and numerous other countries, addresses air emissions, oil and chemical pollution, and sewage and garbage disposal from ships. Annex VI of MARPOL, which specifically concerns air pollution, is incorporated into the Act to Prevent Pollution from Ships.[79] This U.S. law applies to all U.S.-flagged ships and foreign-flagged vessels transiting U.S. navigable waters and is enforced by EPA and the Coast Guard.

While an offshore project developer need not obtain emission permits for particular vessels, the emissions from a project taken as a whole may require a permit. This is the case even if the proposed project is located beyond state waters. CAA section 328 treats emissions from facilities on the outer continental shelf within 25 miles of the seaward boundary of a state as though the source is located within the onshore area of a state. Such offshore emissions are subject to local requirements for emission controls, emission limitations, offsets, permitting, monitoring, testing, and reporting.[80] EPA's outer continental shelf air regulations require new OCS emissions sources to submit a notice of intent containing information about the project and its anticipated emissions to the agency and the adjacent affected state's air pollution regulators.[81] The notice would then be followed by an application filed with EPA or the state agency with delegated authority under section 328 to construct and operate the OCS facility.[82]

Emissions from vessels servicing or associated with an OCS source are considered to be emissions from the OCS source.[83] Vessels that are permanently or temporarily attached to an OCS facility also are considered part of the OCS source itself and would be subject to permit conditions applicable to the OCS source.[84] In addition, the emissions from vessels in transit are included in the project's *potential to emit* (i.e., potential emissions): "Emissions from vessels servicing or associated with an OCS source shall be considered direct emissions from such a source while at the source, and while enroute to or from the source when within 25 miles of the source, and shall be included in the 'potential to emit' for an OCS source."[85] As a result of these rules, an offshore renewable energy project that requires frequent vessel trips could be considered a major pollution source requiring an operating permit, regardless of the fact that the generating technology used in the project is pollution free.

Whether or not an offshore project requires an operating permit, the project will be subject to the EPA's general conformity process. The CAA prohibits federal agencies from licensing or permitting any activity that is not in conformity with a state implementation plan designed to achieve national ambient air quality standards.[86] (The EPA rules for evaluating conformity of federal actions to state implementation plans are in 40 C.F.R.[87]) As a practical matter, this means that the permitting agency may not approve a project that would exacerbate the deficiencies in local air quality addressed in the state implementation plan or otherwise delay the achievement of the state's air quality goals. As stated in CAA section 176(c)(1),

> *Conformity to an implementation plan means [A] conformity to an implementation plan's purpose of eliminating or reducing the severity and number of violations of the national ambient air quality standards and achieving expeditious attainment of such standards; and [B] that such activities will not [i] cause or contribute to any new violation of any standard in any area; [ii] increase the frequency or severity of any existing violation of any standard in any area; or [iii] delay timely attainment of any standard or any required interim emission reductions or other milestones in any area.*[88]

Achieving conformity in a nonattainment area may require a proposed project to offset its projected emissions with emissions reductions achieved elsewhere. It remains to be seen how EPA will apply its offset rules to offshore renewable energy facilities. The agency should allow vessel and other project emissions to be offset by regional air quality improvements from reduced fossil-fueled generation that otherwise would be dispatched to serve load in the absence of the renewable energy project.

Lastly, it is notable that EPA also has a role in the NEPA analysis of an offshore renewable energy project. The CAA requires EPA to review and comment in writing on the environmental impact of proposed legislation, federal construction projects, and certain major federal actions subject to NEPA.[89]

Foreign-Built Construction Vessels

Discussions of offshore construction sooner or later turn to the Merchant Marine Act of 1920, commonly referred to as the *Jones Act*.[90] A problem may arise under the Jones Act if appropriate U.S.-flagged vessels are not available to provide construction services to an offshore project and a developer seeks to hire foreign-flagged vessels.

The Jones Act protects the U.S. ship-building industry by prohibiting the transport of merchandise and passengers between points in the United States subject to the *coastwise laws*, except in a vessel that is *coastwise qualified*.[91] In general, a vessel that is U.S.-built, owned and controlled by U.S. citizens, and documented with a coastwise endorsement is coastwise qualified.[92] There is no statutory definition of *coastwise laws*, but U.S. Customs and Border Protection (CBP) interpretations hold that the term applies to the inland waters and territorial sea (i.e., extending from the coastal baseline 3 nm seaward), as well as to points on the outer continental shelf. The Outer Continental Shelf Lands Act extends U.S. jurisdiction to "the subsoil and seabed of the outer Continental Shelf and to all artificial islands and all installations and other devices permanently or temporarily attached to the seabed, which may be erected thereon for the purpose of exploring for, developing, or producing resources therefrom, or any such installation or other device (other than a ship or vessel) for the purpose of transporting such resources."[93]

The act was intended to extend federal jurisdiction over activities conducted for the purposes of exploring, developing, and producing resources contained within the seabed. In *Guess v. Read*, it was ruled that the Continental Shelf Act was enacted for the purpose, primarily, of asserting ownership of and jurisdiction over the minerals in and under the Continental Shelf.[94] This view would also encompass offshore renewable energy facilities. In contrast, it has been held that the Jones Act does not apply to sludge dumping sites on the outer continental shelf that are wholly unrelated to the development of natural resources in the seabed.[95]

Rulings by U.S. Customs and Border Protection are consistent with this interpretation. Customs has held that the Jones Act applies to *jetfoil vessels* used to transport workers to a drilling platform located on the continental shelf, more than 200 miles off the coast of Alaska.[96] Customs, however, has issued several opinions that transportation to a dumping point off the coast is not within the coastwise laws of the United States. The CBP has held, for example, that an energy production facility located on the OCS would be a point subject to the coastwise laws. In response to an inquiry whether transportation by jetfoil between a point on the Alaska coast and an oil platform located over 200 miles offshore was subject to the Jones Act, CBP held that the OCS Lands Act extended the reach of the Jones Act to the production platform.

> *The United States outer continental shelf extends to "a depth of 200 meters or, beyond that limit, to where the depth of the superjacent waters admits of the exploitation of the natural resources of the said areas" for purposes of the application of the coastwise laws to the outer continental shelf.*
>
> *The drilling platform under consideration is located in less than 200 meters of water and, furthermore, the operation of the drilling platform at its location is evidence that the "depth of the superjacent waters admits of the exploitation of the natural resources" in the area. Accordingly, the drilling platform under consideration, as well as any other drilling platform located in a similar location, whether or not it is more than 200 miles from the coast of Alaska, is a coastwise point.*[97]

Interpretations of the Jones Act, however, often turn on the slimmest of technicalities. A separate CBP ruling concerning the use of a dynamic positioning vessel to install a suction pile by means of a winch and cable concluded that the vessel/piling combination did not constitute a coastwise point because the vessel was not anchored to the seabed but rather maintained its position through its dynamic thrusters. The vessel's connection to the piling through the winch cable was insufficient to create a solid connection to the seabed that would cause jurisdiction under the Jones Act to attach.[98] The CBP clarified that a vessel that was connected to the seabed, such as with anchors, would be subject to the coastwise laws, and that transporting merchandise from shore to a piling connected to the seabed also would be subject to the Jones Act.

In 2004, the Chinese-built, British-owned *MV Resolution* became the "world's first purpose-built vessel for installing offshore wind turbines, foundations and transition pieces."[99] Built at a cost of £60 million (approximately $93 million), the *Resolution* is self-powered and has six retractable legs and two cranes with a capacity up to 300 tons. Capable of transporting and assembling the large components of offshore wind turbines, the vessel uses a dynamic positioning system to hover in place over a site while setting its legs on the seabed and jacking its hull above the water to form a stable working platform. Under current CBP interpretations, the Jones Act would prohibit use of the *Resolution* (or a similar foreign vessel) for the movement of foundations and other turbine components from one point in the United States to a project site on the U.S. outer continental shelf.

Does this mean that developers must rely on a limited selection of coastwise-qualified vessels for U.S. offshore renewable energy projects and sacrifice the efficiencies available through the use of specialized foreign vessels? The answer is yes—at least in part—although foreign vessels like the *Resolution* still could have a central role in U.S. projects. Assume, for example, that the heavy lifting and jack-up capabilities of specialized vessels like the *Resolution* are the key capabilities in short supply in the U.S.-flagged fleet. The *Resolution* could be positioned at the project site while construction materials are transported to the site on coastwise-qualified barges. These barges could pull along side the *Resolution*, which would then lift the turbine components from the barges and place them on the site. Provided that the *Resolution* did not carry wind turbine components from one place to another, but simply placed components on the sites while it was in a fixed position above each turbine site, the *Resolution* should not be engaged in merchandise transportation in violation of the Jones Act. This means that to avoid the "transport" of merchandise between two U.S. points, the *Resolution's* decks should be clear of all wind turbine components when it lifts its legs to move from one turbine installation site to another and when entering any U.S. port.

The Jones Act also requires dredges to be coastwise qualified to work in U.S. waters.[100] CBP defines dredging as "the use of a vessel equipped with excavating machinery for digging up or otherwise removing submarine material" and "the removal of soil from the bottom waters by suction or scooping or other means."[101] Conventional dredging, such as leveling the seabed in preparation for the placement of a concrete gravity base wind turbine foundation, would be included in this definition. It is less clear, however, whether cabling with a jet plow also would constitute dredging. In one simultaneous operation, a jet plow liquefies sediments with high-pressure water jets, places the cable, and then allows the sediments to settle back into the trench. Arguably, there is no removal of seabed materials, only momentary disruption. CBP has held that the temporary displacement of the seafloor as occurs during the installation of a suction pile or cable installation with a jet plow is not dredging under the Jones Act.[102] CBP also has held that laying out pipeline and cable is not coastwise trade because the pipe and cable are laid out in a continual operation while the vessel proceeds. This is distinguishable from the act of unloading merchandise since there is no single identifiable coastwise point involved in the laying of pipe or cable. For example, U.S. Customs ruled in 1998 that a foreign vessel that takes on cable in a foreign country and lays it between two points in the United States is not engaged in coastwise trade in violation of the Jones Act.[103] These holdings imply that a coastwise-qualified vessel would not be required for cable installation by means of a jet plow.

Each offshore renewable energy project will be unique. Developers should anticipate each step in the process—from marshalling components at dockside and moving them into place, to installation at the project site—and understand the tasks required of each vessel participating in the job. While prior CBP rulings are helpful in understanding the Jones Act implications of the project, such rulings are not definitive. A developer should not rely on a ruling issued to another party unless (i) the developer's factual situation is exactly the same as presented in the ruling letter, and (ii) the developer has confirmed with CBP that the ruling has not been modified or withdrawn. Regulations provide that each ruling letter "will be applied only with respect to transactions involving operations identical to those set forth in the ruling letter."[104] If in doubt, a developer's best option for minimizing Jones Act problems is to seek and obtain a ruling from CBP directly related to the particular facts of the developer's proposed project. Procedures related to ruling requests are in 19 C.F.R.[105] Jones Act exemptions or waivers also may be available under conditions of national emergency (as during the aftermath of Hurricane Katrina).

Marine Navigation

Marine renewable energy devices will have markings, lights, and warning horns. It is illegal for any person other than the Coast Guard or the military to "establish and maintain, discontinue, change or transfer ownership of any aid to maritime navigation," such as a buoy, light, beacon, or fog signal, in U.S. waters without the prior authorization of the Coast Guard.[106] (The regulations applicable to private aids to navigation are found at 33 C.F.R. parts 66 and 67. An application to establish a private aid to navigation is also available on the Internet.[107]) To obtain authorization, a project developer would file an application with the commander of the Coast Guard district in which the navigational aid will be located stating, among other items, the position of the aid, details of the technology used (e.g., light color, candle power, and height above water) and the name of the person or entity with responsibility for maintaining the aid. (Coast Guard regulations in 33 C.F.R. describe the function of various aids to navigation in detail.[108]) When a private aid to navigation takes the form of a fixed structure erected in U.S. waters, authorization of the Corps also is required.[109]

Fixed structures, including floating structures moored in place for which a Corps permit is issued, are required to have certain warning devices under part 67 of the Coast Guard regulations. The term *structures* as used in part 67 "shall include all fixed structures, temporary or permanent, for which a Corps of Engineers' permit is issued. It shall include, but is not necessarily limited to, all drilling platforms, Mobile Offshore Drilling Units (MODUs) when attached to the bottom, production platforms, quarters platforms, pipe line riser platforms, manifold platforms, loading platforms, boat landings, caissons, well protective structures, tank battery barges submerged on station, drilling barges submerged on location, breakwater barges submerged on location, artificial islands and all other piles, pile clusters, pipes, or structures erected in the waters."[110] Structures must have at least one obstruction light visible regardless of the angle of approach to the structure, and the lights are generally required to be flashing and red or white.[111] Structures also are generally required to have sound signals, except when they are so close to other structures that have sound signals that they are enveloped by the sound signals of the other structures and do not present a hazard to navigation.[112]

Air Navigation

The builder of any structure that would extend more than 200 ft (61 m) above ground level or water level is required to notify the Federal Aviation Administration (FAA) before commencing construction. (Regulations at 14 C.F.R. part 77 govern notification and other requirements applicable to structures affecting the navigable airspace. FAA form SF 7460-1, Notice of Proposed Construction or Alteration, should be submitted to advise the FAA of a proposed project that triggers the notification requirement.[113]) Notification also is required for the construction of structures to be located near an airport (including seaplane bases) or heliport (table 8–1). The test for structures located near aviation facilities is whether the proposed structure, when built, would penetrate an imaginary cone of airspace extending from the runway or heliport at a given slope for a given distance (fig. 8–1).[114]

Most, if not all, offshore wind turbines would exceed the height threshold for notification. Conversely, marine hydrokinetic devices have a much lower profile, or would be submerged, and would avoid FAA regulation. A wind energy developer would file a notification with the FAA that identifies the project location, height, and configuration of the turbines, among other details. FAA would then either make a preliminary determination of no hazard to navigation or find that the likelihood of a hazard exists. In the latter case, upon request by the project sponsor, the FAA would conduct a more extensive aeronautical study of the obstruction to determine its effect on the safe and efficient use of airspace.[115] FAA's review also will address lighting requirements and radar and radio frequency interference and may entail a public hearing.

The lighting and marking requirements for various structures, including wind turbines, are described in an FAA advisory circular *Obstruction Marking and Lighting*.[116] In general, wind turbines should be marked with flashing red or white lights (red is preferred). Lights should be mounted as high as possible on the turbine nacelle and should be visible from all angles. Not all turbines in a group need to be lighted, but unlighted gaps along the perimeter of a group of turbines, or in a string of turbines, should be no more than 0.5 statute miles in length. Turbines in the interior of a cluster or grid should be lit at intervals if the distance across the grouping is greater than 1 mile.[117] FAA recommends that turbines be painted in a bright white or light off-white color to provide effective daytime visibility. A project using these colors would not be required to have daytime lighting, but if gray, blue, or other colors are employed, daytime lighting would typically be required.

Table 8–1. FAA air hazard notification requirements

Notification to FAA is required	For	That	If	And
(a)	Any construction or alteration	would exceed 200 ft above ground level.		
(b)	Any construction or alteration	would be within 20,000 feet of a public or military airport	any runway is more than 3,200 feet long	the structure would penetrate a 100:1 sloped surface from any point on the airport's runways.
(c)	Any construction or alteration	would be within 10,000 feet of a public or military airport	the longest runway is not more than 3,200 feet long	the structure would penetrate a 50:1 sloped surface from any point on the airport's runways.
(d)	Any construction or alteration	would be within 5,000 feet of a public heliport	the structure would penetrate a 25:1 sloped surface from the heliport takeoff and landing area.	
(e)	Any other activity, as specifically requested by FAA.			

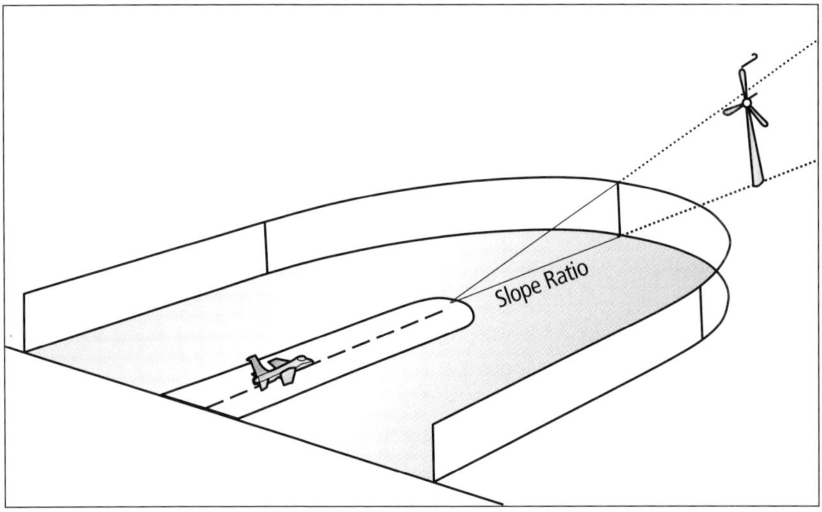

Fig. 8–1. Restricted airspace surrounding aviation facilities

FAA paint color requirements have the potential to conflict with developer efforts to minimize the aesthetic impact of an offshore wind farm. Wind farm lighting also must strike a balance between effectively alerting aviators to the presence of turbines over the water and minimizing bird attraction and disorientation problems.

As noted above, an FAA aeronautical study determines whether the proposed project will present a hazard to air navigation. The determination is based on standards similar to those that trigger the notice requirement involving the height of the structure and imaginary airspace surfaces surrounding aviation facilities. For example, any structure would be deemed an *obstruction to air navigation* if its height is greater than 500 ft (152 m) above ground level. Additional complex measures for ensuring adequate clearance for airport approaches (in some cases extending 50,000 ft from runways) also apply.[118] Suffice it to say that a developer considering a project that would be located in proximity to any aviation facility should consult early with the FAA to determine the extent to which FAA may find an obstruction to exist. In the course of the study process, the FAA may solicit comments from interested persons and hold a hearing to gather facts. The FAA also may suggest revisions to help the project proposal comply with FAA standards and avoid being considered an obstruction to navigation.[119]

An FAA determination that a proposed project presents no hazard to air navigation is valid for 18 months but may be extended. If the project proceeds to the construction stage, the developer must submit a supplemental notice to the FAA within five days after the construction reaches its greatest height.[120] A determination that a project would pose a hazard to air navigation does not, as a matter of law, block another federal agency from licensing or permitting a project. However, as a practical matter, the FAA's ruling would have that effect. Federal permitting agencies are required to base their decisions on the public interest, which encompasses a public safety element. It is difficult to imagine that an agency could find a project to be in the public interest in the face of an FAA air hazard determination. In addition, the Outer Continental Shelf Lands Act expressly requires the Secretary of the Interior to ensure that authorized offshore renewable energy activities are carried out in a manner that provides for "safety" and "coordination with relevant Federal agencies," among other criteria.[121]

Aristocratic Opposition to Renewable Energy Facilities

Developers of offshore renewable energy projects must obtain a raft of federal and state licenses, permits, certifications, and consents prior to placing steel in the ground. (The state laws have only been addressed tangentially, since an extensive treatment of state laws that may be applicable to marine renewable energy projects is site specific and beyond the scope of this work.) Federal permits will typically be contingent upon a developer's receipt of all necessary state permits and clearances that are not preempted by federal laws.

This concludes four chapters on the legal and regulatory challenges of acquiring permission to build. It is easy to imagine that once there was a simpler time, when regulations (and special interests) did not hold progress in a stranglehold. Today's wind generators have their roots in the windmills and waterwheels of eras past, and those early windmills evolved, like today's machines, out of a struggle against dependency. Twelfth-century English lords controlled water rights and monopolized grain grinding, extracting a steady fee income from local farmers and villagers for the use of the manor's waterwheel. One theory holds that windmills evolved from vertical water wheels that share a similar horizontal axis and gear design. Middle-class land owners built windmills to break away from the lords' monopoly over water power. Wind technology challenged the social order of the time and, predictably, the aristocracy sought to block windmill construction, usually by claiming authority over land use. According to wind power historian Robert Righter, "Windmills proved particularly alarming, because they utilized a primary power source that the lords could not oversee. A tiny ruling class, which thrived on monopoly and privilege, began to lose out to rural entrepreneurs and a growing urban middle class."[122]

Developers of offshore renewable energy projects will challenge the expectations of traditional users of the marine environment and incumbent energy providers. This happens as the myriad technical approvals outlined in these chapters are sought and obtained. Yet regulators also operate against a nontechnical social background that developers must shape. A successful strategy emphasizes building a coalition of supporters that will counterbalance entrenched interests and advocate for the project (or renewable energy in general) in both the larger social context and in specific regulatory battles. The next chapter on the public consultation process addresses this point.

9 Renewable Energy Goodwill: Managing the Public Consultation Process

Rotating wind turbines have emerged as popular icons—symbols of progress, modernism, reliability, and environmentalism.

—*Robert W. Righter*

The French are keen on nuclear energy, while neighboring Germans agreed to close all of their country's 19 nuclear power reactors by 2020. What accounts for differences in public attitudes toward a technology?

Germany's strong focus on promoting renewable energy has created more than 150,000 jobs in the sector, and revenues from the renewable energy industry (11.6 billion euros in 2004) have exceeded those of the German pharmaceutical industry. Many Germans view their efforts to expand renewable energy and to use energy more efficiently as critical to building the innovative and sustainable German economy of the future. Renewable energy provides high technology and new industrial jobs, while shifting the economy away from reliance upon declining traditional smokestack industries.[1]

Nuclear power facilities, in contrast, are complex, vulnerable to potentially disastrous human errors, and subject to terrorist threats. Improperly controlled sources of nuclear radiation can cause illness, and nuclear waste will likely persist in the environment beyond the life of the civilization that created it. Moreover, nuclear technology is understood only by experts. Why then do the French so strongly endorse nuclear power?

A Nuclear Community

Need and culture combine to make nuclear energy acceptable to the French—even popular. France has no oil or gas reserves, and little coal. Nuclear power provides energy independence, and even positions France to export electricity to the rest of Europe. Support of nuclear energy is patriotic. Nuclear energy provides jobs and prosperity. Scientists and engineers are held in high regard in France, and the country has a tradition of popular large technical projects such as high-speed trains and the

Concorde supersonic jet. While the French understand the potential for nuclear accidents, they have confidence in their scientists and engineers to manage the risk appropriately. Government officials also are more likely to have a technical background than to have formal legal training.

Ironically, while the French welcomed new nuclear power stations, they fought vigorously to oppose nuclear waste dumps. Christian Bataille, appointed by the French Parliament to find a solution to the nuclear waste impasse, investigated the psychology behind the opposition. He found that rural citizens viewed nuclear dumps as a "desecration of the Earth" and, unlike a nuclear plant, could see no benefit from being host to a dump site.[2] Residents near the proposed dump sites thought that if waste was deposited permanently, it would be forgotten and not properly cared for.

Bataille's proposed solution was to establish several nuclear waste research laboratories to investigate methods of storing and even detoxifying the waste. Eventually, one of the laboratories would become a waste repository—not a dump—where the waste could be managed until technology evolved to control its toxicity. This approach promised stewardship of the waste and would bring to the residents of the rural laboratory sites the added benefit of jobs to staff the research effort.

Building on Common Goals, Understanding Differences

Energy concerns are prominent in the public consciousness, on par with terrorism, access to health care, and education. The public generally supports efforts to promote renewable energy, and ordinary citizens increasingly make the connection between foreign energy dependence and national security. A history of technological progress supports the belief that technology might make energy independence achievable. Bold efforts to increase the supply of energy from renewable sources can attract support even if they require significant investment and result in higher energy costs.[3] The adoption of renewable portfolio standards in many states reflects the public's desire to achieve energy independence.

Renewable energy project developers start with the advantage of generally positive public attitudes. The battle for the public's hearts and minds is theirs to lose. Government agencies and developers may choose to invest in early and inclusive public consultation, which should result in better projects and greater public satisfaction with the outcome. Alternatively, a poorly conceived strategy and clumsy execution may turn a community of likely supporters into opponents. Fumbling the consultation process may result in protracted hearings and lawsuits, expensive studies, and delayed project timelines, while construction and materials costs increase.

Accepting that all conflict cannot be avoided, developers, stakeholders, and the government should promote the conditions under which unnecessary conflict caused by emotionalism and miscommunication can be reduced, because that is the path to forging workable compromises. Government owes project sponsors a review process that progresses steadily to a final decision without interminable delay. And the best developers do not merely rely on the required statutory project review process. They identify numerous stakeholders and are proactive in addressing their often-competing interests.

Sara MacIntyre, who is working with NaiKun Wind Development to build a 1,750-MW offshore wind farm in British Columbia, recommends starting off by listening to stakeholders to hear their concerns. MacIntyre states that earning the "social license to operate… is just as important as anything else in the business model…When you realize that, you approach community outreach a little bit differently than when it is a mandated, regulatory checklist."[4] Building alliances, or at least converting opponents into informed skeptics, is far better than unwittingly stepping on a hornet's nest of uninformed opposition.

The contrast between the perception of nuclear energy in Germany and France is a lesson for the marine renewable energy industry developing in the United States today. Public acceptance of offshore renewable energy technology is at the mercy of broad cultural and political sentiments operating in the United States. Yet each individual project also presents an opportunity to shape the public discourse on the nation's energy future and to move toward sustainable solutions.

Organizing public outreach and identifying stakeholders

The British have a long seafaring tradition. Their marine capabilities were instrumental in developing the United Kingdom's offshore oil and gas industry, which is now on the wane. So it is not surprising that the United Kingdom now supports offshore renewable energy as a logical extension of an existing skill set. Offshore renewables are a way to strengthen the competitive position of U.K. businesses in the world economy. Renewables offer a way to leverage the United Kingdom's competitive advantage in the maritime trades and to challenge Denmark's lead in the global wind energy industry.

The United Kingdom has completed a first round of offshore wind development and is in the midst of a second round of even larger offshore wind projects. Lessons learned from the initial round of project permitting are reflected in a set of best practice guidelines for public consultation. The guidelines were developed by a coalition of renewable energy industry participants, environmental groups, and U.K. government agencies to improve the consideration of offshore wind energy projects.[5] The coalition takes the view that through participation by all interested parties, projects can be improved and more people will be satisfied with the outcome. It is a pragmatic view that acknowledges that the offshore resource is public and that all projects are likely to consume significant public subsidies.

The key contribution of the best practice guidelines is to establish principles of effective consultation:

- Make consultation inclusive—all stakeholders deserve to have their views heard and addressed.
- Treat people equally—judge ideas by their merits, not their source.
- Share responsibility for the process and feedback—stakeholders need to feel that the process meets their needs and that they know how their ideas and opinions have filtered into the decision-making process.
- Consider using an independent facilitator for workshops and public meetings—it may ease stakeholder doubts about impartial treatment.
- Make the process transparent, especially about uncertainties—untruths and omissions ruin trust.[6]

There are three basic types of stakeholders:

1. Government agencies that by law must review a project proposal and participate in aspects of the permitting decision
2. National-level (or international) groups that represent a broad constituency; e.g., environmental or trade groups
3. Community groups representing those most directly affected by a particular project

Listing the existing ocean users is a good place to start when compiling a list of stakeholders affected by a particular project:

- Shipping interests and agencies such as the Coast Guard and Army Corps of Engineers, whose missions include shipping and boating safety
- Commercial and recreational fishers and recreational boaters
- Offshore mining, oil and gas production interests
- The military
- Beachgoers and surfers
- Native American tribes
- Environmentalists and state and federal environmental agencies representing society's interest in protecting marine mammals, fish, and birds

Additional stakeholders include property owners who may be in the viewshed of an offshore project, businesses and local governments concerned with impacts on tourism, and ports and unions who may see potential revenue and employment opportunities associated with offshore construction. Industries that would participate in the supply chain for offshore renewable energy projects also are interested parties. Lastly, government laboratories and universities may play a role in analyzing the impact of offshore projects and developing offshore technologies. Their members may at times advocate as experts for or against particular projects.

This listing of likely stakeholders is consistent with the Outer Continental Shelf Lands Act, which requires MMS to consider issues such as the following:

- Safety
- Environmental protection
- Waste prevention
- Natural resource conservation
- Coordination with other agencies
- The rights of others to use the outer continental shelf for purposes such as fishing or navigation
- The fairness of the return to the United States
- Public participation in the leasing process

A developer that creates a consultation process that addresses these issues has advanced the agenda and made the regulator's job that much easier.

MMS, FERC, and the Corps also encourage broad consultation as essential to the NEPA process. Many of the various statutes mentioned in the preceding chapters triggered by offshore projects also require consultation, often between agencies on issues such as environmental conservation, coastal zoning, and historical preservation (table 9–1). The task of identifying other, voluntary stakeholders is more subjective and challenging. People positively or negatively affected by the proposed project and officials and other persons of influence in the community should be included in the consultation process. Groups with socioeconomic as well as environmental concerns should be represented. Parties that have commented on other projects affecting the marine environment are likely to be interested in offshore renewable energy projects. Project developers also may advertise an informational meeting at a local town hall or college campus and see who attends.

Table 9–1. Key issues, related laws, and recommended actions

Issue	Laws Implicated	Actions
Project may require site-specific environmental analysis to evaluate potential adverse effects.	National Environmental Policy Act (NEPA)	Assist the lead permitting agency with early outreach to solicit stakeholder views, organize scoping, and determine research emphasis. Leverage analysis already conducted in the MMS Programmatic Environmental Impact Statement, or prior FERC and Corps EISs, as applicable, to lessen burden of preparing site-specific EIS. Take advantage of any "categorical exclusions" under NEPA created for activities that would not have a potential for a significant impact (e.g., site characterization, meteorological towers, small scale technology testing).
Project may affect endangered species.	Endangered Species Act	Consult with the National Marine Fisheries Service (NMFS) and the U.S. Fish and Wildlife Service (FWS). Contact appropriate local, regional, and perhaps, national environmental groups.
Project may affect protected marine mammal species and have an adverse effect on essential fish habitat.	Marine Mammal Protection Act; Magnuson-Stevens Fishery Conservation and Management Act; Fish and Wildlife Coordination Act	Consult with NMFS, FWS, and affected state fish and game agencies. Contact appropriate local, regional, and perhaps, national environmental groups.
Project may present a collision risk or otherwise adversely affect birds	Migratory Bird Treaty Act	Consult with FWS. Contact appropriate local, regional, and perhaps, national environmental groups.
Project may be located near a national marine sanctuary and affect sanctuary resources.	National Marine Sanctuaries Act	Consult with National Oceanic and Atmospheric Administration.
States have adopted plans for the development and management of their coastal regions. Offshore projects may not be consistent with these plans.	Coastal Zone Management Act	Consult with state agency that administers the coastal zone management plan. Consult with state agencies and local groups that deal with tourism.
Access to fishing grounds may be limited by a renewable energy project.	Public trust doctrine	Consult with fisher groups, related industry groups (e.g. fish processors), and regional fishery management council to identify most important fishing grounds.
Project involves placement of structures in navigable waters and other water impacts (e.g., dredged material disposal; cable laying within wetlands; stormwater runoff from terrestrial construction).	Rivers and Harbors Act; Clean Water Act; Marine Protection, Research and Sanctuaries Act	Consult with the U.S. Army Corps of Engineers (Corps). Consult with relevant state environmental agency with regard to impacts of project on state waters.

Outreach should start well before particular developers begin mapping out their projects. In his comments to the President's Commission on Ocean Policy on establishing marine protected areas, David White, director of the Ocean Conservancy's Southeastern Regional Office, emphasized that "ALL stakeholders, not just fishing interests and conservation groups, must be involved in the process from

Issue	Laws Implicated	Actions
Emissions from vessels building and servicing project may, in the aggregate, be substantial enough that a permit is required.	Clean Air Act; Act to Prevent Pollution from Ships	Consult with relevant state environmental agency and the EPA.
Project may encroach on shipping lanes, safety fairways, or otherwise present a hazard to navigation. Project also may include private aids to navigation.	Ports and Waterways Safety Act	Consult with the U.S. Coast Guard and the Corps.
Project located near airports may present a hazard to air navigation. Lighting and other markings on tall structures also may be subject to regulation.	Federal Aviation Act	Consult with the Federal Aviation Administration (FAA).
Project may interfere with military and civilian radar, and military training.	Outer Continental Shelf Lands Act (Section 388); Federal Aviation Act	Consult with the U.S. Department of Defense and the Department of Homeland Security through the Long Range Radar Joint Program Office with respect to military issues. Consult with FAA and National Weather Service regarding effects on civilian aviation radar and weather radar, respectively.
Project may require upgrade to terrestrial transmission network to interconnect.	Federal Power Act	Consult with local utility and regional transmission organization to evaluate interconnection options and to secure a place in the interconnection queue.
Project may have a visual or other impact on historic properties.	National Historic Preservation Act; Abandoned Shipwreck Act; Archaeological and Historic Preservation Act	Consult with the Advisory Council on Historic Preservation and the relevant state historic preservation office. The Department of Interior also may be consulted with respect to certain archeological or historic artifacts. Consult with affected parties to incorporate visual design considerations into project.
Project involves oil stored offshore that may leak.	Oil Pollution Act of 1990	Coordinate with first responders, wildlife agencies, and the lead authorizing agency to prepare a spill response plan.
Project may require use of foreign flagged construction vessels.	Jones Act	Consult with U.S. Customs and Border Protection.
New technologies may result in unintended effects.	Many	Employ "adaptive management strategies," including ongoing monitoring of projects to minimize adverse effects. Implement a monitoring program covering the construction, operation, and decommissioning phases to identify and respond promptly to unintended effects.

the earliest stages. This is harder than it sounds, and requires deliberate and targeted education and outreach. But we have learned that it is counterproductive to come to the table with maps and boundaries drawn before the various stakeholders are engaged."[7] It makes little sense to solicit public views if the project is so far along in its planning that it cannot practically be changed, such as after statutory permits have been obtained. Efforts to solicit comments at this stage would be perceived as not genuine and would simply erode the developer's credibility. (Where the developer has already settled on a project design, it is usually better to limit public outreach to informational meetings where the project is explained, but comments are not solicited.)

Part of the groundwork of educating the public about offshore renewable energy involves soliciting general comments about concerns and obstacles to offshore development and prequalifying certain sites as appropriate for development. Ideally, this should be done by entities other than the developer. Some forward-looking states have taken on this task. Hawaii, New Jersey, and Rhode Island have prepared studies that begin the difficult process of balancing competing interests.

Developers will usually require legal counsel to assist with project permitting, but counsel may not have the skills to manage the public consultation process. Project counsel will have researched and prepared memoranda detailing the required statutory approvals. However, a legalistic "check-the-box" approach to project permitting is likely to underemphasize human nature and the fluid, political/social environment in which the permitting process operates.

Legal compulsion is just one way to achieve an end, and it is a poor path to choose when the objective is to earn a "social license to operate." Economic incentives in the form of jobs, construction and maintenance contracts, and tax revenues build support for renewable energy projects. Environmental concerns about the public health effects of burning fossil fuels to generate electricity, and climate change–driven plant and animal extinctions, move people through emotion. The desire to live sustainably and to improve the inheritance of the next generation also is a powerful motivator. Seafaring traditions and an attachment to the physical beauty of a place also create strong feelings that can work for or against an offshore renewable energy project. Effective persuasion combines all of these factors, and its force is unlocked through building trust, presenting a logical case, capturing the public's imagination, and respecting the power of emotion.

Renewable energy machines are an elegant testament to the ingenuity of the mind. They attract tourists and visitors for the same reason that people like to visit airports simply to watch the planes land and depart. Operating in the marine environment requires cutting-edge engineering mixed with a healthy dose of courage. The reward is progress toward a planet where future generations can flourish. This message has the appeal of other great undertakings in the country's past—space exploration, the construction of the Panama Canal, and building the western railroads across the continental divide. Developers should invite stakeholders to join them in one of the great challenges of this generation. When persuasion, not legal compulsion, is the focus of the public consultation process, the public imagination can be captured, and winning popular consent, as well as formal regulatory approval, becomes much easier.

Engaging the stakeholders

Once the developer has identified the stakeholders, the process of collecting information about issues and concerns begins. Local fisherman might be consulted to avoid siting a project on prime fishing grounds. Indian tribes may have customs and beliefs regarding the marine environment, or knowledge of spiritual or historic sites that should be considered in project planning. Only after *listening* can the developer be effective in structuring a project that incorporates, as best as possible, ways of addressing stakeholder views. The listening and project structuring exercise should precede any meetings with stakeholders where detailed project plans are announced and specific comments are solicited.

There are many ways to collect and share information. A developer might start by recruiting an advisory panel consisting of experts and lay people from the community to serve as a sounding board. A workshop is a way to have a more public dialogue, yet still explore issues in depth. An unwieldy group of stakeholders might participate in a workshop indirectly through a core group of representatives who speak for people or groups with similar views. The exchange of ideas and the ability to delve into issues often make a workshop especially productive. A workshop might uncover the need for additional factual development, and study scoping, monitoring, and reporting can be agreed on at these meetings.

Other forums include town meetings, leafleting, focus groups, opinion surveys, and exhibits. Each approach has its merits and demerits (table 9–2). The large numbers of people that may attend a town meeting usually limit the opportunity each person has to express his or her views. Since each person may only get one chance to speak, views tend to be expressed as fixed statements of position—which may tend to polarize the participants. A public exhibition may be erected in a local community center where interested people might visit at their leisure to view photographs and short videos, obtain written project information, and ask questions of exhibit staff.[8] Project Web sites are a popular and effective way to provide accessible information. Web sites may explain the project construction process, provide answers to frequent questions, and illustrate (perhaps with a link to real-time information) how much power would be generated, and pollution avoided, by the project. Such sites also can provide links to project documentation, such as maps and environmental studies.

Table 9–2. Comparison of public outreach methods

Forum	Pros	Cons
Internet	Allows display of a variety of information from non-technical summaries and answers to frequently asked questions to the full text of project permit applications including technical information such as environmental impact statements. Once initial set-up cost incurred, it is an inexpensive form of outreach.	Limited in its ability to be interactive. Participants may not be satisfied that their concerns are heard and they do not have the opportunity to hear other community members' views.
Public meetings	Allow dissemination of high-level information to many people. Permit public voicing of opinions, but depending on attendance some may not have opportunity to speak.	Not ideal for developing nuanced views or for working on solutions to differences among stakeholders. Limited time and number of participants tends to polarize views.
Workshops	Allow for setting an agenda of issues and working through them in depth with a limited number of participants. Have the potential to be most productive in terms of identifying commonality and developing solutions that are supported by stakeholders.	Can be time consuming and costly (e.g., requiring expert attendance and facilitator services).
Leafleting	Inexpensive means of building a basic level of project awareness. Can be used to direct interested persons to website, public meeting, or exhibition for additional information.	Not interactive. Not conducive to explaining complex or controversial issues.
Focus groups	A controlled means of probing deeply into community views and beliefs regarding a proposed project.	May be difficult to explore the full range of stakeholder views. Not appropriate for building consensus within a community or developing solutions to problems or differences.
Opinion surveys	Good initial method of evaluating public perceptions and defining issues. Allows project developer to anticipate issues that may arise in a more public forum so that adequate responses can be prepared in advance.	Not interactive. Those with strongly held views tend to respond, while the majority is likely to remain silent.
Advisory panels	A neutral group of experts and lay people can provide useful insights into complex issues in an informal and confidential setting. Allows for testing ideas before they are made public. The inclusion of respected national or local leaders on an advisory panel may lend credibility to the effort.	May become insular over time and, as "group think" takes effect, stop functioning as an effective mirror of stakeholder views.
Public exhibition	Allows public to attend and gather information at their leisure and according to their level of interest. Permits large numbers to participate over time. Permits participation by people who would not ordinarily speak out at a public meeting.	Exhibition must be staffed with knowledgeable people that can appropriately answer questions and receive public comments. Requires investment in preparation of exhibit items and non-technical explanatory materials for the public to take away.

Source: British Wind Energy Association. 2002. Best Practice Guidelines: Consultation for Offshore Wind Energy Developments: 18-19. http://www.bwea.com/pdf/bwea-bpg-offshore.pdf.

While each of these methods is different, the objectives are largely the same: (a) to collect information from the community and (b) to respond with understandable and trustworthy information. Both of these tasks need not always be accomplished in the same forum. It is helpful to recognize that the offshore environment is unfamiliar to most and that the public record, consisting of applications and studies provided to government agencies, is often technical and difficult to digest. Special attention to making information understandable to a nontechnical audience will help to expose preconceptions, clarify issues, and reduce uncertainties. Shifting out of technical language also helps to encourage a dialogue, allowing a two-way exchange of information rather than a briefing. Dialogue helps both sides to acquire information.

This chapter emphasizes building trust between the developer and the affected community. For this to happen, project proponents must be seen as honestly seeking to protect the environment and other important community interests. Efforts at inclusion and good corporate citizenship should bring tangible benefits to the developer. A notable 1997 study of the NEPA process found that the potential for litigation and delay can be minimized by providing a meaningful opportunity for the public to participate in the evaluation of a project.[9] Fortunately, the marine renewable energy industry can build a compelling case for how its projects can both protect the environment and improve communities.

A Tempest in Nantucket Sound

There is perhaps no better example of the divergence between the generally supportive public opinion of renewable energy and the vigorous public opposition to an actual offshore renewable energy project than the storm that has roiled Nantucket Sound. Cape Wind, the project's developer, failed to conduct sufficient outreach to local opinion makers before announcing the project in Boston's newspapers. Opposition swiftly mounted among wealthy Cape Cod landowners, whose open water vistas and recreational boating would be affected by the project. In just one meeting at a Cape Cod yacht club, the Alliance to Protect Nantucket Sound raised $4 million for its litigation and public relations efforts against the project.[10] The local press also took sides, turning many local residents into opponents. Cape Wind simply lost control from the beginning of how the project was characterized. Seemingly overnight, the busy waters of Nantucket Sound became a "pristine" national treasure that a greedy developer was attempting to capture for his own personal gain.

A survey of 1,500 Cape Cod residents by Jeremy Firestone and Willett Kempton, professors at the University of Delaware College of Marine and Earth Studies, found that an astonishing 99.8% of respondents were aware of the project. This awareness was a good indication of how thoroughly it had been imprinted onto the public consciousness by proponents, opponents, and the media.[11] The survey indicated that 42.4% of the local residents opposed the project, 24.6%

supported it, and 33% had not yet decided their views. People who would see the project from their homes or during their daily routine opposed the project by a 4:1 margin. Although significant, aesthetic concerns were by no means the only source of opposition. Other objections included concerns about the project's impact on community harmony, the fishing industry, recreational boating, property values, birds, marine life, and tourism. The assortment of negative sentiments about the project clearly outweighed the relatively few positive impacts of the project listed by survey respondents, including the impact on electricity rates, job creation, and improved air quality.

In contrast, the developers of a wind energy project proposed off of Long Island's southern shore were public relations masters. They started by structuring the project as a joint public-private effort between the local public power authority and a private developer (FPL Energy). They minimized the viewshed issue by proposing to site the project in offshore waters opposite a state park, rather than offending the scenic views of wealthy and influential landowners elsewhere along the coast. The developers enlisted local celebrities to endorse the project, and they cultivated citizens groups to promote clean offshore power. In the end, however, the project's high cost was its undoing. The developers failed to position the project as advancing important national and local interests and thus worth the extra cost.

The debate over offshore renewable energy, like the debate over wind power in general, has split the environmental community. Firestone and Kempton observe that each side in the Nantucket Sound debate uses environmental arguments to support its cause, but the arguments are based in belief more than fact. The top reason given by respondents for supporting or opposing the project was its impact on marine life and the environment (table 9–3). The second most important reason given by opponents was the perceived aesthetic impact of the project, while supporters cited the impact on electricity rates.

There is clearly a gap between facts and beliefs. Opponents of the Nantucket Sound project who base their objection on concerns about marine life and the environment should prefer clean, renewable electricity to the nearby coal- and oil-fired generating plants that provide power to Cape Cod. Apparently these opponents do not understand the choice, or they actually value other criteria more highly. The fossil-fueled plants serving the region have a significant impact on the health of the local population, contributing to ailments such as asthma and to the destruction of bird and marine life. Substantial numbers of fish are killed by impingement on cooling water intake screens, and millions of fish larvae die when they pass through the screens and are entrained within the cooling systems of conventional power plants. The

Table 9–3. Public opinion of the Nantucket Sound project

Issue	Opponents' Top 3 Reasons (%)	Supporters' Top 3 Reasons (%)
Marine life/environmental impacts	65	48
Aesthetics	51	14
Fishing impacts/boating safety	50	15
Electricity rates	20	47
Tourism	15	4
Private use of public lands	15	5
Property values	14	7
Foreign oil dependence	5	37
Global warming/climate stability	4	4
Air quality	3	23
Jobs/economic concerns	2	18
Alternative/renewable energy	1	36
Other	55	41
Total	**300**	**299**

Source: Firestone, Jeremy and Willett Kempton. 2007. Public Opinion About Large Offshore Wind Power: Underlying Factors. Energy Policy 35. March: 1589, Table 6.

environmental impact of the Nantucket Sound project is modest in comparison. The project's aesthetic impact also would be minimal; at a distance of several miles, the project's turbines would appear quite small when viewed from the beach.

New facts may sometimes soften opposition and even change minds. For example, the Firestone/Kempton survey found that opponents would be more likely to support the project if Cape Cod benefited by receiving the electricity, electricity rates decreased, the local fishing industry benefited, or air quality improved.[12] It was curious, however, that increased property values or having the project less visible from shore were new facts that did not appear to have the potential for converting many opponents. The study found that 48% and 34% of respondents, respectively, would not change their position if their property values increased or the project was not visible from shore. Because this is inconsistent with many opponents' concerns about the aesthetic impact of the project, it may be that respondents were hesitant to admit in a survey that self-interest, rather than the betterment of the community, could change their minds.

Facts are difficult to communicate in an atmosphere of heightened emotion. The study found that opponents were much less likely than supporters to change their minds if presented with new facts. This finding highlights the need to learn the concerns of stakeholders and to dispel misconceptions that may be held about offshore renewable energy generally. This should start before specific projects are identified and people and interest groups solidify their positions. After particular projects are identified, expenditures on newspaper and television advertisements and other efforts to influence public opinion may simply polarize the public and harden opinions, rather than succeed in educating the public with objective information. That opponents of the Nantucket Sound project were reluctant to change their views also may signal mistrust of the developer or the process followed by the regulatory agencies charged with reviewing permit applications. Cape Wind lost the opportunity to establish an environment of trust at the beginning of the project planning process, which limited the company's ability to convey information effectively later on.

Culture, community, and the common good

How can one understand the contradiction inherent in continuing to use dirty fossil-derived electricity while opposing offshore wind power in Nantucket Sound on environmental grounds? The impressive natural beauty of Cape Cod, Martha's Vineyard, Nantucket Island, and Nantucket Sound inspires strong emotions and a spiritual connection to the region. The emotional connection to the sea, like the connection between the French country dweller and the land, are part of the cultural roots of the Nantucket Sound community. Residents resent the intrusion of a for-profit corporation onto a space that is reserved in public trust. That some opponents of the Nantucket Sound project might change their minds if the community received a benefit expresses a desire for reciprocity and fairness.

France's commitment to nuclear energy is only possible because nuclear technology has achieved social support. Similarly, cultivating a broad public acceptance of offshore renewable energy is necessary in the United States if these projects are to become a significant part of the energy future. Firestone and Kempton's study suggests that residents would have been more likely to support the offshore wind project in Nantucket Sound if they believed that the project would be the first of many projects along the Atlantic Coast and that their sacrifice would lead to important national progress in renewable energy and climate stabilization.

Cooperative ownership lessened public opposition in Denmark to the Middelgrunden offshore wind farm by increasing the number of people that perceived a benefit from the project. Middelgrunden, a 20-turbine farm near Copenhagen's harbor, is owned 50% by the local municipally owned utility and 50% by a cooperative with 8,650 members. Each share in the cooperative entitles its holder to 1,000 kWh/yr of the wind farm's production. Cooperative ownership gave members a personal stake in the project, allowing them to view it as "their" project and not that of an outside entity.

The Middelgrunden site was initially identified by a Danish government action plan to promote offshore wind power. Copenhagen's Energy and Environment Office acted as the catalyst to form the cooperative and establish the connections between the utility, the cooperative, and the city. Because the project was established by municipal entities, it began with the general support of local politicians. Use of an abandoned Copenhagen shipyard for foundation construction and marshalling of turbine components provided an additional community economic development benefit.

Despite the initial support, permitting the Middelgrunden project required three public hearings transpiring over three years from the submission of the initial application. Public views influenced the project design. Project promoters had initially proposed to arrange the turbines in a grid of three rows of 9 turbines each. Due to aesthetic objections, the final design featured 20 turbines arranged in a gentle curve roughly parallel to the shore. Noise from the offshore turbines also was presented as a concern. These concerns subsided after project promoters arranged a tour of an onshore wind farm to demonstrate noise levels. A wind farm at a distance of one-quarter mile is comparable in volume to typical rural nighttime background noise and is not nearly as noisy as roadway traffic or train traffic (table 9–4). The noise emitted by an offshore wind project is likely to be too far from the nearest residence to be heard above the background of wind and wave noise at the shore.

Table 9–4. Noise levels from various sources

Source	Noise level dB(A)
Jet aircraft at 250m	105
Pneumatic drill at 7m	95
Truck at 48 kph at 100m	65
Busy general office	60
Car at 64 kph at 100m	55
Wind farm at 350m	35–45
Quiet bedroom	35
Rural night-time background	20–40

Source: Sustainable Development Commission. 2005. Wind Power in the UK.

The process of educating potential shareholders in the cooperative paid benefits later as those who subscribed for shares became the project's advocates. "At the final hearing a large number of local groups and committees, not mentioning several thousand shareholders, recommended and supported the project—only a relatively small group of yachtsmen, fishermen, individuals and politicians remained in opposition."[13] Public involvement is credited with the willingness of the local population to accept the project, even though it is located only a few kilometers from Copenhagen's popular waterfront.

There is nothing uniquely European to this approach. Wind farm developers building projects in the U.S. heartland often do many of the same things. Successful developers meet with local politicians and the municipal staffs of the communities affected by a project, assemble advisory committees of locally influential people, and try to make farmers, landowners, and anyone else who might benefit from a wind farm advocates for the project.

In the view of Susan Nickerson, the executive director of the Alliance to Protect Nantucket Sound, an opponent of Cape Wind's project, Denmark's development process starts with a strong national commitment to energy conservation and environmental protection and then adds genuine democratic involvement in siting offshore facilities. In her view, the process in the United States has operated backwards, with the developer making the site selection and presenting it to the community as nearly a fait accompli. "Neither the public nor the government was consulted on the acceptability of this site from its perspective…Leaving the task of site selection purely to developers like Cape Wind effectively severs the public partnership that must exist before offshore wind development can succeed here in the United States."[14]

The opposition to development in Nantucket Sound could leave the impression that Massachusetts residents simply will not tolerate offshore wind projects. Ironically, another Massachusetts project proves otherwise. The municipal utility in the town of Hull (near Boston) is proposing a small, four-turbine offshore project that appears to be moving rapidly towards regulatory approval. The town has had good results from two existing shore-based turbines, which has fostered public acceptance of the offshore proposal.

Even the cooperative ownership aspect of the Middelgrunden project could be replicated in the United States, after a fashion, by involving public power authorities, municipal utilities, and rural electric cooperatives in a project. These entities generally have broad community support and can play an important role in helping a project attain the political support for site permitting and local infrastructure improvements. Cooperative and municipal utilities tend to be risk averse, and their access to capital may be limited. They also may not be large enough to build big offshore projects on their own. For these reasons, an approach that combines public and private investors may improve the chances of a successful offshore development.

Europe's success with offshore wind might suggest that European attitudes about the technology are dramatically different from those in the United States, but that is probably not the case. The locations of existing and planned European offshore wind projects tend to be geographically separate from areas of high coastal tourist use. Prime resort areas such as the Mediterranean coast have yet to sprout offshore wind farms. Current offshore wind projects located in the North Sea and Baltic Sea do not share the same characteristics as proposed offshore wind projects in the United States such as Cape Cod, with high recreational use, expensive vacation homes, and high levels of tourism. In short, the current European sites have different stakeholders than might be typical at potential sites in the United States, which may lead to different outcomes.[15]

Developers of the proposed Long Island offshore wind farms clearly learned some lessons from the Nantucket Sound experience. FPL Energy, an investor-owned utility and the project's developer, and the Long Island Power Authority (LIPA), the power purchaser, both played a significant role in planning the project. FPL Energy and LIPA avoided commercial fishing grounds and areas with high concentrations of birds. They also declined to encroach on the natural beauty of eastern Long Island's Montauk Point or its nearby summer communities. Fighting the wealthy and politically connected landowners in the Hamptons, where it is not unusual for an oceanfront home to sell for $25 million, was something to avoid. The preferred site was off the shore of Jones Beach State Park and adjacent Robert Moses State Park, which together have almost 12 miles (19.3 km) of broad, sandy beach. These parks are frequented during the summer months by millions of day-trippers. There are few if any homes or businesses along this stretch of coast that would be affected by an offshore wind farm and few people to raise alarms about lowered property values.

Although the Long Island project received a supportive public reception, it was not wholly without opposition. The ad hoc Committee to Save Jones Beach arose to represent the views of certain taxpayers and LIPA customers. They objected to the project's "industrial" character, its effect on "some of the most pristine beaches and parks in the world" and the lack of sufficient public meetings to vet the proposal. Blunting this criticism was the fact that LIPA, a public agency, was a key proponent of the project. The visual intrusion on the public's offshore domain would at least be offset by benefits accruing to the public at large. Moreover, LIPA's process was transparent. It had engaged in a multiyear public scoping process to evaluate potential environmental impacts and identify the best sites.

Advocacy through proxies can be more effective than messages that flow directly from a developer. WindWorks Long Island was formed early in the LIPA project scoping process as a partnership of local, regional, and national environmental, civic, health, business, and faith-based groups to promote offshore wind power on Long Island. The group educated stakeholders throughout the community on the need for renewable energy and the negative impacts of traditional energy sources derived from fossil fuels. WindWorks even hosted a guest from Denmark to answer questions about the Danish offshore wind experience. WindWorks advanced public acceptance of the project through its credible, independent perspective and its representation of multiple stakeholders.

In addition to the usual public forums, newspaper opinion pieces, and advertisements, the LIPA project's public relations efforts extended to seeking endorsements from prominent community leaders, scientific experts, trade unions, and celebrities. The developers sponsored student research projects and science fairs and became involved in the local Audubon Bird Survey and beach cleanup efforts. FPL Energy's strategy was to employ a multipronged approach to penetrate the cultural consciousness of Long Islanders: "In addition to standard outreach, we will incorporate subliminal marketing to make this project part of the Long Island identity and jargon. This tactic could include the use of local faces/celebrities in support of the project."[16] Actor Alec Baldwin, a Long Island native, was featured in radio commercials supportive of the project. FPL and LIPA understood that they were competing for a favorable station in the public's consciousness and, on an even deeper level, for residence in the local cultural identity.

The Effect on Tourism

Arguments about the impact of offshore wind turbines on coastal tourism generally assert that visitors will object to the view of wind turbines on the horizon and will go elsewhere to vacation, or cut their visit short, spending less on local goods and services. The reality is probably more complex. Some unspoiled natural coastal areas may attract visitors looking for beauty, while other coastal zones may already be commercialized, industrialized, or adjacent to densely developed metropolitan areas. The impact of a wind farm on each area will vary.

Middelgrunden proves that it is possible to reach a compromise on wind turbine aesthetics. By arranging the turbines off of Copenhagen's harbor in a gently curved line, visitors to the shore could gaze out on modern kinetic sculpture, rather than a jumble of towers jutting from the sea. Visitors to Atlantic City, New Jersey, a place once known as

America's playground, seem to be drawn less by nature than by the diversions in the town's many casinos. A row of turbines in the ocean, visible from the casino hotels and the boardwalk promenade, would not be out of place in this environment. In fact, the presence of several large wind turbines located on a marsh behind Atlantic City's coastal barrier island have not detracted at all from the character of the place.

When faced with a project in Delaware, members of the tourism industry were split on whether the wind turbines would draw or repulse tourists. One Delaware businessman observed, "Its not like Rehoboth in the summertime is a peaceful, pristine experience…the ocean's here… which means the tourists will come. They put up with crowds, they put up with all the things they have to [to] be here. I can't believe they'd stay away because of a bunch of little windmills."[17] Injecting some levity into the debate, the project's developers offered that to the casual observer, "the giant turbines would look smaller than a boardwalk French fry."[18]

Visitors to seaside locations such as New Jersey's Atlantic City, Maryland's Ocean City, or Delaware's Rehoboth Beach will experience a coast that is fully developed with high-rise hotels, apartment buildings, and amusements. Tourists to these locations do not expect to commune with nature but rather simply to vacation by the water. Tourists may consider wind turbines on the horizon an interesting curiosity rather than an intrusion.

Wind power and tourism can coexist. Residents and visitors to Tehachapi celebrate the town's leading role in California's wind industry with an annual wind fair featuring educational exhibits, lectures and seminars, concessions, and music on outdoor stages. The town has identified a site for a wind energy museum and is in negotiations with the local community college to establish an educational center to train the labor force needed to support the growing wind energy industry.[19]

Property Values

Objections on visual grounds also are related to a fear of decreased property values. That problem does not seem to have occurred near terrestrial wind farms. Property values in the vicinity of 10 large U.S. land-based wind projects that came online from 1998 to 2001 were examined before and after the construction of the wind farms. The study reviewed over 24,000 real estate transactions both in the viewshed of the wind farms (within a 5-mile radius) and in nearby comparable areas. No statistical evidence was found for the "contention that property values within the view shed of wind developments suffer or perform poorer than in a comparable region."[20]

Fishing Industry Objections

The U.S. fishing industry generally takes a dim view of offshore renewable energy development. The industry is concerned about exclusion from productive fishing grounds, damage to fishing gear, construction impacts, and the cumulative effects of development. Factors such as the spacing between renewable energy devices, whether the devices have moorings or moving parts that could become entangled with fishing gear, and the type of fishing gear employed in the fishery will determine the extent of any exclusion of fishers from the facility site. Wave energy devices that float and are cable-moored would most likely present more significant obstacles to fishing within the boundaries of the wave farm due to the danger of entangling fishing gear with mooring lines and electrical cables. Wave energy devices also would typically be spaced more closely than wind turbines. Lobster and crab harvesting, which uses traps placed on the seabed, would likely coexist well with most offshore renewable energy technologies. Trawling, in which bottom-dwelling species are caught by dragging a weighted net along the seabed, however, may be restricted in the immediate area of a wind or wave energy farm to avoid damage to undersea cables and entanglement in moorings.

Fishing restrictions within the boundaries of a wind or wave farm may not necessarily reduce fish yields. A wind or wave farm may serve as a protected spawning area, much like a marine reserve. Turbine foundations and rubble scour protection provide a hard surface to which benthic organisms may attach. This community then attracts other sea life seeking food and/or shelter. The phenomenon has been observed around offshore oil and gas platforms that have become popular recreational fishing sites. MMS created the Rigs-to-Reefs program when it found "a profound and pervasive connection between fish, fishing, and oil and gas structures in the marine environment."[21] Under the program, obsolete rigs are submerged in designated areas to promote reef development.

The popularity of oil and gas rigs as recreational fishing sites and differences in the gear and techniques used by recreational and commercial fishers argues against treating recreational and commercial fishers as a single interest group. Each group is likely to have unique concerns and views, and outreach to these groups should, accordingly, be separate.

Limited fishing within the confines of an offshore energy park would create valuable protected fish habitat. The Merritt Island Wildlife Refuge, a marine reserve located in the waters off of Florida's Atlantic coast, has populations of snook, drum, trout, and mullet that are 5 to 12 times higher than in waters adjacent to the reserve.[22] Fishers that initially opposed the creation of marine reserves have found that these protected areas actually increase the size and amount of fish caught outside the reserve, despite the decrease in area available for fishing. As a result, fishers have become some of the strongest advocates for the creation of such reserves. Rather than losing out, fishers stand to gain when offshore renewable energy facilities become de facto fishing reserves.

Job Creation and Economic Development

Building a terrestrial wind energy project is akin to retail politics. Developers knock on farmhouse doors and talk at town meetings. Arguments based on environmental stewardship and economics are usually received well by farmers and ranchers. Taking just this approach, Bill Moore, the developer of the 197-MW Maple Ridge Wind Farm in northern New York state, worked to convince dairy farmers and town people that wind energy could bring prosperity. Moore promised jobs, enhanced tax revenues, and better schools. He encouraged farmers considering leasing turbine sites to visit wind farms in nearby counties to learn from other farmers' experiences.

Farmers participating in the Maple Ridge project would earn approximately $6,000 per year per turbine in the form of a guaranteed minimum payment, adjusted annually for inflation. The community also would benefit tangibly, since local tax revenues of approximately $8.5 million per year would help to build new schools. The sale of gravel, concrete, and other construction supplies, and food and lodging for 400 temporary construction workers, would lift the local economy. As many as 20 operations and maintenance jobs would provide permanent employment; an important benefit, given that the county's alternatives for economic development consisted of a landfill and a low-level radioactive waste facility. Participating in the project also would help farmers retire without leaving the land, provide a legacy for the farmers' children, and help to rescue rural towns from slow decline.[23]

Similar benefits would accompany offshore renewable energy projects. Tourism in coastal communities is generally seasonal. Small commercial fishers are struggling against heavy debts, high fuel costs, and competition. The capacity of fishing fleets plying many fisheries is greater than the available catch. Employment opportunities in offshore renewable energy could benefit fishers and provide an alternate source of economic growth for coastal towns.

Some New Jersey fishers certainly see it this way. Fishermen's Energy of New Jersey (FERN), a consortium of commercial fishers, wants to build a 350-MW offshore wind park in south New Jersey. FERN's president, Daniel Cohen, believes that fishermen are best positioned to move this project forward, due to their experience working in a harsh marine environment, knowledge of the seabed geography, and access to key dock facilities. "FERN and its fishing industry supporters and investors have chosen to view offshore energy as an opportunity, rather than as a threat. FERN intends to be the agent for change, instead of a victim of changes going on around it."[24]

The renewable energy sector provides more jobs per unit of energy delivered than the fossil-fuel-based energy sector, according to an examination of the impact of renewable energy jobs by the University of California.[25] The study considered jobs in construction, manufacturing, and installation, as well as O&M and fuel processing. Direct employment and indirect employment of suppliers and others through the multiplier effect also were factored into the analysis. Manufacturing and construction jobs represent the majority of renewable energy sector employment, except in biomass energy production, which employs many agricultural and forestry workers in fuel gathering. A separate study by the New York State Energy Research and Development Authority found that wind energy produces 27% more jobs per kilowatt-hour than coal plants and 66% more jobs than natural gas plants.[26] (Developers of offshore renewable energy projects may use the Job and Economic Development Impact [JEDI] model created by the Department of Energy to estimate the job creation potential of proposed projects.[27])

Offshore renewable energy has the potential to rejuvenate a declining American manufacturing base and to provide employment that pays living wages. George Sterzinger, executive director of the Renewable Energy Policy Project, argues that broad public support for renewable energy and actions to address climate change and energy insecurity will come when the public and politicians understand the job potential of renewable energy technologies. "Unlike fossil energy, which is discovered, renewable energy is conceived and created in labs and universities, brought to commercial readiness by developers, manufactured as component parts, and assembled into finished products. In the end, renewable energy is manufactured energy."[28] Sterzinger estimates that there are in excess of 42,000 firms active in manufacturing components for renewable energy systems. Many of these companies are located in states that have experienced losses in the manufacturing sector.

Royalties for offshore leases in state waters are another source of benefit to local economies. Offshore lease payments in Texas support public education. More than $6.8 billion has been deposited in the permanent school fund since the Texas General Land Office was established in 1854. Much of this money has come from oil and gas leases on state land, but offshore wind power is poised to make a contribution. A project sited in federal waters also may produce a local benefit. Federal law requires MMS to share 27% of the revenues it receives for offshore renewable energy leases with any state located within 15 miles of the geographic center of a project, any part of which is located within 3 nm of the most seaward extent of a state's submerged lands.[29] Developers also may make a payment in lieu of taxes (PILOT) as a gesture of goodwill to compensate a community affected by an offshore renewable energy project when little project property lies within the community's taxing authority.

Comparing Offshore Renewable Energy to Traditional Energy Facilities

When Delaware residents were asked to choose from a variety of energy options, more than 90% supported offshore wind, and fewer than 10% supported expanding coal or natural gas generation. Public consultation on offshore renewable energy can be made more effective by contrasting the benefits and detriments of offshore renewables with other available energy options.

Pollution related to combustion is the most obvious detriment of using traditional forms of energy. The public tends to disregard other significant life cycle costs. Mining, transport, combustion, cooling, and waste disposal are all part of the life cycle of a fossil fuel plant. Costs include habitat destruction and pollution from mining, security costs from protecting foreign fuel sources, dangers associated with the importation of explosive LNG, and environmental damage from pipeline construction and oil spills. Marine renewable energy largely avoids these costs. An offshore renewable energy facility can be removed once it has served its purpose, and the natural environment will quickly return to its undisturbed state. Appalachian mountain streams buried by mountaintop removal coal mining will never be restored.

Developers of offshore renewable energy should ensure that the public understands the risks of traditional energy sources that they have come to take for granted. Natural gas from domestic supplies and imported LNG will be relied upon to support future electrical load growth. A General Accountability Office evaluation of the risks of transporting LNG found that people as far as a mile away could be burned if an LNG spill was ignited, as might happen in a terrorist attack on an LNG tanker.[31] Air pollution has immediate health effects on vulnerable populations such as children, asthmatics, and the elderly, and the long-term effects of greenhouse gas emissions caused by fossil fuel combustion are increasingly a concern.

Some Web sites for offshore wind projects feature a rolling counter that shows how much pollution and carbon emissions could be avoided if the proposed project were running and displacing conventionally generated energy. Graphic comparisons of this kind communicate effectively the significant benefits of offshore renewable energy.

When improperly disposed of, toxic coal ash, laden with heavy metals, may contaminate wells and drinking water. Mercury emissions from coal combustion taint waterways, placing them off-limits to fishing. Thermal pollution from conventional power plants changes fish habitat, and fish and their larvae are trapped and killed in plant cooling water systems.

The 1,456-MW Kingston Fossil Plant near Knoxville, Tennessee, consumes about 14,000 tons of coal a day and, according to the *New York Times*, in just one year, "The plant's byproducts included 45,000 pounds of arsenic, 49,000 pounds of lead, 1.4 million pounds of barium, 91,000 pounds of chromium and 140,000 pounds of manganese. Those metals can cause cancer, liver damage and neurological complications, among other health problems."[32] In December 2008, an earthen dike containing a coal ash pile at the plant gave way, and 5.4 million cubic yards of toxic coal ash slid into the Emory River, jeopardizing drinking water supplies and the health of local residents.

The hazards of the traditional energy infrastructure in the United States impose real economic and social costs. These costs range from increased hospitalizations from pollution-related illnesses to increased educational expenses for developmentally handicapped children. Recreational and fishery values are decreased. Each of the issues mentioned above has the potential to engage a constituency that may support and advocate for offshore renewable energy.

Though the life cycle of renewable energy technologies is not completely benign, it is much less destructive than traditional energy sources. Renewable energy devices must be manufactured, transported, maintained, and eventually decommissioned. The materials used are principally steel and other metals used for wind turbine towers, gearboxes, generators, cabling, and other machinery and glass-reinforced plastics used in wind turbine rotor blades. These materials are the principal determinants of the life cycle cost. Where concrete foundations are used, the energy impact of concrete and its steel reinforcement also must be considered. ExternE, a European Commission–funded project to consider the life cycle costs of numerous energy alternatives, found that wind was the least expensive source of power (fig. 9–1). Although the study considered terrestrial wind power, the life cycle costs of offshore wind power and other marine renewable energy technologies should be similar.

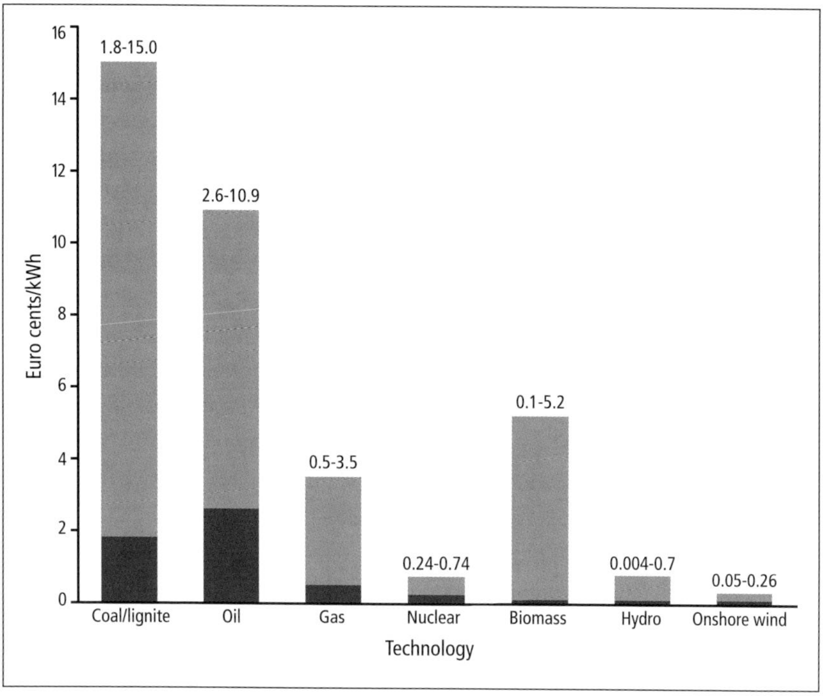

Fig. 9–1. ExternE life cycle cost estimates
Source: Garrad Hassan and Partners Ltd. 2004. Sea Wind Europe: 38, Table 6.3

MMS found substantial pollution, land use, and fuel cycle impacts when it examined the environmental and socioeconomic consequences of traditional energy production, with electricity from coal being the most detrimental. Offshore renewable energy development was clearly less harmful from an overall environmental perspective. Most renewable energy impacts were limited to the construction and decommissioning phases, when there is significant site activity, noise, and sediment disturbance. MMS also has considered the climate change issue.

> *Alternative energy development would result in only very minor greenhouse gas emissions...[and] has the potential to provide significant benefits in terms of reducing greenhouse gas emissions from energy consumption. Alternative energy development could be used as part of a strategy to manage carbon emissions by providing energy with very low emissions. If such a strategy includes a cap-and-trade program, a carbon tax, or various incentives for cleaner technology, alternative energy could become more competitive and could provide a meaningful contribution to efforts to achieve national carbon emission reduction goals.*[33]

Lastly, comparisons between offshore renewable energy and traditional energy sources should emphasize the value of a resilient energy system. A market or regulatory system that produces the lowest priced energy to the exclusion of other considerations will produce an energy infrastructure and economy that is vulnerable to instability in the long run. A mix of generating technologies provides increased security to consumers through reduced exposure to price volatility and supply disruption. Including offshore renewable energy as one component of a diverse supply also provides indirect benefits through domestic employment and investment, strengthening the social and economic fabric of communities.

Regulators influence utility power supply decisions by allowing utilities to recover costs for purchased power, utility-owned generation, and related fuel costs. Regulators should promote electrical system resiliency by encouraging choices that result in a diverse supply, including offshore renewable energy, even when the short-term cost of renewable energy seems greater than traditional, ostensibly lower-priced options.

Bird Deaths and Other Harm to Nature

Bird deaths are a fact of the industrial landscape. Astonishingly, about 1 billion birds a year are estimated to be killed by anthropogenic sources in the United States. Given the ecological importance of birds and the popularity of bird watching, public concern about bird deaths from wind turbine collisions is understandable. Unfortunately, these concerns are largely out of proportion to the real impact of wind turbines. The principal causes of bird mortality are collisions with buildings and power lines, and domestic cat predation. Other significant causes of bird deaths include vehicle collisions, pesticide exposure, and impacts with communications towers. When properly sited, wind turbines have been shown to be a minimal cause of bird mortality (fig. 9–2). Wind turbines account for only 0.003% of all bird fatalities.

The wind energy industry's vulnerability to bird deaths is a symptom of public attitudes nurtured by an assortment of man-made environmental catastrophes. These attitudes continue to shape decisions about the use of the offshore environment today. One accident that was seared into the public consciousness occurred in 1969 on Union Oil's platform Alpha, located in the Pacific Ocean off of La Conchita, California. Oil and gas reservoirs are often under great pressure and, when not successfully contained, the result may be disastrous. In a blowout, high-pressure gas trapped in an underground rock formation expands and rises through the well bore at up to supersonic speeds. The blowout at platform Alpha was so forceful it fractured the seafloor surrounding the wellhead, and oil and gas began to pour from multiple

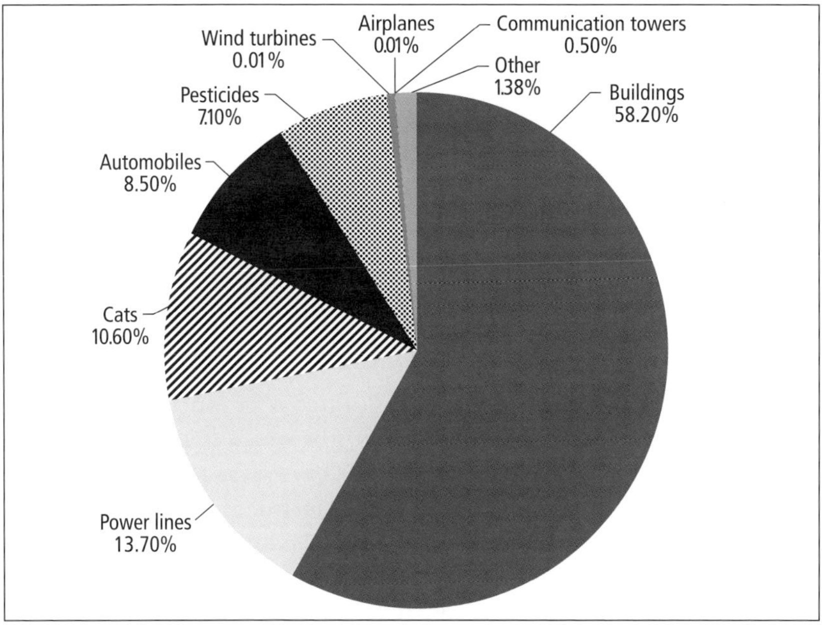

Fig. 9–2. Causes of bird deaths in the man-made environment
Source: Erickson, Wallace, Gregory Johnson, David Young Jr. 2005. A Summary and Comparison of Bird Mortality from Anthropogenic Causes with an Emphasis on Collisions. USDA Forest Service Gen. Tech. Rep., PSW-GTR-191 at 1039.

underwater fissures. Over 10 days, approximately 3 million gallons of oil escaped into the Santa Barbara channel before the well was brought under control. Oil coated the nearby coast up to six inches thick in some places, poisoned seals and dolphins, and killed at least 3,600 seabirds.

The environmental movement in the United States was gaining momentum at the time. Rachel Carson's *Silent Spring* had been published several years earlier in 1962. The Santa Barbara spill spurred the public's wariness of the offshore resource extraction industry. President Nixon, himself a Californian, appreciated the depth of emotion stirred by the event. "The Santa Barbara incident has frankly touched the conscience of the American people," he said. Union Oil's president was deaf to popular sentiment and remarked, "I am amazed at the publicity for the loss of a few birds."[34]

Two decades after the Santa Barbara spill, the single-hulled tanker *Exxon Valdez* ran aground in Alaska's Prince William Sound, spilling approximately 11 million gallons of crude oil. Thousands of birds and sea mammals died, and the expense of the cleanup, litigation, and damages ran into the billions of dollars. Technology being imperfect, oil spills continue to be regular occurrences, and media coverage provides a powerful reminder that, when technology falters, vast damage to nature may ensue. Oil exploration technology and the industry's record have since improved. Far more oil enters the nation's waterways today through natural seepage and roadway runoff than from oil industry mishaps. Yet the memories of catastrophic failures have been etched into the public consciousness, and the mistrust many feel towards big oil extends to the proponents of other large projects on the public commons.

The ramifications of the oil industry's failures go beyond the cost of cleaning polluted beaches; they affect the industry to this day. Drilling for oil and gas along the coasts of the North Atlantic, most of the Pacific, parts of Alaska, and most of the eastern Gulf of Mexico was prohibited.[35] The first moratorium, adopted in 1982 in connection with a U.S. Department of Interior appropriations bill, covered only 736,000 acres off the California coast. Over the years, the areas covered by oil and gas exploration and production moratoria have expanded. Moratorium supporters argue that fishing, recreation, tourism, and other benefits of the protected areas exceed the value to be gained by drilling. When pump prices rise, so do claims by opponents of the moratoria that the United States must secure its energy supply with increased domestic exploration and production. So far, the industry remains confined mostly to the western Gulf and parts of the Alaska coast.

Offshore renewable energy facilities pose little potential for catastrophic environmental damage, yet project developers will encounter a fearful, distrustful, and cautious public. Oil industry missteps have created the combustible precursor for public opposition. The wind energy industry's own early actions have reignited the public's concern and fanned the flames that singed the wind industry's green credentials. Well-publicized bird kills at wind farms located in California's Altamont Pass were particularly to blame. Between November 1984 and April 1988, for instance, 99 raptor deaths were recorded at Altamont Pass. For comparison, in California's Tehachapi Pass, another area of intensive wind energy production, only 9 bird deaths were measured during the same period. Bird mortality at Altamont Pass was largely attributed to collisions with turbines, but about one-third of the deaths occurred due to electrocution from bird contact with uninsulated conductors and wires on transmission poles. The electrocution deaths probably could have been prevented with foresight and a small, incremental investment in insulators.[36]

Experiences such as these color the perceptions of the public, regulators, and politicians and raise the bar for even the most benign offshore projects. Developers proposing to build on the public commons will be expected to demonstrate the highest sensitivity to environmental and economic concerns. At times this will seem quite strange. The marine environment has for decades been the victim of a variety of insults perpetrated by its entrenched users. Polluted urban and agricultural runoff, sewer outflows, over fishing and destructive fishing methods, industrial uses, and vessel discharges are just some of the causes of ocean degradation. Yet the newcomer to the public commons, the renewable energy developer, is held to a higher standard. Many will expect developers to prove that their renewable energy technologies will have almost no detriment.

Opponents of the Nantucket Sound project have pointed to the 40,000 gallons of dielectric cooling oil that would be housed in Cape Wind's transformer substation and the 1,000 gallons of diesel fuel stored for backup generators. They have raised legitimate questions about the likelihood that an oil tanker or conventional ship could collide with a wind turbine tower and spill diesel fuel or heavy crude oil into the environment. (MMS required Cape Wind to address these concerns in part with an oil spill trajectory analysis that would show offshore and onshore areas that could potentially be affected by a spill.) Yet in nearby Buzzards Bay, while the debate on Cape Wind's project raged, an oil tanker bound for a local oil-fired power plant ran aground, split open its tanks, and spilled almost 100,000 gallons of thick No. 6 fuel oil. The ecological impact of that spill was many times greater than what might possibly happen should all the oil in the wind farm transformer station spill. The irony was lost on most who accept familiar technology largely without question.

Developers of offshore renewable energy projects should help to foster public perceptions that are grounded in fact. Collisions at a four-smokestack Florida coal-fired power plant killed more than 3,000 birds during just one night of their fall migration. The increasing prevalence of communication towers is also taking a toll. One 1,000-ft (305-m) television tower in Eau Claire, Wisconsin is estimated to kill 3,199 birds per year. Several factors contribute to these fatalities. Tall, lighted towers may attract birds that migrate at night, particularly when the weather is cloudy or foggy. Birds may circle the towers and collide with the guy wires that support the towers, hit other birds, or become exhausted and fall to the ground. Solid and pulsating red lights appear to be more attractive to birds than white strobe lights.

Studies of bird deaths at wind farms show an average of about 2.11 birds are killed per turbine per year.[37] The comparatively low rate is attributable to tubular towers that, unlike lattice towers, do not provide places for birds to nest or perch, and the diminished use of guy wires. Better wind farm siting, chiefly avoiding areas frequented by numerous birds, such as migration corridors and nesting areas, also have reduced the opportunity for collisions. Minimizing tower lighting within a wind farm and using lights principally to mark the perimeter of a wind farm also has helped to reduce the lighting distraction to nocturnal migrants, while keeping the farm visible to aviators and navigators.

Rediscovering the "Can-Do" Spirit

Shared values allow diverse stakeholders to become invested in finding solutions despite their differences. In 1973, Richard Nixon called for shared sacrifice to change the nation's energy habits.

> *Let me conclude by restating our overall objective. It can be summed up in one word that best characterizes this Nation and its essential nature. That word is 'independence.' From its beginning 200 years ago, throughout its history, America has made great sacrifices of blood and also of treasure to achieve and maintain its independence. In the last third of this century, our independence will depend on maintaining and achieving self-sufficiency in energy.*[38]

Unfortunately, Mr. Nixon's call went unanswered by the nation through a succession of presidential administrations. The country seemed to have lost its can-do spirit. Perhaps nowhere has this been more evident than in the context of the Nantucket Sound project. Jim Gordon, president of Cape Wind, has persevered in pursuit of his visionary project. However, in the words of Robert Whitcomb and Wendy Williams, authors of *Cape Wind*, "It was getting harder to introduce such ideas in a nation that had become increasingly dominated by an entrenched plutocracy that had little, if any, sense of national or global responsibility."[39]

Almost three decades after President Nixon's call for energy independence, President George W. Bush found himself the leader of an America addicted to oil, seemingly with no recourse but to expend blood and treasure to pacify the Middle East. Now, more than ever before, people are beginning to understand the cost of the nation's complete failure to address energy dependence and the increasing costs of delaying the transformation to a sustainable economy. Once afraid to get too far ahead of the electorate, some politicians now are emboldened to support alternatives to business as usual. A cultural shift appears to be under way.

Developers of offshore renewable energy projects in the United States have an unparalleled opportunity to reinvigorate the can-do spirit within community, business, political, environmental, regulatory, and opinion leaders at all levels. Developers that listen to stakeholders, assemble well-conceived projects, and build coalitions of supporters can, through force of vision, advance the offshore renewable energy industry.

The following chapter addresses the specific challenge of connecting an offshore renewable energy project to the transmission grid. Transmission is a major project cost and a source of significant delay when upgrades to the terrestrial grid are required to accept offshore power. The chapter also considers, and refutes, the charge that intermittent offshore renewable energy supply will decrease grid reliability.

10 Connecting to the Grid: Adapting to Intermittent Renewable Energy Supplies

There are no overriding technical reasons why wind energy could not achieve close to 100% penetration— it is economics that will determine the ultimate mix.

—*Garrad Hassan and Partners, Sea Wind Europe*

Connecting an offshore intermittent renewable energy source to the electric transmission grid involves addressing two broad challenges. The first is to minimize the cost in both time and money required to build the transmission infrastructure that feeds power from the facility to the grid. This includes the cost of upgrading the transmission grid to accept offshore generated electricity without overloading circuits, transformers, and such. The second broad challenge is operating the transmission system on an ongoing basis to maintain balance between the fluctuations in load and generation. This requires both long-term planning and split-second responses to meet demand and regulate system voltage and frequency to a high standard of reliability.

Electrical transmission systems are the largest and most complex machines in the world. They are dynamic systems, and adding new technology requires care. When the machine breaks and outages cascade throughout the system, the consequences extend beyond melted ice cream in the home freezer. The Northeast Blackout of August 2003 cost $7 to $10 billion.[1] Businesses as diverse as Broadway theaters and steel mills lost millions from the forced shutdown.[2] It is in this context that renewable project developers, engineers, and regulators are learning to manage the challenges of integrating renewable power and the grid.

Making the Grid Connection

A typical offshore wind farm has a low-voltage collection system linking the individual turbines and a high-voltage transmission link from a central collection point offshore to a substation located on land. In a large offshore farm, the collection system transmits the power to one or more offshore transformer platforms, where the power is stepped up to a higher voltage and transmitted to shore via high-voltage transmission cables. A small group of offshore turbines may connect directly to the local transmission grid without an offshore transformer station. The collection system and transmission system cables are usually placed into a trench on the seabed. Directional drilling may be used to bring the cable onshore with minimal disturbance to shoreline habitat. Once onshore, the transmission line terminates at a point of interconnection with the regional high-voltage grid, generally at a substation.

The layout of the inside-the-farm collection system should take into account the possibility that undersea cables may be damaged by boat anchors and fishing gear and will sometimes fail. Because bad weather may delay repairs for a considerable period, a layout that isolates the fewest number of turbines from the central collection point in the event of cable failure will result in the lowest loss of revenue while the turbines are off-line awaiting cable repair. A looped network collection system in which a series of turbines are interconnected by a loop of cable that begins and ends at the transformer station provides two alternate paths for energy to reach the transformer. In the event of a cable failure, a looped network would isolate fewer turbines than a radial collection system where several turbines share one line that extends from the transformer station to the outermost turbine on the line. If a spoke of a radial collection system fails, only the turbines between the break and the transformer platform will stay connected, while the production from all turbines beyond the break will be lost until repairs can be made. Similarly, a multicable, high-voltage connection from the transmission platform to shore provides a redundant transmission path if one cable suffers a fault. Of course, the desire for redundancy must be balanced against the considerable cost of undersea cabling.

Most offshore wind farms built today have used *high-voltage alternating current (HVAC)* transmission systems to bring the power to shore. This mature technology is the most cost-effective for moderate distances and for projects where the voltage is lower than 175 kV. As distances increase, AC systems experience significant losses and generate large reactive currents that require compensation. Real current, which drives loads such as motors, and reactive current, which stores and releases energy during every cycle, work best when they are in phase. The task of managing the resonance between the cable, shore-based transmission systems, and the turbine generators becomes complex when long AC cable spans are involved.

As a result of the challenges of using HVAC, developers of large projects far from shore are exploring the practical applications of *high-voltage direct current (HVDC)* technology. HVDC is more suitable for long-distance transmission than HVAC, because it permits much higher current flows with minimal power losses. HVDC also allows for the control of both active and reactive power independently and almost simultaneously, which greatly improves the ability to manage the transmission system. Unfortunately, HVDC systems require a converter station at each end of the transmission line, which adds significant cost to the system. Improvements in converter electronics are expected to decrease HVDC system costs over time.[3]

At what distance from the shore does the advantage shift from HVAC to HVDC? Based on technological and economic considerations, HVAC technology starts to lose its appeal at distances in excess of 25 miles (40 km) from shore. A study by a consortium of German utilities that considered the requirements for integrating a large quantity of wind power into the grid by 2020 provides an example.[4] They considered 6 offshore wind projects in the Baltic Sea aggregating 1,711 MW, and 13 projects in the North Sea aggregating 18,640 MW, located from 20 to 200 km (12 to 124 miles) from shore. The configuration of the offshore transmission system, available technology, and cost were all considered.

The study determined that HVAC transmission technology is the most cost-effective and technically straightforward approach to connecting all but the most distant offshore plants. HVDC systems were the choice for offshore wind farms over 150 km (93 miles) from shore with a capacity of 250 MW or more. At this distance, the initial higher cost of HVDC technology becomes cost-effective because power losses are lower than with AC technology.[5] Regardless of the transmission technology employed, the study found that connecting the offshore farms in groups reduced the cost of interconnecting with the onshore grid and provided enhanced electrical stability to the system.

The proposed U.S. offshore renewable energy projects all are located relatively close to shore in shallow water to minimize transmission and foundation costs. As demonstrated in the German study, grouping offshore projects so that they can share an optimally sized transmission path to shore is another way to lower transmission costs. Unfortunately, the present system of state renewable portfolio standard (RPS) programs discourages grouping, which could lead to an oversupply of renewable energy credits (RECs) in one state and a drop in the REC subsidy earned by the grouped projects. A federal RPS, however, would eliminate the disadvantage faced by grouped projects through the increased REC price stability that can be expected from a larger market. (Chapter 11 describes RECs and their use in renewable energy project finance in greater detail.)

Apportioning Interconnection Costs among Project Developers and Transmission System Customers

The high-voltage transmission system can be compared loosely to a hub-and-spoke network. Radial lines leading from a generator to a point of interconnection with the grid (i.e., the hub) are referred to as *generator tie lines*, *interconnection facilities*, or just *interties*. These are considered *sole use facilities* because they have the singular purpose of establishing an interconnection for one supplier. *Network facilities*, comprising the web of high-voltage lines and substations beyond the interties, can be considered the hub. FERC generally requires generators to bear the full cost of interconnection facilities under a theory of causation. The generation project uniquely benefits from the interconnection infrastructure and, but for the project, the construction of the transmission facility would not have been required.[6] For the most part, this rule encourages generators to site near transmission facilities with adequate available capacity and promotes the efficient use of existing transmission resources.

Upgrades to network facilities at or beyond the point of interconnection that are required to accommodate generation additions are generally paid for by the generator. However, because the network upgrades benefit customers broadly, the generator may receive a transmission credit that reimburses the generator for all or part of its investment in network upgrades over several years. Expenditures for transmission interties and network upgrades can be substantial, running into the tens of millions of dollars. Cost apportionment can significantly affect project economics and frequently gives rise to disputes among interconnection customers, transmission providers, and load serving entities that would bear the transmission system costs not imposed on the interconnection customer.

Transmission policy that increases the availability of low-cost interconnections will promote the expanded use of renewable energy technology. In Denmark, where wind penetration is approximately 60% of the system peak load, large wind facilities including offshore projects have been encouraged by interconnection policies that share costs broadly. The cost of an intertie and any network upgrades in Denmark are paid by the transmission system operator and are borne by all transmission system customers.[7]

California's new approach to building renewable energy transmission

To state the obvious, renewable energy projects must be built where renewable resources are found. In many cases this may not be convenient to existing transmission. The operator of California's high-voltage grid, the California Independent System Operator (CAISO), found that many renewable energy projects were discouraged by the transmission system cost allocation rules. The theoretical basis for allocating interconnection costs to the project developer—to encourage projects to locate near existing transmission facilities—breaks down in the context of renewable energy projects. These projects can be located near existing transmission only if, by accident, those places also happen to be (in the case of wind energy) windy and otherwise suitable sites. Renewable energy projects also usually come in small increments. The developer of an individual project does not have an incentive to build an intertie sized to accommodate the various renewable energy projects that could be expected to be constructed in a given area.

These barriers to adequate transmission for renewable energy projects were impeding the development of renewables at the same time that utilities were under increasing pressure to supply load with renewable energy under California's RPS (20% renewable energy by 2010). To correct this problem, CAISO asked FERC for permission to establish a new category of transmission facilities supported by a unique financing mechanism. Conceived as trunk lines to regions of the state with significant potential for renewable energy development, the facilities would initially be paid for by network customers under the standard transmission access charge. As developers connected their renewable (or traditional) energy projects to the trunk line, each would pay a proportionate share for the line capacity used by the project. The shared burden would significantly reduce the entry barrier for renewable energy projects in California.[8]

FERC approved CAISO's novel proposal. The agency's chairman, Joseph Kelliher, noted:

> *Commission policy with respect to allocating interconnection costs is designed to encourage a generator to locate where interconnection costs are lower. In my view, this policy is sound and should be maintained. However, there is inherent tension between this policy and the unique characteristics of renewable energy projects. Unlike fossil fuel and nuclear generation projects, renewable projects are location constrained. A fossil fuel generation project can choose more easily a location that minimizes interconnection costs. The location of renewable energy potential is dictated by nature, not by proximity to the transmission grid. A strict application of our current policy will curtail development of our renewable energy potential, by limiting development to those areas where interconnection costs happen to be lowest, not where renewable energy potential is greatest. Many areas of greatest renewable potential are in remote areas, where interconnection costs are relatively high.*[9]

FERC commissioners Philip Moeller and Jon Wellinghoff encouraged other transmission owners and operators to bring forward similar proposals to reduce the transmission investment barriers to renewable energy projects. Renewable energy transmission corridors in other states have since been proposed as a first step in unlocking wind resources for development. Plans for *competitive renewable energy zones (CREZ)* in West Texas, for example, contemplate construction of new transmission facilities and upgrades between remote, but windy, West Texas and eastern Texas load centers. These projects would support the development of additional wind energy that Texas needs to meet its renewable portfolio standard of 5,000 MW by 2015.[10]

Will the transmission operators in the Northeast and Mid-Atlantic regions determine that transmission barriers to developing offshore renewable energy are too high without a similar cost allocation mechanism? Participation in MMS offshore site auctions might be encouraged by a proposal to build an offshore high-voltage transmission loop to serve leased tracts. Each developer of an offshore tract along the transmission path would pay only its proportionate share of the trunk line's capacity cost. An arrangement of this kind that integrates plant and transmission siting makes for more efficient offshore transmission investments and more affordable offshore renewable energy.

Obtaining Permission to Interconnect with Existing Transmission Facilities

Negotiating a transmission connection can be a contentious and protracted dialogue between project developer and transmitting utility. If the transmitting utility's system is constrained, the process will require a system impact study. The transmitting utility will evaluate the effect the project will have on grid reliability and power quality and propose system modifications to make service possible. The proposed terms of interconnection may be costly if severe constraints must be relieved. Offshore project developers should begin informal discussions with their transmission provider at an early stage to understand the transmission environment before the project design proceeds too far.

Assuming that a proposed marine renewable energy project cannot connect with an existing offshore trunk line, the developer must build a tie-line to connect the project to a substation on land. Alternatively, transmission interconnections built by pioneering offshore project developers may have extra capacity that could accommodate the power from a second or third offshore farm. As has happened with land-based projects, developers that follow the pioneer into an area may seek access to the interconnection facilities built by the pioneer. The pioneering developer, however, may have its own plans for a new or expanded wind farm in the area and may desire to save any available capacity for its own projects.

This was the case in California's Tehachapi region, an area that has experienced substantial wind development. Early wind developers financed the Sagebrush transmission line, a 46-mile, 230-kV transmission line extending from the Tehachapi Mountains to load centers in Southern California. Sagebrush Partnership, an entity formed by several partners, each holding a proportionate share of the line's capacity to transmit power from the partners' wind projects in the Tehachapi area, owns the Sagebrush line. The wind farms connected to the line are currently using most of its 420-MW capacity. With adequate transmission capacity, the area's strong winds could support construction of an additional 4,500 MW of wind capacity.

One wind developer, Aero Energy, LLC, sought to connect its 50- to 120-MW project to the Sagebrush line and was rejected by the Sagebrush Partnership and the line's operator, Eurus Toyowest Management LLC. They claimed that they could not make available more than 3 MW of unused capacity on the line. It is telling that the owner and operator declined to conduct a study of the amount of capacity available or to state what improvements or reinforcements to the line would be required to accommodate Aero Energy's request. They also refused to provide Aero Energy with information that it could use to prepare an independent study. Faced with this roadblock, Aero Energy brought the matter to FERC.[11]

Under sections 210 and 211 of the Federal Power Act, FERC has authority to require a transmitting utility, such as the Sagebrush Partnership, to interconnect with a generating utility and provide transmission services. This includes authority to require expansion of capacity as necessary to provide transmission service. The commission may provide this relief if it is in the public interest, which in this context generally means that requiring transmission service would "encourage [the] overall conservation of energy or capital," "improve the reliability of any electric utility system," and "promote the economically efficient transmission and generation of electricity."[12] The commission's order must provide the transmitting utility with recovery of its costs for providing service, including any facility additions or improvements. These costs should be allocated to the party seeking interconnection and not the transmitting utility's existing customers.

To obtain FERC relief, the applicant has to have first made a good faith request to the transmitting utility at least 60 days prior to asking the commission to act.[13] This requirement is intended to open a constructive dialogue between the transmitting utility and the generator to reduce the need for FERC intervention. In its request, the project developer should specify the transmission services it will require in sufficient detail that the transmitting utility can model the transaction. The explanation should include delivery and receipt points, the total amount of transmission capacity requested, and load factor data such as the hourly quantities of energy that the generator would expect to deliver to the points of interconnection with the transmitting utility. If nonfirm service is requested, the generator should specify the conditions under which it is willing to accept interruption or curtailment. Curtailment is an important issue to a renewable generator, because it can have a significant impact on project profitability. When the transmission operator curtails the output of a renewable energy facility, the project owner loses three sources of revenue: (1) the commodity sale; (2) the value of any RECs that would have been generated; and (3) the PTC, which is only earned when energy is produced and sold. The developer's request also should discuss the rates, terms, and conditions for service, typically by describing a rate methodology or referencing an existing transmission tariff.

Commission rules require a transmitting utility that has received a good faith request for service to reply promptly with an offer of transmission service in the form of an executable service agreement that explains the basis of the charges for each component of transmission service to be provided. Alternatively, where the transmitting utility identifies constraints, it should describe them and provide supporting documentation. The transmitting utility must offer to conduct a system impact study to determine what changes to the grid are necessary to remedy the constraint, estimating the cost of the study and the time necessary to complete it. If the project developer still wants to proceed, it can then enter into an agreement to reimburse the transmitting utility for the costs of the study. The transmitting utility is required to proceed diligently to complete the system impact study within 60 days. (Compliance is encouraged by penalty provisions in section 19.9 of the pro forma *open access transmission tariffs [OATTs]*.[14])

Under FERC rules, the study requester should be given access to documents and work papers used to perform the study to facilitate an independent evaluation of any constraints and remedies.[15] The system impact study may indicate that the transmission system must be fortified or upgraded to provide the requested service. In this case, the transmitting utility is required to offer to conduct a facilities study to obtain a detailed picture of the needed improvements and their related cost. A completed facilities study should include "a good faith estimate of (i) the cost of Direct Assignment Facilities to be charged to the

Transmission Customer, (ii) the Transmission Customer's appropriate share of the cost of any required Network Upgrades as determined pursuant to the provisions of Part II of the Tariff, and (iii) the time required to complete such construction and initiate the requested service."[16]

If the dialogue between project developer and transmitting utility outlined in FERC's regulations breaks down, the developer can seek a FERC ruling. In some instances, FERC finds adequate basis to issue an order requiring the transmitting utility to provide service. In such cases FERC will as a procedural matter first issue a proposed order that sets a reasonable time for the parties to agree to the terms and conditions of service, such as rates, cost allocation method, and the length of the transmission service agreement to be imposed by the order. This gives the parties one more opportunity to negotiate a transmission agreement that the commission will adopt in its final order. If the parties cannot agree, the commission will independently set the terms and conditions by order.

In *Aero Energy*, FERC found that if Sagebrush determined that it could not, at a minimum, provide nonfirm service for Aero Energy's project, Sagebrush and Aero Energy must work together to complete a study of transmission constraints and negotiate a transmission services agreement. The commission gave the parties 28 days to come to agreement or FERC would evaluate the positions of the parties, apportion the costs, and set the other terms and conditions of service.

Although FERC's transmission interconnection process seems well-defined and time limited, it still can take an inordinately long period to conclude an interconnection agreement. Aero Energy claimed to be in discussions with Sagebrush for three years before it sent Sagebrush its good faith request for transmission service under FERC's regulations. Litigation, more negotiation, and a flurry of FERC decisions then ensued, culminating in an executed transmission service agreement between the parties. In all, four years and eight months transpired between the initial overture to Sagebrush and the parties' eventual agreement on the terms of transmission service.

FERC has recognized that wind plant developers may need data from the transmission provider, particularly about conditions on the grid at possible points of interconnection, to self-study the feasibility of a project's proposed interconnection. Access to this information permits the developer to plan the electrical design of its proposed plant to the level of detail required in a formal interconnection request. The timing of the formal interconnection request determines the position of the developer's project in the transmission provider's queue of interconnection requests, which is usually managed on a first-come, first-served basis. A good queue position can shorten the time it takes to reach an interconnection agreement. It can mean the difference between obtaining firm transmission at a time when there is adequate available transmission capacity versus settling for nonfirm service and possible curtailments. A spot back in the queue may allow transmission service requests with priority in time to absorb some or all of the available capacity, subjecting projects later in the queue to a greater risk of having to pay for grid reinforcements.

Critical Energy Infrastructure Information

Detailed engineering information about the local transmission infrastructure is more difficult to obtain after the terrorist attacks of September 11, 2001. FERC restricts the availability of critical energy infrastructure information (CEII), which FERC defines as specific engineering, vulnerability, or detailed design information that: (i) relates details about the production, generation, transportation, transmission, or distribution of energy; (ii) could be useful to a person in planning an attack on critical infrastructure; (iii) is exempt from mandatory disclosure under the Freedom of Information Act; and (iv) does not simply give the general location of the critical infrastructure.

A developer that has been refused access to CEII by a transmission system operator may make a formal request to FERC. The request should explain the need for and intended use of the information (e.g., to study the feasibility of connecting an offshore project) and should include an executed nondisclosure agreement. FERC will balance the sensitivity of the information against the developer's need and may place conditions on the release of information beyond those in the standard nondisclosure agreement.

Developers looking to step into the queue as early as possible can take advantage of FERC's policy of permitting wind project developers to satisfy the requirements of an interconnection request by submitting preliminary electric design specifications that represent all the turbines comprising a wind farm in the aggregate as a single generator. Once this information is submitted, the transmission provider should release the base case transmission information to the developer. The developer then has up to six months to complete detailed design specifications and submit them to the transmission provider so that it can begin the system impact study.

Aero Energy is a cautionary tale about the difficulty of obtaining transmission access for location-constrained renewable energy projects. The case also provides a roadmap of sorts for offshore renewable energy developers that seek to reduce transmission costs. If other factors permit, it may be helpful to consider locating a project next to an existing offshore facility and asking for permission to connect to its transmission tie-line. Even if the connecting project must absorb costs associated with reinforcing the existing intertie, those costs may be less than the expense of building a new intertie. The case also illustrates a potential trajectory for offshore development. The ability to share transmission interties and thereby lower interconnection costs suggests that, where practical (and where the REC markets permit), offshore renewable energy facilities should be clustered. Wind and wave energy facilities may be sited together, for example. Clustering of projects also could help to simplify and accelerate the interconnection queue process and reduce the costs of onshore upgrades to the grid. As FERC has noted,

> *We believe clustering request studies offers potential benefits as the needed transmission upgrades are frequently large enough that the upgrade can accommodate more than one transmission service request. In addition, jointly modeling transmission service requests can allow the transmission provider to more efficiently design transmission system upgrades. Clustering also allows the transmission provider to include, to the extent it is consistent with good utility practice, the potential counterflows created by the clustered requests.*[17]

PJM and NYISO, for example, may group generation projects proposed within semi-annual or annual periods for study. Clustering benefits project developers by spreading the cost of needed transmission system upgrades proportionately among the projects in the study group.[18]

MMS should consider structuring its offshore leasing program to promote the clustering of facilities to minimize the per-megawatt cost of providing offshore transmission links and land-based grid reinforcements. Since clustering projects may impact REC values, MMS should recommend the adoption of a federal RPS or the creation of regional REC markets in coastal states. MMS also might promote the construction of offshore transmission corridors, perhaps by seeking authority to allocate a portion of lease royalties to funding these projects. Investments in offshore transmission could accelerate offshore renewable energy development, generating royalties to repay the government's initial transmission investment.

Accommodating Intermittent Supplies

Wind power's bane has been its intermittency and unpredictability. Other renewable energy technologies share this problem to a greater or lesser extent (fig. 10–1). Like wind energy, solar power depends on the weather. Tidal current energy, although intermittent, is as predictable as the tides. Astronomers know years in advance how high the tide will be on a daily and hourly basis, based on their ability to predict the earth's position relative to the moon and sun. With this information, a transmission grid operator is able to schedule a tidal energy plant for dispatch secure in the knowledge that, absent a mechanical fault, the plant will contribute to meeting the forecast load. Wave energy falls midway along the predictability spectrum. Ocean buoy systems monitor and report wave direction, height, and frequency information. This data is input into models that predict with reasonable accuracy when a distant wave train will arrive to energize a wave farm.

Fig. 10–1. Variability of renewable energy sources

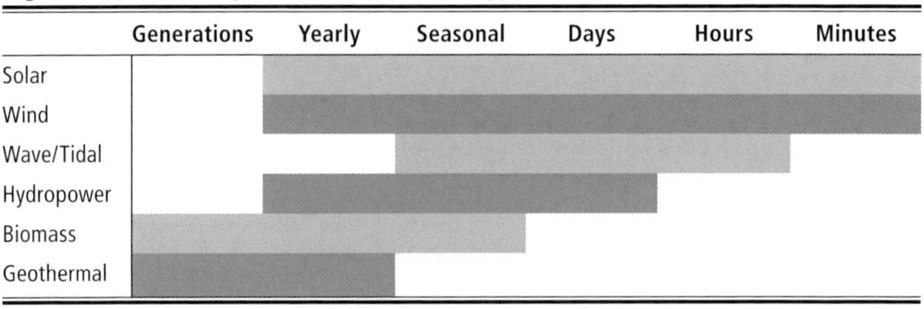

Source: International Energy Agency. 2005. Variability of Wind Power and Other Renewables; Management Options and Strategies: Fig. 2.

A common misconception is that electrical system reliability requires one unit of traditional dispatchable generation capacity to back up each unit of so-called unpredictable renewable energy generating capacity. In reality, the ratio is much lower. Reserve capacity amounting to only a fraction of the installed renewable energy capacity is required, since the pooling of resources across the grid minimizes the impact of the variability of any particular resource.[19] Loads on the electrical system also are variable, so a system operator's task is to balance the combined variability of loads and resources, more so than the variability of a single resource such as a wind turbine or wind farm. The larger the grid in terms of its geographic dispersion and loads, the greater the benefit from pooling, and the easier it is to manage the combined variability of resources and loads.

Studies show that ancillary service costs—the cost of fine-tuning or balancing the grid to accommodate wind power's natural variability—are small at low levels of wind penetration. While ancillary service costs increase as penetration rises, costs remain moderate even at higher penetration levels. A review of U.S. studies on the additional operational and capacity reserve costs of wind power found that at wind penetration levels ranging from 3.5% to 29%, additional system costs ranged between $1.47/MWh and $5.50/MWh.[20] A study of U.K. plans to increase biomass and wind power to a 20% market share by 2020 found that this goal would result in total additional transmission system costs of approximately EUR 4.9/MWh, or about 10% of current wholesale electricity prices.[21] Interestingly, the study found that significantly higher system costs (on the order of EUR 14/MWh) would come from a future renewable energy scenario that relied predominantly on wind power from Scotland and offshore. This scenario resulted in the highest costs since the wind resources in Scotland and offshore are remote from the load centers in southern England and require expensive transmission construction and upgrades. In this respect, the additional transmission system investments required to develop offshore renewable energy sources in the United States are likely to be significantly less than predicted in the United Kingdom. The planned U.S. offshore wind projects all are located close to the nation's populated coasts, where electricity loads are large and the transmission infrastructure is reasonably robust. U.S. coastal areas should be able to support higher levels of renewable energy penetration than remote parts of the United Kingdom, or for that matter, areas of the sparsely populated U.S. interior.

A collaborative effort to investigate wind energy technology, led by the U.S. Department of Energy (DOE), considered manufacturing, transmission and integration, markets, environmental impacts, and siting issues to determine whether the United States could obtain 20% of its electricity from wind by 2030. DOE explored how to rapidly expand wind energy while keeping delivered costs, which include both generation and transmission, low. In the DOE scenario, the nation would add 54 GW (i.e., 54,000 MW) of shallow-water offshore wind, located primarily along the northeastern and southeastern coast, and 241 GW of land-based wind farms. Significant investments in additional transmission capacity would be required to move the wind energy from the resource to the load. Although the needed transmission investment is large, it represents only a fraction of the value of the potential wind energy production. Furthermore, even if additional wind capacity is not built, much of the investment in building a more robust electric transmission highway would be needed in any event to move traditional energy resources to satisfy the nation's expanding electricity appetite. The technical challenges are surmountable. As the DOE report noted, "Utility engineers in some parts of the United States now have extensive experience with wind plant impacts... Wind's variability is being accommodated, and given optimistic assumptions, studies suggest the cost impact could be as little as the current level—10% or less of the value of the wind energy generated."[22]

Supporters of renewable energy emphasize that traditional fossil-fueled (and nuclear) power plants also have their intermittency and predictability issues and require reserves to maintain system reliability. Increasing U.S. reliance on renewable energy technologies is really about learning to deal with a different kind of intermittency and unpredictability. Amory Lovins, lead scientist at the Rocky Mountain Institute, an environmental policy and research organization, observed that large traditional power plants also experience unpredicted outages. When they do go down, the loss of a big chunk of generating capacity can be harder to manage than fluctuations in smaller, dispersed renewable generation. "*All* sources of electricity are unreliable—for different reasons, with different degrees of predictability and different scales and durations of outage. If you properly diversify variable renewable technologies and disperse them over hundreds of kilometers, and forecast them and integrate them with your supply-side and demand-side resources, their variability is not a significant issue, even at very high penetrations."[23]

How much wind energy can transmission system operators accommodate in the energy mix? Some U.S. utilities, such as Xcel Energy and MidAmerican Energy, already have wind energy penetration in excess of 10%, and in some windy parts of the country, penetration could reach the high levels currently found in Denmark. Denmark derives an annual average of 30% of its energy from wind, and during some hours of the year, wind supplies the country's full electrical load.[24] *Penetration* is usually defined in terms of the percentage a particular energy resource (in this case, wind power) is of total peak electricity demand on a system. A utility with 10,000 MW of peak demand with 1,400 MW of wind generating capacity installed on its system has a penetration of 14%. This measure is more technically termed *capacity penetration*. Other ways to measure penetration include *energy penetration*, which is the ratio of energy delivered from wind generation compared to the total energy delivered. The term *instantaneous penetration* also may be used, which refers to wind plant output at a point in time, or over a short period, as compared to total system load at that point or over the same period.

Large regional grids connecting many generation resources and loads have a greater resiliency and ability to accept renewable sources of generation. The depth of reserves in a large interconnected area enhances reliability and cushions the intermittency of wind power. In 2000, western Denmark received more than 80% of its off-peak energy from wind. Denmark can support such a large wind penetration in part because it is interconnected with Sweden, Germany, and Norway.

Norway's large hydroelectric capacity can be shut in when wind is plentiful; in effect, the hydro plants operate as a battery to store excess wind power.

As Denmark's neighbors build out their wind capacity, the Danes have had to become more creative in balancing the grid. In addition to large amounts of wind power, Denmark has a significant base of combined heat and power (CHP) plants that are dispatched according to the demand for heat and are not generally controllable for electricity production. Particularly during off-peak, windy weekends or holidays, the aggregate CHP and wind plant output can result in excess supply. Eltra, the transmission system operator for western Denmark, exports the excess on its high-voltage interties with Sweden, Germany, and Norway. To better manage the wind capacity, and to avoid shutting down wind turbines during times of low demand or restricted export capacity, Eltra is exploring other options. Two options include improving its dispatch control over local CHP plants and introducing flexible loads such as heat pumps, which would use electricity to generate heat during periods of electricity surplus.[25]

Denmark's experience illustrates that although integrating wind power into the grid is not without challenges, and requires a willingness to adapt, significant penetration of renewable energy is possible. Loads on the populated U.S. East Coast where offshore wind is likely to first be developed are much larger than in western Denmark. U.S. offshore wind resources could make a large contribution to serving this load before the challenges of grid integration rise to the levels experienced in Denmark.

Terrestrial and offshore wind energy differences

How do terrestrial and offshore wind energy compare in terms of their effects on the transmission system? A study commissioned by the New York Independent System Operator (NYISO) and the New York State Energy Research and Development Authority (NYSERDA) answers this question. The study assumed that 3,300 MW of wind generating capacity was distributed over 37 sites throughout New York, including 600 MW off of Long Island's south shore. With a peak load of 33,000 MW, the New York grid has a robust capacity to absorb wind power. Like the DOE study, the New York wind integration study found that the impact on grid reliability of increasing wind penetration to 10% of New York's peak load was both minor and manageable. The New York wind integration study commented that for the most part, small increases in measures such as forecasting error and variability can be accommodated by existing processes and resources.[26] As in the country as a whole, the distribution of wind resources in New York is uneven. Rural, windy regions of the state might have a local penetration exceeding 20% or 30%, while in other regions of the state, there was little or no installed wind capacity. In addition, when viewed over time, wind power's effective penetration at some times was much greater than 10%. At times of low load and high winds, the instantaneous penetration of wind power could be up to 35% in some regions of the state, including Long Island, where the offshore wind capacity was modeled.[27] Even at high penetration wind power did not cause significant adverse effects, which bodes well for the development of offshore renewables.

Integrating wind power works due to a combination of grid management changes by grid operators and technical solutions such as better weather forecasting. A grid operator traditionally must balance uncertain and varying load levels with unexpected changes in the output of conventional generating resources. When wind power is added to the mix, the complexity of the grid operator's tasks increases. At times the variability of wind plant output and load, which are generally not correlated, will tend to cancel each other and smooth fluctuations in system load, as adjusted for wind plant contributions to system supply. Balancing can be particularly challenging, however, when load rises while wind generation sinks, or vice versa. Grid operations and planning must function well on various levels to meet such challenges. A grid operator must plan years ahead to ensure that adequate generating capacity is built to satisfy projected loads (*planning*). In the short-term, an operator prepares forecasts to establish the day-ahead schedule in which generating units are committed to meet the next day's anticipated load (*scheduling*). In the very short-term, an operator manages generation on an hourly basis and at intervals of minutes to follow the load (*load following*). Lastly, on intervals of only seconds, a grid operator engages in *regulation*, typically using automated generator controls, to respond to changes in load and to maintain the frequency and voltage of the system within acceptable parameters. Users of the electric grid benefit from each of these functions or services, the cost of which is included in the rates charged to wind energy generators and other grid users.

Wind energy output assumes seasonal patterns and is variable in both time and space. Output can vary in the following ways:

1. From one region to another within the grid operator's control area
2. From one farm to another according to local conditions
3. Among the turbines within a particular farm

The distributed nature of wind power, however, helps to smooth out this variability. The New York wind integration study found that the wind will not stop blowing at one time all across the state of New York. According to another report, "The longest continuous time of calm weather in a 21-year hourly wind data set from 13 different locations in England and Wales was 11 hours, incidentally over night at a time of low demand."[28] So while grid operators should run scenarios to determine how to respond if a wind farm should go off-line, or a substation serving two or more wind farms experiences a fault, there is no need to plan for the contingency of all wind plants on the system being out at the same time. When individual turbines are grouped in a farm, their individual variability is absorbed by the overall average output of the farm. Similarly, when wind farms are widely distributed throughout a grid control area, the variability of individual farms has a less marked impact on the grid as a whole. In dispersed systems, the regional fluctuations in wind output occur over hours rather than minutes and are more easily managed by grid operators.

Offshore wind farms are likely to have a less variable output than their land-based counterparts. Typically, large offshore projects spread the variability of individual turbines over more machines. Offshore winds also tend to be steadier than the wind at onshore sites. When winds are gusty, the increased inertia of offshore wind machines, which tend to be larger than terrestrial wind machines, helps to smooth power production.[29]

Wind turbine capacity ratings

As a consequence of its intermittent nature, many think of wind energy as a source of energy, not capacity. In other words, wind may be good for power, but it has little to no value for purposes of supporting grid reliability. This distinction is important, because a project's capacity value affects revenues. The weather-related variability of each wind project influences the capacity value granted to the project by the grid operator.

Traditional fossil-fueled generating facilities can produce energy on demand at their full capacity for an extended period of time. Like all machines, traditional generators require scheduled maintenance and occasionally break down unexpectedly. Thus grid operators must discount the installed capacity of the unit for the probability of a forced outage. A 100-MW unit with a 10% expected forced outage rate has a 90% probability of being available to serve load when needed. Grid operators would give this unit a 90-MW *unforced capacity (UCAP)* rating.

Wind generator capacity factors and generating plant *capacity ratings* are easily confused. Terrestrial wind turbine generators are highly reliable machines that are available to generate electricity approximately 97% to 99% of the time. However, their actual energy output is much less; capacity factors in the 30% range are typical. It would be wrong to conclude that a 100-MW wind farm (with a 30% capacity factor) has a 30% probability of being available to provide 100 MW of capacity when called upon. A plant's capacity factor measures the total actual energy output over a year compared to the plant's potential maximum energy output. In other words, this particular wind farm's output is about 30% of its total installed capacity on average. Since a wind farm's output can vary significantly due to the weather, and that variability takes on seasonal and daily patterns, it would not be appropriate to assign a UCAP rating to a wind farm that is 30% of its installed capacity. The capacity value of a wind farm, and hence its capacity rating, relate instead to how well the farm's wind energy production pattern aligns with load patterns because, as described below, capacity ratings represent each plant's contribution to electrical system reliability.[30]

The system modeled in the New York wind integration study exhibited wind turbine capacity factors ranging from 50% during the winter months to as low as 20% during the summer peak load time for the system. During daytime peak hours (10 a.m. to 6 p.m.), the wind turbine capacity factor was usually less than 25%, but during the evening and night hours, 40% or more was typical.[31] Because wind power production generally is not coincident with peak loads, it does not have a high capacity value. In fact, during the summer, wind power production is likely to be out of phase with system load, which means that wind power contributes little to system reliability (fig. 10–2). The phasing problem also has an economic impact, because the power produced by wind plants generally competes with the lowest value (and most difficult to cycle) baseload generating plants, not expensive gas- and oil-fired peaking generators.

The grid operator evaluates a wind power plant's effective capacity by comparing wind generators to traditional generators in terms of their contribution to grid reliability (i.e., avoiding a blackout). The likelihood of a blackout is known as the *loss of load probability (LOLP)*, and this is highest at midday during the summer.[32] (As a technical matter, LOLP is defined as the probability that the load will exceed the available generation.) Generally, grid operators aim to keep the LOLP down to about 0.1 days per year (a 1-day blackout every 10 years). Each generator added or system modification has an effect on the LOLP. The grid operator may determine, for example, that adding a traditional generator with a UCAP of, say 100 MW, would lower the LOLP by the same amount as adding 500 MW of wind generation. In that case, the wind generation should be assigned an effective capacity rating (i.e., UCAP) of 100 MW. Based on comparisons such as this, the New York wind integration study found that, due to its poor summer peak period performance, terrestrial wind farms should be assigned a UCAP rating equal to 10% of installed capacity.

Offshore wind, however, breaks the general rule (fig. 10–3). The New York wind integration study found that "the offshore site has a much different daily pattern that peaks several hours earlier in the day and is much more in line with the load patterns."[33] Offshore wind's 36% to 42% capacity factor, both on an annual basis and during the peak summer period, also is higher than typical inland wind farms. For this reason, the study concluded that offshore wind should be given an effective capacity rating equal to about 40% of installed capacity.[34] Since part of project revenues come from capacity payments, the difference between terrestrial and offshore wind UCAP ratings represents a significant economic benefit that helps to justify the higher cost of building projects offshore. The value of the energy produced by offshore wind facilities also should be higher than onshore wind power, because more offshore wind power is produced during expensive peak periods (fig. 10–4).

Other transmission operators also have recognized that offshore wind performs differently. PJM, for example, grants offshore projects an initial UCAP rating of 20% of installed capacity, which may be adjusted upwards over three years if actual experience proves that the facility operates at a higher effective capacity.

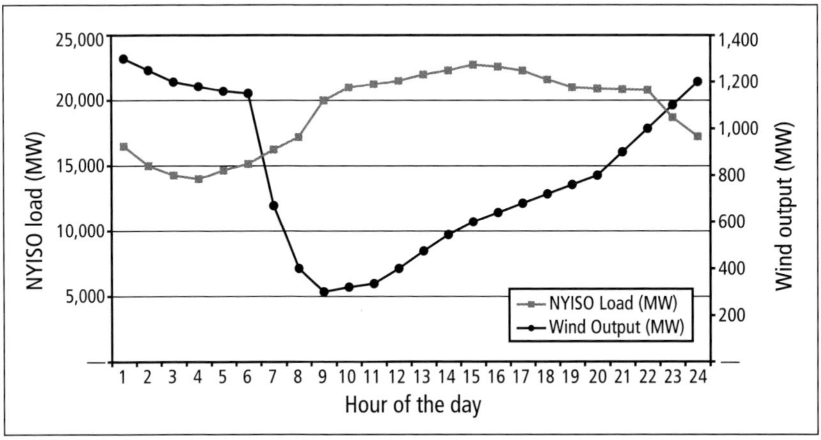

Fig. 10–2. Approximate summer (July) load and projected wind turbine output
Source: GE Energy Consulting. 2005. The Effects of Integrating Wind Power on Transmission System Planning, Reliability and Operations

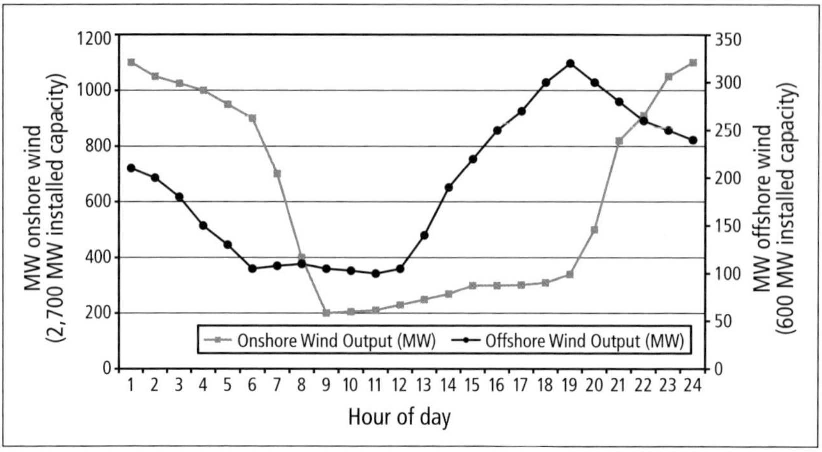

Fig. 10–3. July onshore and offshore wind power production profiles compared
Source: GE Energy Consulting. 2005. The Effects of Integrating Wind Power on Transmission System Planning, Reliability and Operations

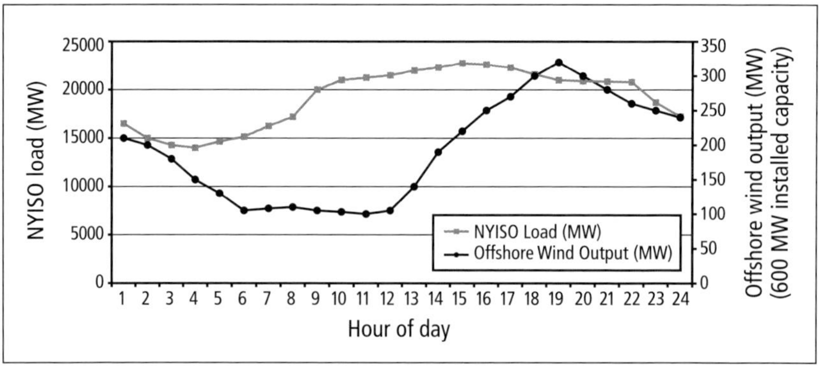

Fig. 10–4. July load and offshore wind production profiles compared
Source: GE Energy Consulting. 2005. The Effects of Integrating Wind Power on Transmission System Planning, Reliability and Operations

Forecasting the wind and scheduling wind energy production

Improved wind forecasting methods are another tool for grid operators trying to juggle shifting loads and resources. As one might expect, near-term forecasts are more accurate than long-term forecasts. The New York wind integration study found that day-ahead wind generation forecasts had a standard deviation of approximately 400 MW (12%) of the aggregate rating of all the wind generators (3,300 MW) in the study scenario. Standard deviation is a measure of the dispersion of values about the mean value of a distribution. In other words, the actual wind generation was within 400 MW, higher or lower, than the day-ahead forecast most of the time. If the forecast horizon is decreased to only one hour ahead, the standard deviation drops to about 145 MW, or only 4.2% of the aggregate wind generating capacity.

Despite the lower accuracy of day-ahead forecasting, the study concluded that such forecasts are valuable because they allow the system operator to better manage the commitment of other generation and lower the costs of energy systemwide. For example, shutting down a baseload plant with a long shutdown and restart time is expensive and can reduce system reliability. A system operator may respond to a wind forecast that indicates strong wind power output during a period of expected low demand by asking the wind generators to curtail output during the low demand period to avoid requiring critical baseload generators to shut down. Alternatively, following the example of Denmark described above, the system operator may choose to schedule wind generation before hydroelectric plant output and impound water for future hydro generation when wind power is unavailable. The study found that accurate forecasting and creative scheduling of wind energy could result in an average of $1.80/MWh reduction in the spot price of electricity in New York. All users of the grid, not just wind plant owners, benefit from such an integrated and coordinated approach.

Coordinating weather data collected by wind farms, weather stations, and other sources permits forecasters to have better regionwide information about fast-moving weather systems. A more accurate regional perspective should improve the prediction of local conditions. Denmark's Risø National Laboratory has produced a weather model that takes national weather forecasts up to 36 hours ahead and then applies information about local conditions at the site of a wind farm to estimate wind speeds and wind farm energy output. The model is called *WAsP (Wind Atlas Analysis and Application Program)*, and it adjusts the free flow of wind high in the atmosphere for conditions closer to the surface, where wind is affected by mountains, forests, and open water. A separate model developed by Risø called PARK recognizes that the energy production of a wind farm (i.e., a wind park) is not simply the aggregate of the generating capacity of the individual turbines under a given set of predicted wind conditions. Wind turbulence within a wind farm and sheltering of some turbines by others influences the aggregate output of the farm.[35]

WAsP exemplifies the generic category of *numerical weather prediction (NWP) models* that have increased the ability to schedule wind power. Using expected average, minimum, and maximum wind speeds to estimate energy output, transmission system operators can reduce the need to schedule expensive quick-start or spinning reserves for purposes of maintaining grid stability. An NWP model might predict, for example, that a 100-MW wind farm would produce 85 MW on average tomorrow, but actual production might vary from 70 to 100 MW given a typical range of error.[36] Accurate predictions are especially critical when dramatic weather changes cause wind farms to ramp up or down quickly, requiring a swift response on the part of the grid operator to balance the other generating resources on the system. Consider, for example, how the grid operator should schedule resources when a thunderstorm is predicted to move through a region bringing strong winds and driving up wind farm power production. The NWP model may forecast winds of 50 mph, with an error of ±5 mph, under circumstances where turbines at a wind farm in the affected area have a cutout speed of 52 mph. A conservative grid operator would probably schedule additional spinning reserve just in case winds reach the cutout speed and the wind turbines ramp down. The system operator might have avoided the expense of the spinning reserve if there was a greater certainty that the winds would stay below the cutout speed.

The sea breeze effect

Differences in weather patterns over water and land cause variations in the energy production of offshore and terrestrial wind farms. Meteorological measurements in Nantucket Sound show a positive correlation between the New England region's peak energy use periods and the proposed offshore farm's expected energy production. On hot days when energy use in New England is greatest, differential heating of the land and water produce strong winds. On average during New England's 10 highest demand days, Cape Wind estimated that its project would have produced at 76% of its capacity, adding 321 MW to the grid.[37]

The phenomenon of alternating offshore and onshore winds is called the *sea breeze effect*. As water and land heat and cool at different rates, winds regularly alternate between blowing offshore in the mornings when the land is cold and the ocean comparatively warm, and onshore in the evenings when air warmed by the land rises and cool offshore air rushes in to take its place. This recurring dynamic contributes to the high capacity factor of offshore wind turbines as compared to terrestrial wind farms. The timing of alternating offshore and onshore winds also corresponds nicely with daily energy peaks.

The sea breeze effect also operates during cold winter days. The DOE noted that if Cape Wind's project had been in operation during a severe three-day cold spell in January 2004, it would have contributed nearly its full capacity during most of that period and provided a much-needed alternative to natural gas generation in New England. In fact, gas supply constraints almost led to a blackout that was avoided only by the narrowest margin. Limited gas pipeline and storage capacity conspired with high gas use for heating to restrict gas available for electric generation. Gas demand increased so dramatically during the cold snap that the wholesale price of natural gas reached \$63.23/MMBtu, and electricity prices were driven as high as \$375/MWh.[38] Rather than contributing to lessened grid reliability, offshore wind would make the electrical system more robust through increased supply diversity.

It is reasonable to expect improved meteorological data collection and weather models to continue the trend of improved forecast accuracy. Detailed information on atmospheric conditions in the vicinity of a wind farm leads to better predictions. Meteorological towers placed upstream of a wind farm and ground-based Doppler radar can supplement the data available for terrestrial wind farm predictions. Offshore, however, advances in remote satellite sensing are likely to hold the most promise. Satellite-based sensors can scan the atmosphere above the oceans and distinguish variations in electromagnetic radiation that signify moisture and temperature differences. Visual observations of cloud movement also provide information about wind speeds. Increasingly sophisticated forecasting is improving scheduling and helping make better use of the wind.

Improving Grid Reliability with Wind Turbine Technology

Wind turbine technology has improved substantially over the last two decades, and many of the new features improve the machine/grid interface. Modern turbines have *ride-through* capability that permits continued operation when grid voltage drops unexpectedly. They also allow grid operators improved control over real and reactive power output, which helps to maintain stable grid conditions. Operators now can manage the rate at which a turbine's output ramps up to allow other generators to ramp down in an orderly fashion. Remote communications and control capabilities allow operators to oversee turbine performance and manage these functions from a central station. Each of the new features as applied to wind turbines is described further below. Wave, tidal, and ocean current energy devices also will probably use similar technologies to limit adverse impacts on the transmission grid.

As wind power has expanded, transmission operators have developed unique interconnection requirements for wind energy generators seeking to connect to the grid. The variety of requirements has imposed costs for special turbine and wind farm designs, technology, and equipment. At the urging of the American Wind Energy Association, which argued that it would be more efficient to standardize the requirements for wind interconnection, FERC adopted standard procedures and technical requirements for the interconnection of large wind projects that transmission owners and operators must include in their transmission tariffs.[39] Transmission providers seeking to manage wind power's impact on the grid may require that a wind project comply with certain FERC-authorized interconnection specifications included in a schedule to the transmission operators' terms of service (i.e., their OATTs).[40] The large wind generator interconnection procedures supplement existing standard procedures for the interconnection of traditional large generators and tailor them to intermittent, nonsynchronous generators. Special rules are justified because traditional fossil-fueled generators operate at a speed that matches the grid frequency, while wind turbine generators are nonsynchronous, operating at various speeds determined by wind speed and the gearing between the rotor and the generator.

When the wind farms in a region are small and diffuse, their impact on the transmission grid is generally easily managed. Offshore projects, however, will usually be large. If such large projects connect to a weak onshore grid (e.g., in a lightly populated coastal area), these projects may have a significant impact on the transmission system. To avoid problems, it should be expected that transmission operators will require most offshore wind projects to adopt the full assortment of reliability-enhancing fixes described in the following discussion.

Low-voltage ride-through capability allows a wind turbine to continue operating when the grid experiences a disturbance that results in a voltage drop. A generator that trips off-line during a grid disturbance may exacerbate the instability of the grid and contribute to a cascading failure by further depressing system voltage. FERC's standards require all new wind plants to have low-voltage ride-through capability, which permits them to stay online for as long as nine cycles after a fault, with the actual time determined by the normal fault clearing time of the substation where the wind facility is interconnected.

The induction generators typically used in wind turbines require reactive power from the grid. As a generator's output increases, during wind gusts for example, so does its reactive power consumption. The varying consumption of reactive power can cause voltage fluctuations on the grid. Long underground cables, such as those connecting offshore wind projects to land, also tend to generate reactive power, further complicating efforts to balance reactive power levels. If a system impact study indicates that a proposed wind farm could cause voltage instability threatening safety and reliability, the transmission provider may require that the wind project maintain a power factor within a certain range (0.95 leading to 0.95 lagging). (The *power factor* is a number between 0 and 1 that represents the ratio between real power and apparent power. A circuit with only reactive power would have a power factor of 0, while in a circuit with a power factor of 1, all the energy supplied is consumed by the load.) Equipment such as capacitor banks or static or dynamic VAR compensators may be installed by the project developer to meet the required power factor standards.

Wind generators "are often remote, unmanned, and characterized by an unpredictable rate of change of output, thus making it difficult for the Transmission Provider to limit the output of the wind plant when necessary for system reliability."[41] To address this problem, FERC's standards require wind farms to have remote system operating and control capabilities known as SCADA, an acronym for *supervisory control and data acquisition*. SCADA communications equipment permits wind turbines to communicate with the grid operator and allows the operator to provide instructions to the turbines to alter their operation as necessary to maintain the safety and reliability of the transmission system. A wind turbine equipped with SCADA and blade pitch control could be instructed by the grid operator to feather the blades to temporarily reduce output during certain grid conditions. When might this ability be used? A weather front could move into the area of the wind farm causing the farm to rapidly reach maximum output. The wind farm's output could increase much faster than conventional generators can ramp down their output in response to the grid operator's instructions. SCADA capability allows the grid operator to adjust the ramp up rate of the wind farm to match the ramp down rate of the conventional generators.

Information transmitted by a SCADA system may include real-time real power and reactive power output, as well as information about the status of the interconnection facility. The SCADA system also can communicate data about wind conditions at the wind farm. This data could be fed into weather models to assist the grid operator's near-term forecasting of the output of the wind farm and other wind energy plants in the area.

Wind farm owners are hesitant to relinquish control over their generation facilities to transmission providers, particularly when the transmission provider could use the SCADA system to curtail a wind farm's output and the owner's revenues. However, most wind energy developers understand that failing to work with grid operators to manage reliability issues also does not serve their interests. A moratorium on new wind plant interconnections was imposed in Ireland in 2003 when the lack of a workable grid interconnection code threatened stable grid operations. FERC's requirement for wind farms to have SCADA capabilities does not authorize the transmission provider to control the wind plant. Any control provisions must be separately negotiated as part of the interconnection agreement. The type and extent of SCADA information communicated to the transmission provider also is subject to agreement between the parties. Factors such as the size of the wind farm and its impact on maintaining resource adequacy and transmission system reliability in the area where it is located will determine the appropriate level of information sharing.

Transmission and Ancillary Services

A buyer or seller of energy must purchase transmission and ancillary services to move energy from the point of interconnection to the load, or to move energy to a node on the transmission system where competitive market sales can be made. One reason for wind energy's lower market value, as compared to traditional dispatchable energy supplies, is that its intermittent nature requires more ancillary service support. Wind farm operators must manage the cost of transmission service and ancillary services, while avoiding penalties for unscheduled power deliveries, in an environment of constantly varying wind farm output.

Transmission service can be provided point-to-point on a firm or nonfirm (interruptible) basis. Point-to-point service specifies a transmission path from a specific generator to a specific load. Network service is another type of service used by a utility to economically coordinate multiple generating sources owned, purchased, or leased by the utility to serve the utility's load on the network. Network service is firm and has a higher priority than nonfirm point-to-point service. A network service customer also may obtain energy from generators that have not been designated as part of the customer's network resources by using a nonfirm service called *secondary service.*

A large offshore renewable energy project would generally need firm point-to-point transmission service to assure the ability to deliver power under the terms of a long-term power purchase agreement. However, if the offshore facility is committed to serve a utility through a power purchase agreement or lease, for example, it could qualify as part of the utility's network resources and be delivered to the utility's load under network service.[42] Firm point-to-point service, which reserves a discrete portion of capacity along a transmission path and gives a generator priority rights to ship power, is more expensive than nonfirm service.

Operators of an intermittent resource would typically prefer firm point-to-point service or network service so that energy can be sold when the wind is blowing. But the nature of an intermittent resource also means that periods of low power production will result in excess transmission system capacity that can be made available for other system users. FERC permits point-to-point transmission customers to reassign transmission capacity at market prices.[43] Network service customers have no specific capacity rights and therefore no ability to reassign service.[44] The ability to reassign capacity depends on accurate wind forecasting. Nevertheless, a wind project's financial model may include revenues earned by selling the project's unused transmission capacity.

Firm point-to-point transmission service is more appropriately viewed as a variety of possible services residing along a spectrum extending from the highest transmission priority to service that could be curtailed in the event of certain system constraints. This distinction may be important to owners of offshore renewable energy projects who are seeking to more closely tailor their transmission service to energy production forecasts. When fully firm service is not available, service that is almost firm could still be adequate and also could make it easier for a renewable energy project to obtain transmission service without paying for network upgrades. This category of less-than-firm service, called *conditional firm*, is firm transmission service modified to permit the transmission provider to impose certain hours of possible curtailment during specified times of the year, or curtailment only during defined system conditions. Conditional firm service can be curtailed for reliability reasons but not economic reasons.[45]

Prior to the development of the conditional firm service category, when a generator requested firm service but there was a constraint for a few hours during the year, the transmission provider would reject the request. A system impact study was then required to evaluate system upgrade options and costs and to determine whether redispatching other generation during certain hours could relieve the transmission constraint at a lower cost. This latter option, known as *planning redispatch service*, may sometimes be arranged if the transmission provider determines that dispatching generation out of the order of normal economic dispatch would relieve the constraint while maintaining system reliability. The marginally increased cost of the redispatched generation would typically be borne by the transmission customer. The cost of planning redispatch service is: (1) the higher of (a) actual incremental costs of redispatch calculated monthly or (b) the applicable embedded cost transmission rate on file with FERC or (2) a fixed rate for redispatch to be negotiated by the transmission provider and customer and subject to a cap representing the total fixed and variable costs of the resources expected to provide the service.[46]

Conditional firm service may allow a wind project developer to avoid upgrade and redispatch costs. Conditional firm service also is useful when a project's timeline does not agree with the timeline for transmission upgrades. The generator may take conditional firm service on an interim basis until the system upgrades necessary to allow for firm service have been constructed. A system impact study still would be required prior to initiating conditional firm service, however, to evaluate the system conditions and hours during the year when constraints would require curtailment.

Conditional firm transmission service is not required to be available in systems managed by regional transmission organizations (RTOs) or independent system operators (ISOs) that have real-time energy markets. In these markets, transmission customers can use *financial transmission rights (FTRs)* to *buy through* transmission congestion and do not need to make a reservation to obtain transmission capacity. FTRs are a hedge against transmission congestion costs that pay the holder the difference between the price of energy at the *source point* (where the generator interconnects) and the *sink point* (where the power is delivered).

In addition to electing a type of transmission service, sales of renewable energy will require ancillary services, the cost of which may be borne by either the load (the purchaser of the energy) or the generator. Some ancillary services may be provided only by the transmission provider on the terms stated in its tariff, while others may be provided by the generator itself or other competitive suppliers. The transmission provider performs the following services:

1. Scheduling, system control, and dispatch service
2. Reactive supply and voltage control service

Other services, which may be self-provided or provided by the transmission provider or a third party, include the following:

1. Regulation and frequency response
2. Energy imbalance
3. Operating reserve—spinning
4. Operating reserve—supplemental
5. Generator imbalance services

The preceding section described scheduling, voltage control, reactive power, and frequency maintenance issues. When scheduled production does not match actual production, the system operator must intervene to maintain the balance between generation and load. The operator does this by calling on generators that can ramp up and down rapidly to supply energy, regulate grid voltage, and maintain cycle frequency. Regulation and imbalance services are expensive, since some generation is kept spinning to allow it to respond on a moment's notice.

Generator imbalance service is provided by a transmission provider or other service provider to make up for the overproduction or underproduction of a generating facility in comparison to its scheduled delivery. Providers of spinning and supplemental operating reserves maintain capacity that is ready to respond within seconds (spinning) and within longer lead times (supplemental) to the need for additional generating capacity. *Generator imbalance service* differs from *energy imbalance service*, which is concerned with the behavior of the load. The energy imbalance service provider makes up the difference between the energy that was scheduled for delivery during any particular hour and the actual energy delivered to the load.

Underproduction or overproduction compared to the scheduled output can result in penalties imposed by transmission providers that may significantly affect renewable energy project economics. The Oklahoma Municipal Power Authority, for example, paid generator imbalance penalties of $3.4 million for its 50% share of a 102-MW wind farm. Over the course of a year, the wind farm underproduced more than 40,000 MWh and overproduced 14,000 MWh as compared to its scheduled output. Imbalance penalties are intended to encourage good utility operating practice, discourage gaming, and minimize the need for the transmission system operator to provide regulation service. Although penalty provisions vary among transmission providers, they typically apply when a generator's day-ahead output forecast varies from its real-time production by a defined margin.

Applying an imbalance penalty to wind or other intermittent renewable generating facilities makes little sense if the producer has only a limited ability to forecast and control output. As previously discussed, improved wind forecasting has made production schedules more accurate, but significant margins of error remain. Underproduction usually cannot be controlled, and overproduction can be controlled by intervening through a SCADA system to ramp down wind farm output. But wind energy's marginal cost is zero. Once the wind turbine is installed, it costs no more to produce 2 kWh than it does to produce 1 kWh. It seems against the public interest in low-cost power to ramp wind energy production down unless there is a reliability constraint. Imposing a penalty for generating more low-cost power rather than less merely adds insult to injury. To its credit, FERC recognized that the penalties discourage renewable energy development, and the agency changed the form of imbalance charges as they apply to wind power. The imbalance charges that transmission providers may now include in their OATTs increase as the magnitude of deviation from scheduled production increases. They are designed to allow the transmission providers to recover the cost of managing the system imbalance, rather than to be punitive. Wind and other intermittent renewable generation are exempted from the highest tier of imbalance charges:

> *Specifically, imbalances of less than or equal to 1.5 percent of the scheduled energy (or two megawatts, whichever is larger) will be netted on a monthly basis and settled financially at 100 percent of incremental or decremental cost at the end of each month. Imbalances between 1.5 and 7.5 percent of the scheduled amounts (or two to ten megawatts, whichever is larger) will be settled financially at 90 percent of the transmission provider's system decremental cost for overscheduling imbalances that require the transmission provider to decrease generation or 110 percent of the incremental cost for underscheduling imbalances that require increased generation in the control area. Imbalances greater than 7.5 percent of the scheduled amounts (or 10 megawatts, whichever is larger) will be settled at 75 percent of the system decremental cost for overscheduling imbalances or 125 percent of the incremental cost for underscheduling imbalances… Intermittent resources are exempt from the third-tier deviation band and would pay the second-tier deviation band charges for all deviations greater than the larger of 1.5 percent or two megawatts.*[47]

An *intermittent resource* is "an electric generator that is not dispatchable and cannot store its fuel source and therefore cannot respond to changes in system demand or respond to transmission security constraints."[48]

Even before FERC's changes, most RTOs and ISOs had already done away with severe penalties for generator imbalances. These entities are not required to implement FERC's new standard for imbalance charges if they demonstrate that they have alternate tariff arrangements that are consistent with or superior to FERC's requirement, and the following conditions apply:

1. They are related to the cost of correcting the imbalance.
2. They are tailored to encourage accurate scheduling behavior.
3. They account for the special circumstances presented by intermittent generators, such as by waiving the higher ends of the deviation penalties.[49]

The California ISO, for example, nets positive and negative wind plant imbalances over the course of a month, which minimizes the net imbalance. PJM has done away with generator imbalance charges entirely. Any imbalances are resolved in the real-time energy market based on the difference between the day-ahead market and the final real-time position, which reflects the demand for operating reserves to correct the imbalance.[50] New York's grid operator, NYISO, has waived imbalance penalties for up to 500 MW of new wind power to encourage the location of wind power in New York. ERCOT, the Texas grid operator, permits wind projects to deviate up to 50% from their schedules before imposing imbalance penalties. Further revisions to the generator imbalance tariff provisions may occur in response to FERC's recent guidance.

Another charge that affects the level of transmission service selected by intermittent generators is the unreserved use penalty. This applies when a generator uses more transmission capacity than it has reserved. To explore how this penalty works, assume that wind models and production forecasts indicate that a hypothetical 600-MW offshore wind farm would produce its maximum output only 20% of the time. The project's owner might consider whether it would be more economical to reserve only 500 MW of firm point-to-point transmission capacity and count on the transmission provider to find flexibility in the network to accommodate the occasional full 600-MW peak output as and when needed. The unreserved use penalty is designed to discourage this behavior, which frustrates the orderly allocation of transmission capacity. FERC expressly declined to waive unreserved use penalties for intermittent resources because they can control their output and reserve a corresponding amount of transmission capacity. "Unreserved use penalties are based on the transmission capacity reserved rather than the transmission service scheduled, so an intermittent resource's inability to precisely schedule power in future periods is irrelevant, as long as the resource has reserved sufficient transmission capacity to deliver the resource's full output."[51]

Transmission providers have discretion in setting the unreserved use penalty provided that it is based on the period of unreserved use. For example, one hour of unreserved use could result in a penalty based on the daily rate for firm point-to-point service, and several unreserved uses during a week would result in a penalty based on the weekly rate for firm point-to-point service. FERC interpreted *based on* in the sentence above to mean generally that the penalty should not exceed twice the transmission service rate for the applicable penalty period.[52] "In addition, transmission customers will continue to be subject to civil penalties on a case-by-case basis, so attempts to game this penalty regime could result in additional penalties depending on the specific facts at issue."[53] Details regarding the level of penalties and their method of calculation can be found in the transmission provider's tariff.

Transmission System Congestion and Locational Pricing

Thomas Edison built the first commercial electric station on Pearl Street in downtown Manhattan to serve an area of 1 mi^2. Edison's power plant had to be very close to the load because power could not be transmitted over any significant distance. As transmission technology advanced, the scale of power plants could increase, and the large central generating station model of power supply developed. Much of the nation's coal-fired generating capacity, for example, is now located in the Midwest and rural South (fig. 10–5). Power from these remote plants is transported to population centers on high-voltage transmission lines, which have become increasingly congested as electricity demand grows. These plants are often located along rivers for access to cooling water and to minimize the expense of transporting coal (barge transport being significantly less expensive than rail). To keep pollution away from the cities, the plants are sited in rural areas.

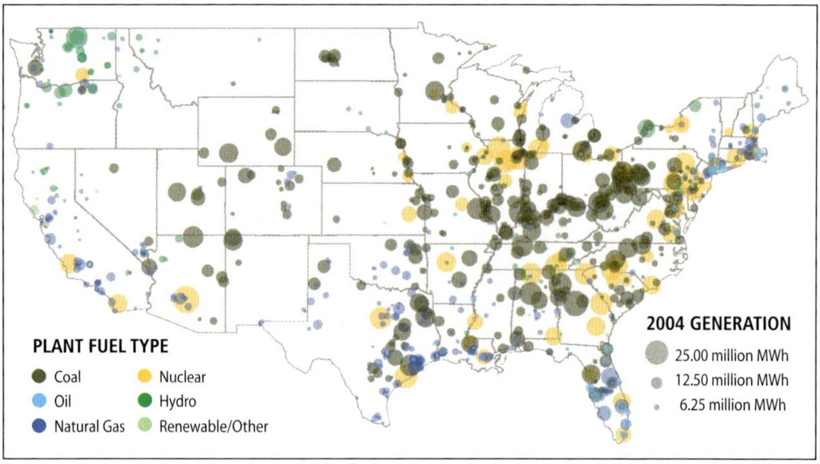

Fig. 10–5. Location, type, and size of U.S. power production
Courtesy of M.J. Bradley & Associates LLC

Advocates of building more transmission lines argue that increasing energy usage and underinvestment in high-voltage transmission facilities has left parts of the country, particularly along the coasts, with transmission bottlenecks. During peak periods, high-voltage transmission capacity is insufficient to bring inexpensive power generated by rural coal-fired plants into the congested areas. As a result, more expensive local gas- and oil-fired generating units must run. Some advocates of new transmission corridors say that they must be built to move terrestrial wind power generated in the nation's heartland to its population centers, again mainly along the coasts. Supporters of wind energy are not alone in calling for new transmission investment. Some utilities with conventional generating resources also would benefit greatly from an expanded transmission highway that could be used to move their low-cost power from coal, nuclear, and hydroelectric plants in the U.S. interior to higher priced urban and coastal markets.

Building new transmission capacity is an enormously expensive and time-consuming undertaking. Allegheny Energy would like to build a 330-mile, 500-kV transmission line that would increase west-to-east transfers of power by 3,800 MW. The project would cost approximately $1.4 billion and may be in service by 2013.[54] American Electric Power (AEP) wants to invest $3 billion to build a 550-mile, 765-kV transmission line starting in West Virginia that would cross Maryland, Pennsylvania, and New Jersey. AEP claims that its proposed line will save $1 billion annually.[55] The company would accomplish this primarily by substituting Midwest coal power for local East Coast gas-fired generation. AEP is one of the largest U.S. utilities, with 80 generating stations (73% of which are coal fired) and an equally impressive fleet of 7,500 rail coal cars, 1,800 coal barges, and 37 tug boats. The company is essentially in the business of bringing coal from the western to the eastern United States. Coal mined in Wyoming's Powder River Basin travels by rail to the Mississippi River and then floats east by barge to AEP's Midwest generating plants. In the final stage, the coal—now transformed to electricity—makes its way to eastern cities by wire.[56]

Given the availability of a clean offshore energy resource located close to major U.S. population centers, it is reasonable to question the case for certain proposed investments in new high-voltage transmission lines. Some of the proposed new west-to-east transmission projects might be avoided if substantial marine renewable energy projects are built near East Coast loads.

Transmission providers are increasingly seeking to make the scarcity of transmission capacity transparent to grid users by implementing transmission congestion charges and generation capacity auctions. Consider the following hypothetical transaction. A load serving utility in Philadelphia has entered a power purchase agreement with a generator in western Pennsylvania, near Pittsburgh, that calls for the sale of electricity at $35/MWh. Assume, as is generally the case, that the transmission grid is more congested in the densely populated eastern part of the state than in the west. PJM, the transmission operator for the region that includes Pennsylvania, imposes a transmission congestion charge related to the hypothetical energy flow based on the difference in the locational marginal price of energy at the source near Pittsburgh and the sink in Philadelphia. The *locational marginal price* is the market clearing price of energy at the time of the transaction at issue, determined through the bids of energy sellers and buyers. At any point in time there are different locational marginal prices throughout PJM—in Pittsburgh, Philadelphia, Atlantic City, Baltimore, and other locations—where generators actually able to deliver power to those areas are matched with buyers needing to serve load. Note that the locational marginal price is not affected by the price terms of the power purchase agreement between the supplier in Pittsburgh and the customer in Philadelphia.

If, at the time of the hypothetical transaction, the locational marginal prices are $80/MWh in Pittsburgh, $85/MWh in Philadelphia, and $92/MWh in Atlantic City, PJM will add a $5/MWh congestion charge to the transmission customer's bill for transmitting the energy from Pittsburgh to Philadelphia. Note that if the Philadelphia customer had instead contracted for offshore wind power supplied to the grid at Atlantic City, PJM would make a payment to the transmission customer of $7/MWh for relieving congestion.

This hypothetical example shows how offshore renewables may provide value by relieving congestion near dense urban areas. The example also illustrates how current proposals to relieve congestion in the East Coast region by building high-voltage links to western coal-fired generation could adversely affect the economics of offshore renewable energy. Additional transmission capacity would lessen the energy price differential between western and eastern markets, reducing the financial incentive to site new generation in the eastern areas, including offshore renewable energy.

Capacity auctions are another approach to managing congestion. They help increase the value of generating capacity located in constrained areas. Utilities in PJM are required to procure sufficient capacity to serve their load, plus a reserve margin. A utility meets its capacity needs through resources owned by the utility, resources procured in bilateral contracts, and lastly, resources procured in periodic capacity auctions arranged by PJM. The auctions result in market clearing prices that vary by region (fig. 10–6). High capacity prices prevailing in congested areas encourage owners that had intended to retire plants to keep them online and also help support the economic case for the construction of new generating facilities.

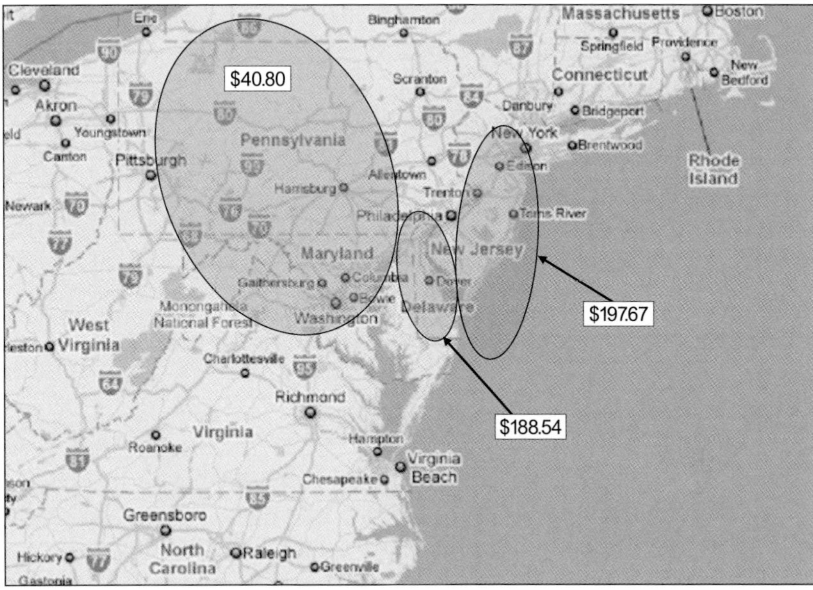

Fig. 10–6. Capacity prices in PJM ($/MW-day)

Offshore projects stand to benefit more than terrestrial renewable energy projects from a system that values generating capacity, as illustrated using prices from a recent PJM capacity auction. Consider two wind farms, one located in the mountains of western Pennsylvania, and the other off of Delaware's coast. Both wind farms have installed capacity of 400 MW, but given the differences in power production characteristics between onshore and offshore wind, the Pennsylvania plant is assigned a 15% capacity rating, while the Delaware plant earns a 30% capacity rating. The PJM capacity auction results in a capacity price of $102.04/MW-day for the western Pennsylvania region, while the capacity auction for the capacity-constrained Delmarva Peninsula (the area of Delaware, Maryland, and Virginia bordered by the Atlantic Ocean and the Chesapeake Bay) results in a market clearing price of $186.12/MW-day.[57] Under these capacity prices, the western Pennsylvania wind farm could expect capacity payments on the order of $6,122/day (400 × 0.15 × $102.04), while the Delaware offshore farm could expect payments of $22,334/day (400 × 0.30 × $186.12). In this example, offshore wind capacity earns roughly 3.5 times the capacity payment of onshore wind.

Transmission access is critical to the success or failure of an offshore renewable energy project. Transmission issues affect project timing and economics; it is not unusual for transmission upgrades to take much longer to get approved and built than the renewable energy facility itself. Transmission capacity constraints also may cause dramatic geographic differences in energy and capacity prices. These issues make transmission a key factor in site selection and project financial modeling. An offshore project proposal must clearly address transmission issues and effectively manage the related risks.

Actual experience with high wind penetration levels in Denmark and certain parts of the United States, and the system models examined in the New York wind integration study, prove that the grid can accommodate large amounts of wind power. In fact, the output of offshore wind energy projects is likely to match well with the typical load profile, contributing to grid reliability. In most of the United States, developers fund interconnections, grid upgrades, and the technical improvements to wind turbines necessary to make a project acceptable from a grid reliability perspective. If the cost of the intertie and upgrades is too high, the developer will abandon that project and seek another less costly place to locate and interconnect. Developers, transmission operators, and government can promote offshore renewable energy with policies that remove transmission barriers. FERC's approval of the novel CAISO renewable energy transmission trunkline financing method is an example of smart transmission policy that promotes renewables through better cost allocation. A wider view of the other elements that must come together to make a financially viable project is offered in the next chapter.

11 Working the Numbers: Financing and Insuring Offshore Renewable Energy Projects

Either you rigorously extrapolate current trends, in which case you get rigorous nonsense, or you adopt an imaginative, blue-skies approach, in which case you get imaginative nonsense.

—*Peter Jay*

Building an offshore wind farm requires a site with strong, steady winds. Financing an offshore wind farm requires strong, steady cash flow. Board members, bankers, insurers, investors, and other skeptics will need to be convinced that a project is commercially feasible. A project is viable if the technology works, the project can be built and operated within budget, and market demand for the product (i.e., clean electricity) would generate revenues sufficient to repay project debt and provide a return to investors commensurate with risk.

Establishing a project's viability starts with a financial model that is adequately robust when tested for sensitivity to various risks. Of course, the most important variables affecting the model are the estimates of energy production and revenues under a power purchase agreement (PPA) or through merchant sales. Capacity value also is important, as is the value of other benefits such as renewable energy credits (RECs) and production tax credits (PTCs). Project capital costs and operations and maintenance (O&M) expenses, depreciation, refurbishment expenses, royalty payments, and decommissioning expenses are other significant factors. The magnitude and variability of all these elements must be understood and modeled before the search for financing begins.

Building the Financial Model

Identifying risks and variables

When the discounted value of the ebb and flow of cash over the life of a project (i.e., *net cash flow*) exceeds its initial cost, the project should produce value and is worthy of consideration among other investment options. Risks, which involve elements of probability and magnitude, will impact a project depending on how they have been apportioned among the participants. Mapping out the expected cash flows involves all of these steps—risk identification, quantification, allocation, and discounting.

Financial theory, however, quickly runs up against the difficult reality of estimating the timing, amount, and variability of cash flows for new technologies like marine renewables. Project risks extend beyond technology and resource issues. Renewable energy project viability depends on government subsidy programs that are subject to political risk. Current subsidies, such as tax credits, may not be continued, and REC subsidy rules may change, altering REC values. The subsidies enjoyed by fossil fuel energy projects, such as the ability to freely emit carbon into the atmosphere, also could change, shifting energy market prices. If so-called pollution subsidies are withdrawn, then carbon emitters will face additional costs, particularly if the greenhouse gas regulatory regime is structured as an auction of emissions certificates. Under such a scenario, the energy market would place greater value on low or zero emissions resources like natural gas, nuclear, and renewable energy.

A disciplined approach to risk analysis should begin with an examination of several categories of risk identified by Standard and Poor's and described in detail below.[1]

Output sales contracts. A long-term PPA is the most important factor in establishing the credit character of an offshore renewable energy project. To minimize cash flow variability, investors usually prefer that a project have a fixed price PPA at least equal in duration to the project debt term. Project debt with a 20-year term would ideally be matched by a 25-year PPA to give the lender some room to restructure in case the project hits a financial rough spot. Industry trends are towards shorter agreements, however. Utility industry restructuring has made power marketers hesitant to enter into long-term supply agreements. Power marketers want to manage their exposure to price changes by matching supply contract inventory with the tenure of sales contracts. In other words, the length of a marketer's wholesale supply agreements should generally match the length of its commitments to sell to retail purchasers. The latter are generally of a short to medium duration (3 to 7 years). A PPA of that duration would unfortunately leave a long-term project lender exposed to the risk that when the initial supply agreement expires, a new supply agreement might not be negotiated on favorable terms.

Since many power marketers must purchase RECs to cover a portion of their total sales, the market for RECs also favors short-term contracts. Renewable portfolio standard (RPS) programs are relatively new, and REC values may vary over a wide range, influenced largely by the state law that creates and implements each state's program. RECs are intended to be a market-based incentive, and REC values also depend on the behavior of other renewable energy project developers who affect REC supply. Numerous projects will add to the supply of RECs and depress prices unless demand is maintained through continually higher renewable energy obligations.

Will the developer or the power purchaser bear the risk that a project does not qualify for a tax credit such as the PTC when the project comes online because Congress has failed to renew this incentive? That governments may disallow or limit certain payments is an especially important risk factor for renewable energy projects, according to a U.N. Environment Programme study:

> *Any investment made under a policy regime is exposed to the numerous reviews and potential changes which may take place between the time the investment is made and the time at which invested capital is fully repaid from project cash flows. Such regulatory risk is common to many infrastructure deals and particularly affects utilities. However, many fickle subsidy and price support regimes are RE-specific as governments find their way with energy policy and security issues.*[2]

Power purchaser's credit strength. Whether the power purchaser will stand behind its obligation under a long-term PPA may be judged by its bond rating. Lenders may insist that the PPA include credit maintenance provisions binding the purchaser to maintain a credit rating above a certain level or else post collateral or a guarantee from a creditworthy entity. A reciprocal credit maintenance provision that places an obligation on the part of the project owner/seller to maintain a credit rating above a certain level may not be viewed favorably by the lender. The lender is likely to see the requirement as increasing the risk of default under the PPA and creating a situation in which the purchaser could terminate the PPA. When the purchaser is a regulated utility, approval of the PPA by a state utility commission helps ensure that the purchaser will recover its costs and continue to honor the contract. The problem of weak credit among utility power purchasers in Nevada was addressed through the establishment of a state fund. Nevada's Temporary Renewable Energy Development Program was funded by setting aside a portion of Nevada utility rates. The funds, held in a trust, provide an assured source of funding for renewable energy power purchases.

When power is sold on a merchant basis, there is no fixed obligation on the part of any buyer that lenders can look to for assurance that project revenues would be sufficient to service debt. Market price projections over the next 20 years are not "bankable." Before the banking meltdown, some developers had success negotiating financial hedges that allowed merchant projects to be financed, since the hedge provider assumed the market price risk. In the absence of such an arrangement, it would probably be impossible for a one-project company with fixed debt service obligations to rely on merchant sales in a regional power market, either in the form of short-term bilateral contracts or for the daily market clearing price. Merchant projects, however, continue to be an option for large corporations with strong balance sheets.

Resource variability risk and estimated annual energy production. Fuel, a key input for fossil-fueled plants, is subject to supply and price risks. Renewable energy projects face a comparable problem known as *resource variability risk*. Disappointing capacity factors and low production at some terrestrial wind farms illustrate the consequences of hasty resource evaluations. Developers need adequate site data to prepare both optimistic and pessimistic project revenue scenarios. Annual average wind speed typically varies over a standard deviation of ±10%, though a maximum deviation of ±20% from the long-term average is possible according to a study of 20 years of wind data in Denmark.[3] Ideally, several years of resource data should be collected, though in areas where research has already been conducted, the period of data collection may be shortened. Weather derivatives, discussed later in this chapter, also could be used to manage resource variability risk.

A wind farm's annual energy production depends on the efficiency with which the wind turbines chosen for the site convert the available wind into energy. Efficiency is affected by array design, distance from shore (i.e., transmission losses), reliability, and rate of performance deterioration. Shadowing and wake interference affect the output of a turbine array as a whole. Transmission line losses increase with transmission distance. Device reliability affects downtime and availability, and device operating efficiency gradually degrades as components wear, corrode, and accumulate dirt or marine growth. Consequently, various adjustments must be made to initial energy production estimates based on resource measurements to arrive at reasonable long-term energy output estimates.

A Weibull distribution, typically derived from on-site measurements, provides the starting point for a wind energy projection. The distribution illustrates the percentage of time during a year that the wind blows at various speeds (fig. 11–1). Various wind turbine models perform differently depending on wind conditions. Some turbines may be more efficient at harvesting lighter winds, while others may be designed to maximize output during heavy winds. These differences are apparent when comparing the power curves of various turbine models. A power curve shows the relationship between a turbine's designed power output and wind speed (fig. 11–2). Offshore turbine models are typically

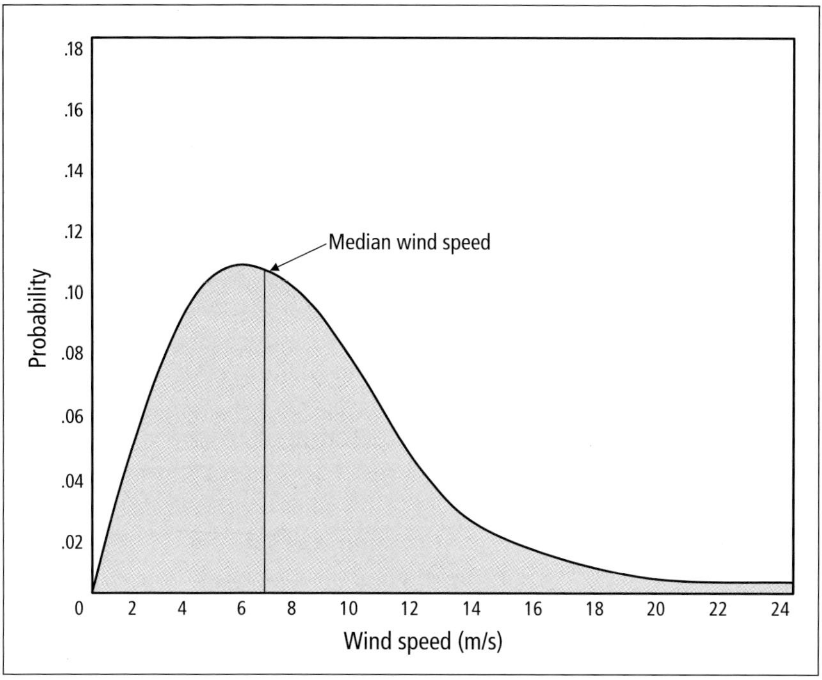

Fig. 11–1. Weibull distribution example

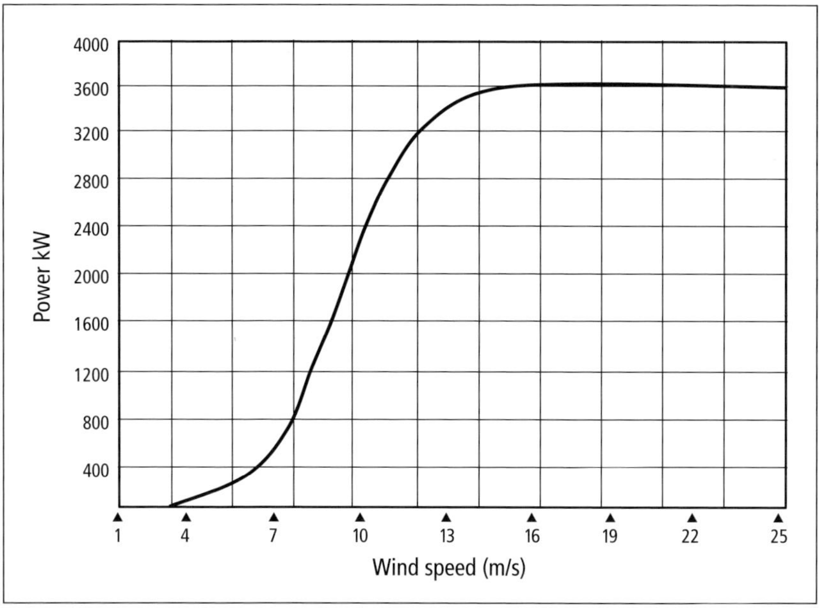

Fig. 11–2. Sample wind turbine power curve

optimized for stronger offshore winds. Manufacturers publish the power curve for each turbine model and are generally expected to guarantee performance at that level for some period. To maximize energy production, a developer will want to find the best match between wind turbine performance characteristics and the wind regime prevalent at the site. In other words, each wind speed probability shown on the Weibull distribution for the area of interest must be weighed with the corresponding amount of power from the various turbine models under consideration to find the model that produces the highest power across the full spectrum of wind probabilities.

Seemingly small differences in wind speed may have a large effect on energy output. A study of Rhode Island's wind energy potential estimated annual energy production at 10 offshore sites where the average annual wind speed varied from 7.75 m/s to 9.25 m/s. Using the power curve from GE's 3.6-MW offshore wind turbine, the study found that at the low wind speed sites, each turbine would produce an estimated 12,720 MWh annually. At the highest wind speed sites, an annual production of 16,105 MWh was possible. Given a difference of only 1.5 m/s in average wind speed, the best sites were expected to produce 27% more energy than the worst sites using the same turbines.[4]

The energy output of a wind farm is not simply the aggregate of each individual turbine's theoretical production based on the expected wind conditions. Each turbine creates a wake that disturbs the air striking downwind turbines in the same farm, reducing the efficiency of downwind turbines. As a rule of thumb, wind turbines should be separated by 7 to 10 rotor diameter lengths in the direction of the prevailing wind, and 3 turbine diameters or more in the direction perpendicular to the prevailing wind. Offshore turbine spacing may be equal in all directions in the absence of a predominant wind direction. The spacing guidelines are intended to promote efficient use of the site area and minimize array losses without incurring too much of a penalty in extra transmission cable costs.[5] Wind farm modeling software can be used to analyze and predict the wake effect and design a wind farm layout that minimizes array losses.

Wind turbine output also degrades over time. In its critique of Cape Wind's offshore project, the Beacon Hill Institute at Suffolk University reduced its estimate of the farm's annual production by 0.8% each year to account for the gradual degradation in equipment performance. This loss in operating efficiency was assumed to be restored with overhauls of the drive train and blades after 10 and 15 years of operation, respectively.[6]

The turbine array effect and other system losses require that gross energy output based on the turbine power curve and Weibull distribution be corrected downwards. The Rhode Island study decreased the gross output prediction by 18% to correct for transmission system losses, turbine availability losses due to scheduled and unscheduled maintenance, and losses due to turbulence within the farm array. A lower loss factor of approximately 15% might be used for a terrestrial wind turbine array, with the difference primarily attributable to lower transmission losses and fewer hours of maintenance downtime.[7]

Projecting revenues. Electricity prices vary by time of year and time of day. The correlation between a project's expected energy output and prices in the local energy market will determine the project's revenue potential. Developers cannot assume, however, that the power produced by a project will make it to the point of delivery agreed to by the purchaser. Deliverability risk arises when a transmission system is constrained and unable to move all the power that generators and purchasers would like to schedule. Either the project owner or the purchaser will suffer the consequences of transmission curtailment. If the PPA places the risk with the purchaser, the result is in effect a *take-or-pay contract*. In cases where a transmission constraint prevents delivery, the purchaser must nevertheless pay for any power that could have been produced and delivered. If, on the other hand, the curtailment risk stays with the generator, any transmission limitations will directly affect the project's bottom line.

When transmission is unconstrained, generating units with low variable costs are dispatched first by the transmission system operator. Unit dispatch continues up the generating cost ladder (also referred to as the *stack*) until enough capacity is online to satisfy demand. Wind turbines are generally bid into the market at a price at or near zero, given that they have virtually no variable operating costs. Fuel-burning generating plants, however, bid into the market at prices based on the cost of fuel used and the efficiency with which a unit converts fuel into electricity. This measure of efficiency, known as the *heat rate*, is stated in terms of heat input required to produce one unit of electrical energy output (typically British thermal units per kilowatt-hour). In an auction-based market, the price of the last (marginal) unit required to satisfy demand during any hour of the day sets the price (the *market clearing price*) that all generating units selling into the market at that time receive. Wind and other renewable generators that have no variable costs, and coal and nuclear plants that have limited ability to ramp up and down to follow load, are *price takers*—they receive the market clearing price. In many parts of the country, and particularly at times of peak system load, the market clearing price is set by gas-fired generators dispatched as the marginal units. These units are operationally flexible and inexpensive to build, but their operating cost is high.

Rhode Island's wind energy study provides an example of revenue forecasting based on educated guesses about future energy and capacity prices, as well as REC and PTC values. The study's authors attempted to forecast energy prices for the 20-year assumed life of a wind farm project, which is the first step in either setting the price under a PPA or projecting revenues that could be earned through merchant sales. Yet a lot can happen over two decades. Imported LNG or newly found gas in shale formations could increase gas supplies. Several large nuclear plants could be built, changing the price-setting role of gas at the margin. Both examples could lead to lower future power prices. Carbon emissions restrictions, on the other hand, could encourage massive fuel switching from coal to gas internationally, driving gas and electricity prices up. While forecasts will always be imperfect guesses, the Rhode Island study used the collective wisdom of many gas industry participants, as expressed in gas futures market prices, and supplemented this data with the predictions in the Energy Information Administration's *Annual Energy Outlook*. The study then used the average heat rate of gas-fired generating plants in New England (8,200 Btu/kWh) to determine the cost of generating 1 MWh of power. This method provided a rough estimate of the future marginal cost of electricity.

Life Expectancy of an Offshore Wind Turbine

The life expectancy of an offshore turbine is an open question, since none of today's offshore turbines have been in place for an extended service life. Bluewater Wind and Delmarva Power entered into a PPA with a 25-year term, implying that their project would last at least that long. Reduced wind turbulence offshore may reduce fatigue stresses on turbine blades, drive train components, and towers, leading to longer-lived machines. It is also possible that corrosion (and other hazards such as lightning strikes) may prove more difficult to control than anticipated, and turbines could experience a shorter life offshore. The financial value of an extended project life merits consideration. The extra value from prolonged operation in the "out" years could be substantial given that the project's debt will have been fully amortized in later years and decommissioning expenses may be deferred, particularly if expensive overhauls are not required. Increased maintenance outlays and a greater initial capital investment in more durable components may, however, be required to achieve longer life.

The Rhode Island study accounted for the cost of carbon by increasing power market prices by $1.30/MWh in 2009, trending upwards to $3.50/MWh in 2028. The impact of carbon emissions restrictions is just beginning to be felt in the New England and the Mid-Atlantic regions under the Regional Greenhouse Gas Initiative and will likely broaden under any future federal greenhouse gas legislation. The study also corrected for wind's intermittency by discounting the expected average market energy price for wind power by $2.50/MWh compared to the overall market price. Table 11–1 below summarizes the revenues that a Rhode Island offshore wind farm might earn per megawatt-hour generated over a 20-year life.

Table 11–1. Rhode Island wind power revenue projections

($/MWhr at the plant busbar)					REC Values		Estimated Total Revenue Range	
	Year	Energy*	Carbon Adder*	Capacity	Low	High	Low	High
1	2007	66	0	2.8	20	56.51	89	125
2	2008	72	0	3.2	20	57.92	95	133
3	2009	70	1.3	3.6	20	59.37	95	134
4	2010	66	1.4	5.3	20	60.85	93	134
5	2011	62	1.5	6.4	20	62.37	90	132
6	2012	59	1.6	6.4	20	63.93	87	131
7	2013	55	1.7	6.4	20	65.53	83	129
8	2014	56	1.8	6.4	20	67.17	84	131
9	2015	56	1.9	6.4	20	68.85	84	133
10	2016	58	2.1	6.4	20	70.57	87	137
11	2017	62	2.2	6.4	20	72.34	91	143
12	2018	62	2.4	6.4	20	74.14	91	145
13	2019	62	2.5	6.4	20	76.00	91	147
14	2020	65	2.7	6.4	20	77.90	94	152
15	2021	66	2.9	6.4	20	79.84	95	155
16	2022	69	3.1	6.4	20	81.84	99	160
17	2023	72	3.3	6.4	20	83.89	102	166
18	2024	76	3.5	6.4	20	85.98	106	172
19	2025	77	3.5	6.4	20	88.13	107	175
20	2026	79	3.5	6.4	20	90.34	109	179

* "Energy"—equals the projected spot market energy price adjusted downwards to reflect intermittent production stream.

"Carbon Adder"—an increase in the spot market energy price due to likely impact of carbon emissions restrictions. Figures may include small rounding errors.

Source: Applied Technology and Management, et al. 2007. Final Report RIWinds Phase I: Wind Energy Siting Study: tables 6-1 and 6-2 (RI Winds).

Estimated energy costs and government incentives are discussed further below.

Project structure. Traditional energy projects such as gas-fired generators benefit from a fortunate match between revenues and expenses, which makes them easy to finance. Revenues for such projects are generally separated into two distinct streams: energy payments and capacity payments. The energy payment for a gas-fired plant usually tracks the plant's variable fuel cost, while the capacity payment is designed to cover fixed costs.

In contrast, renewable energy PPAs typically set a payment for energy per megawatt-hour delivered, and a much smaller capacity payment compensates for a project's availability to serve load. (Generators are compensated for maintaining capacity available to serve load, independent of the actual sale of energy into the market.) Purchasers see little value in renewable capacity, since most renewable energy sources are intermittent. Accordingly, the fixed capacity payments are small.

The vast majority of a renewable energy project's cash flow—from power sales and under government incentive programs—is tied to variable energy output, yet renewables have almost no variable cost of generation. The traditional hydroelectric dam is the exception that proves the rule. The reliability and flexibility of hydroelectric power is why there are very few rivers left to dam. The ability to impound water and open sluice gates as necessary to generate power means that hydroelectric plants can command substantial fixed capacity payments. Capital-intensive renewable energy projects have a cost structure dominated by fixed costs, but project revenues are dominated by variable output payments. Under these conditions, balancing the demands of regular debt service with a mostly variable cash flow is a major challenge. When fixed debt expenses are not well matched by fixed contract (or incentive) payments, a renewable energy project's borrowing capacity suffers.

Financing strategies such as prepaid power contracts have the effect of shifting revenues from a variable to a fixed basis, thus lowering project risks. (However, a prepaid power purchase contract may raise the issue of whether the power purchaser is the owner of a project for tax purposes, complicating efforts to structure a project to comply with the production tax credit rules.) As noted in chapter 10, auction-based locational capacity markets provide an incentive to locate new generating capacity in transmission constrained areas, which may help offshore wind power to earn greater fixed revenues through capacity payments. Offshore renewable energy projects that can be located closer to coastal load pockets than new coal-fired or nuclear power plants have a comparative advantage under locational capacity markets. Transmission operators also are recognizing that offshore wind power production more closely matches load profiles than onshore wind. If the wind resource evaluation at a particular site shows a good correlation between energy output and demand, developers may earn a stepped-up capacity rating (and higher fixed capacity payments) from the transmission system operator.

Lenders will require owners of projects with high cash flow variability to maintain substantial reserves for debt service and other expenses. Maintenance of reserves sufficient to cover at least six months of debt service in the event of temporary project problems, such as higher than anticipated O&M expenses or low-wind conditions, is typical. High reserve requirements also will be imposed on technologies lacking a proven track record. Lenders will insist on equity levels sufficient to keep project owners committed to the project and require debt covenants that restrict the owners' ability to sell assets or withdraw cash until appropriate reserves are established.

A covenant that borrowers maintain a minimum *debt service coverage ratio (DSCR)* is one way that lenders guard against unsound leverage. The DSCR helps to ensure that a sufficient income cushion is available to make regular debt payments. Generally, the DSCR is calculated with the numerator equal to a project's pretax income minus operating expenses, and the denominator equal to the sum of interest and principal payments. Since the DSCR must be met on a regular basis, and coverage at the point of lowest cash flow must be adequate, highly variable cash flows reduce a project's debt carrying capacity. In such cases, expensive equity capital must replace debt, thus reducing equity returns and increasing the cost of energy.

Managing project cash flows to make them more level and predictable may help a developer to meet a set DSCR and carry more leverage. Revenues may be front-loaded and expenses may be back-loaded. Power purchase agreements also could be structured with level payments for energy, rather than gradually increasing payments over time. This effectively results in higher revenues during the early years of a project. Debt amortization also may be structured so that, instead of level amortization, the amortization increases over time. Longer-term debt has a similar effect in reducing the annual amortization burden.

Technology risk. Is the technology tested and proven? Gearbox failures on terrestrial turbines are expensive. In the offshore environment, where rough seas may keep a project offline for weeks until repairs can be done safely, accurate predictions of turbine availability are even more crucial to anticipating energy output, revenues, and the cost of unscheduled maintenance. Novel technologies also increase the risk that contractors may not build the project on time and within budget. The reputation and credit standing of technology vendors and construction contractors are important, as are any performance guarantees.

Performance guarantees may take several forms. A *mechanical availability guarantee* ensures that the machinery will be ready to produce power under the conditions for which it was designed. Generally, if the machine falls below a stated availability level, liquidated damages calculated with reference to the cost of replacement power and the amount of lost production arising out of failure to meet the availability standard would be due. An *output guarantee*, in contrast, is an assurance that the machine will produce a given amount of electrical power (megawatt-hours) over a period of time. The guarantee might be expressed as a percentage (e.g., 75%) of the expected mean annual output. Output guarantees are substantially riskier for manufacturers to issue. They require not only mechanical availability, and a machine that generates electricity according to its design, but also the cooperation of a resource that by its nature cannot be controlled. An alternative to the output guarantee that does not require the manufacturer to assume weather risk is the *power curve warranty*. This warranty assures that the machine will produce energy as intended by meeting designed output levels at specific wind speeds or wave states.

Finally, delay in its many forms jeopardizes a project's financial results. Cape Wind's difficulty in permitting the Nantucket Sound site for its planned offshore wind park illustrates the risk associated with obtaining control of a project site and the necessary permits for its development. Transmission upgrades may take four to seven years to implement, while wind projects have generally been planned and executed within narrow windows provided by successive PTC authorizations. The delay inherent in obtaining necessary transmission upgrades exposes a project to increases in the price of components and materials. Developers may incorporate a commodities cost escalator into a PPA to shift this risk to the power purchaser. Bluewater Wind initially tried this approach with Delmarva Power but ultimately withdrew the provision. Financial models must account for potential permitting delays and the attendant increased costs.

Failure to manage these and other renewable energy project risks caused the default, delay, or cancellation of just under one-half of the projects from which California utilities had hoped to purchase renewable energy, according to a study prepared for the California Energy Commission.[8] Utilities that issue requests for proposals for renewable energy supply are trying to improve the likelihood of successful projects by screening developers and their projects more rigorously. Inexperienced developers tend to propose speculative projects at unrealistically low prices. The utilities may conduct due diligence on projects and their developers, impose fees and credit requirements on developers to weed out insubstantial operators, and require preoperation deposits tied to milestones and performance guarantees.[9]

Estimating capital costs

The range of offshore renewable energy capital costs varies widely depending on one's source of information. In 2008, the National Renewable Energy Laboratory estimated installed offshore wind turbine costs of $2,500 to $3,000/kW, as compared to onshore wind turbine costs of approximately $1,800/kW.[10] A study of the proposed, but cancelled, 144-MW Long Island offshore wind project by the consultants Pace Global Energy Services estimated that the full project cost, including construction, transmission interconnection, and finance charges, was $811 million ($5,634/kW). Pace estimated that a generic European offshore wind project would cost about $4,000/kW. The large gap between European and North American costs was attributed to installation differences in North America, contingencies in construction costs, and a premium of approximately $500/kW that GE was charging on its 3.6-MW offshore turbine manufactured for the North American market.

European turbines require modification to be used in the U.S. market. Pace Global Energy Services noted that it is difficult to determine whether the turbine-cost differential between Europe and the U.S. "is attributable to an appropriate assessment of the risk of the new technology, or conversion of European technology to a North American application, or simply GE extracting higher profits from an illiquid marketplace."[11] Europe's electrical grid operates at 50 Hz, while in North America the standard is 60 Hz. U.S.-based construction contractors have no experience building offshore turbine foundations, and until they do, this work will likely entail greater risks, time, and expense. Specialized vessels used to lift turbine towers, blades, and nacelles into place, and to hammer monopiles into the seabed, must be brought on-site, sometimes from as far as Europe. Turbine prices also are likely to remain high until sufficient manufacturing capacity becomes available to satisfy both European and North American demand for offshore turbines. The large and complex castings required for these machines can be manufactured by relatively few forges and require significant investment in molds. Manufacturers will hesitate to build new production capacity until they are convinced that prolonged demand for large offshore wind machines will develop.

A wide assortment of suppliers, contractors, consultants, and specialized equipment providers has already coalesced in Europe to form an offshore wind power industry. The United States, in contrast, must overcome start-up friction to prepare the industry supply chain needed to build the first few offshore wind projects. The U.S. offshore oil and gas industry is an obvious source of related experience, equipment, and suppliers. The oil and gas industry supply chain may eventually expand to serve offshore renewable energy. However, costs that are higher than current European projects will likely influence U.S. project budgets during the early stages of U.S. supply chain expansion.

Rhode Island's wind study evaluated the costs of developing small (30-MW) and large (200-MW) projects in 10 offshore areas, each with a unique combination of average annual wind speed, water depth, and distance from shore. Project capital costs were divided into several categories: turbine, electrical collection and transmission system, foundations and towers, and mobilization and construction costs. The GE 3.6-MW offshore turbine selected for the Rhode Island study was at the time still in development and was estimated to cost a unit rate of $1,100/kW or approximately $4 million per machine.

Electrical transmission systems represent a large portion of project capital costs. Recent high-voltage subsea cable projects in North America have cost $10,000 to $40,000 per kilometer for each megawatt of transmission capacity. The wide range is attributable to depth, length, cable type, seabed conditions, and other factors. The cost of cable alone ranges from $90/m to $130/m for medium-voltage cables, and $200/m or more for high-voltage cables.

Installation costs vary from 1 to 3 times the cost of the cable depending in part on seabed conditions. Sedimentary deposits characteristic of much of the East Coast seabed should make for relatively quick and inexpensive cable installation using a jet plow. This device uses a high-pressure stream of water to fluidize the sediment as it digs a trench 1 to 4 m (3 to 13 ft) deep. The cable is laid as the trench is plowed, and the sediment falls back into the trench to cover the cable. Remotely operated rock cutting chain saws may be used to dig cable trenches in rocky seabed environments.[12]

Scale economies dilute the high fixed costs of developing distant offshore sites. The Rhode Island wind study found a combined electrical system and transmission cost of approximately $700 to $1,000/kW for a 30-MW offshore project, while a 200-MW offshore project had a cost of approximately $500/kW. Danish researchers considered the cost of energy from various sized offshore wind farms located at varying distances from the shore. A small wind farm of less than 10 MW located

5 km from shore produced electricity at approximately $0.05/kWh, while the same size farm located 30 km from shore had a generation cost of approximately $0.07/kWh. Increasing the wind farm size to 200 MW lowered the price of electricity generated to the narrow range of $0.040/kWh to $0.045/kWh, with the farm closer to shore residing at the bottom of the cost range and the farm located 30 km from shore at the top.[13]

Most offshore turbines built to date have used steel monopile foundations driven into the seabed. As water depth increases, bottom-fixed structures, as compared to floating devices, must be built larger and stronger and at increased cost. The Rhode Island wind study assumed a typical monopile foundation using a steel tube with walls 2 in. thick and 16 ft (4.9 m) in diameter, driven up to 90 ft (27 m) into a seabed consisting of moderately dense sand. The water depth for monopile towers was limited to approximately 75 ft (23 m), and a design wave of 20 ft (6 m) was used to estimate forces acting on the foundation. Monopiles are usually the least expensive foundation alternative due to the ease with which they can be installed with specialized equipment. The Rhode Island study estimated civil/structural costs of $600/kW for foundations and towers for a 200-MW offshore project. A small 30-MW project carried an additional $200/kW cost burden, thus bringing the total cost to $800/kW. Development costs such as engineering and design, permitting, utility interconnection studies, and legal fees were estimated at between 3% and 5% of construction costs.

At least in Rhode Island, a comparison of the total capital costs of various wind projects, including small and large projects, and projects located onshore, offshore (close in), and offshore (distant) demonstrates that the best performer in terms of energy cost is the large project located at the distant site (fig. 11–3). Although the high cost of cabling and foundations in deep water farther from shore argues in favor of nearshore sites, expensive cable and foundation systems may be economical if they provide access to areas of strong winds. Despite higher capital costs, the distant offshore sites benefit from stronger winds that allow greater energy production to offset the marginally higher cost of foundations and transmission.

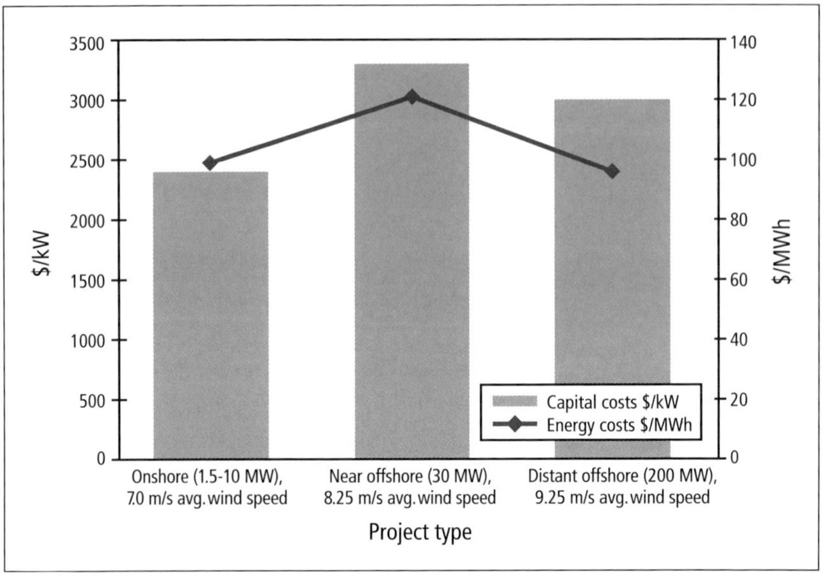

Fig. 11–3. Capital cost and energy cost comparison

Operations and maintenance expenses

Another aspect of project cash flow modeling is estimating recurring expenses, including O&M, depreciation, taxes, royalties, lease payments, and insurance. Debt service, typically the largest recurring expense, is considered separately in the text that follows.

As a rule of thumb, the O&M allowance for an offshore wind project should be approximately $0.02/kWh, which is about twice the O&M costs of a terrestrial wind farm.[14] The estimate covers the cost of long-term service agreements, property and liability insurance premiums, taxes, lease costs, and project administration. Inaccessibility for maintenance and lack of long-term operating experience contribute to the higher offshore O&M cost estimate.

Experience with onshore turbines indicates that O&M costs will increase, perhaps doubling or tripling, over the course of a turbine's useful life. O&M costs for a new turbine may be on the order of 10% to 15% of the levelized cost per kilowatt-hour produced, which can increase to as much as 30% by the end of the turbine's lifetime.[15] These costs are driven by substantial refurbishment expenses for turbine blades, gearboxes, and other components as the project ages. Although newer turbine models have benefited substantially from improved designs and are expected to have lower O&M costs, there is limited long-term experience with turbines in the offshore environment. How these machines will perform over 20 years or longer is unknown. When an assortment of problems developed at the Horns Rev offshore wind farm after only 2 years of operation, Vestas, the wind turbine manufacturer, had to remove 80 wind turbine nacelles from their towers for repairs. In the offshore environment, lessons such as these are especially costly.

Memories of California's wind energy boom, and subsequent crash, will lead cautious investors to prefer well-proven technology. Driven by generous tax incentives, the industry expanded rapidly in California, but haste led to problems, particularly with respect to turbine quality. At one point in the late 1980s, more than 900 turbines were shut down due to faulty blades and safety concerns. A repair bill in the millions, lost revenue, and litigation ensued.[16]

Storm and collision damage, corrosion, biofouling, debris fouling, and other issues will affect O&M costs for marine hydrokinetic energy devices. Cost estimates will be highly device-specific, and little reliable information is publicly available at this time. The hydrokinetic energy industry should take the time to pursue a deliberate development and testing process to avoid repeating the wind industry's early errors.

O&M best practices can maximize the life of capital-intensive assets, minimize cost, and enhance project profitability. In an article for *EnergyPulse*, economist Erik Andersen draws parallels between wind industry maintenance practices and those in the aviation industry. Andersen observed that by following best practice maintenance and upgrading engines, it was possible to extend aircraft life from approximately 15 years and 37,000 hours of operation to 30 years and 50,000 hours. "This experience and knowledge enabled the original depreciation schedules to be extended, resulting in real financial benefits to owners of these aircraft, since, among other benefits, residual values were higher than anticipated."[17] In another example illustrating the difference between operators, two airlines flew identical aircraft, but one taught its flight crews to optimize the use of the plane's engines during flight, while the other did not. The airline that had spent the time and money to train crews on engine optimization was able to substantially increase the duration between engine overhauls, improving its financial performance in comparison to the other airline.

Andersen argues that the wind industry needs a "sophisticated technical appraisal and risk rating system" to improve wind turbine operating and maintenance practices. Better O&M practices can lead to higher project values, lower finance and insurance costs, and more profitable wind farms. Guidelines for the maintenance of different classes of offshore facilities, including wind turbines, have been developed by the risk management consulting firm Det Norske Veritas

(DNV). Investors in offshore renewable energy projects may require operators to demonstrate that they are maintaining their facilities consistent with the DNV guidelines or their equivalent, such as those of Germanischer Lloyd.[18]

The best operators are implementing state-of-the-art maintenance practices with the help of specialists in condition monitoring and fault prediction. Using a combination of hardware (e.g., vibration, sound, and strain sensors, thermal imaging, and remote communications equipment) and software to detect anomalies in machine performance, these experts monitor components such as the rotor, tower, bearings, axles, and gear wheels. A component that is performing outside its normal parameters is scheduled for inspection, preventive maintenance, or repair at a time when turbine downtime would have the least impact on energy production.[19] Since access to offshore wind turbines can be restricted during bad weather, engineers may opt for redundancy with respect to a turbine's critical systems—two oil filtration systems, for example. Condition monitoring and fault prediction systems, which signal developing problems before expensive damage occurs, may permit the elimination of redundant systems, lowering the capital cost of offshore turbines.

A less desirable maintenance strategy is to cyclically replace a component based simply on time in service without reliable information about the part's remaining useful life. This approach typically results in premature replacement of expensive components. Yet another alternative is to replace parts only after they have failed. This can cause both expensive turbine downtime and secondary costs as the failure of a small part may affect larger components.

Royalties and lease expenses

MMS lease and FERC license charges are described in chapter 5. Use of state lands also will typically require payment of a fee and/or royalties, and municipalities adjacent to a project may request and receive compensation as well. Cape Wind entered into a community host agreement with the town of Yarmouth in which the company would pay $350,000 per year to the town, indexed for inflation. Where substantial operations are conducted in a particular jurisdiction, but the business is not subject to local taxation, a business may negotiate a payment-in-lieu-of-taxes (PILOT) with the locality. Businesses often are willing to pay a PILOT to express their intentions as a "good neighbor" and compensate the locality for the benefits the business receives, such as public safety services. Above all else, PILOTs build goodwill in favor of a project by providing a tangible benefit to the community.

Salvage value and decommissioning costs

A developer should expect to post a bond or other financial security to provide assurance that the expense of decommissioning an offshore renewable energy project will not become a government burden. According to FERC,

> *The licensee shall, at least 90 days before commencing project construction and installation, file proof of the purchase of a surety bond, or equivalent financial assurance instrument (e.g., insurance, corporate guarantee, letter of credit, fully funded trust fund, etc.), to cover the entirety of the costs of removing the project in accordance with the Project Removal Plan required by this pilot license. Thereafter during the term of the license, the licensee shall maintain the bond, or equivalent financial assurance. By January 1 of each license year, or as otherwise directed by the Commission or its authorized representative, the licensee shall file proof of the maintenance of the bond, or equivalent financial assurance.*[20]

Decommissioning costs about $300,000 for an offshore wind turbine. This is consistent with a U.K. government study that estimated offshore turbine decommissioning costs at approximately £40,000/MW ($81,400/MW), or 2% of operating costs if spread over the lifetime of a project. This estimate assumes complete removal of the device, including foundations and cables to a depth of 3 to 7 ft (1 to 2 m) below the seabed.[21]

The decommissioning process is essentially the reverse of construction. Cranes remove the rotors, nacelle, and tower. The foundation is cut and removed or toppled in place. Disposal in place may be preferred where an artificial reef community has become established on a foundation.

MMS requires lessees to provide at least $100,000 of lease-specific financial assurance, plus additional financial assurance in an amount determined by MMS based on the complexity, number, and location of the facilities involved in the planned activities.[22] Removal of all facilities to a depth of 15 ft (4.6 m) below the mudline is required, unless otherwise permitted.

Some owners of terrestrial wind projects have found that *repowering* (i.e., replacing old turbines and perhaps towers, and reusing the transmission and other project infrastructure) can be cost-effective. With appropriate corrosion control measures, offshore towers and foundations may achieve a useful life decades longer than the turbines initially mounted on them. Major turbine components also may have significant additional life left after the initial 20- to 25-year term of a PPA if they are replaced at 10- or 15-year intervals. In sum, today's decommissioning estimates may ultimately prove to be conservative if technological progress provides new and productive ways to give a used facility a second birth.

Government Incentives

Renewable energy credits

Renewable energy incentive payments are a substantial component of a project's economic return. Sometimes derided as subsidies, these incentives address the free-rider problem, providing compensation to project owners for environmental benefits consumed by the public at large. At this time, the absence of a federal RPS means that those states without RPS programs are free riding at the expense of states that have implemented their own programs. (At present, over one-half of the states plus the District of Columbia have adopted an RPS.) Some state RPS programs have mandatory renewable energy purchase requirements, while other states have renewable portfolio goals. The program rules vary from state to state. All this regulatory experimentation, while allowing state programs to emphasize certain local priorities, is a poor way to advance renewable energy nationally.

RPS requirements start low and trend upwards to require increasing amounts of renewable energy as a portion of a state's total energy mix over several years. For example, New York's requirement of 1,360,424 MWh in 2006 rises to 11,988,888 MWh (or about 25% of total load) by 2013. Eligible resources include biogas, biomass, liquid biofuel, fuel cells, hydroelectric, photovoltaics, ocean or tidal power, and wind. New York chose to exclude municipal solid waste from eligibility due to concerns about the emission of mercury and other heavy metals but left the door open to future consideration if technologies such as gasification could be shown to reduce emissions. Unlike most other states, New York's program is funded by a volume-based charge on the state's investor-owned utilities' delivery customers. The funds are disbursed to project owners by the New York State Energy Research and Development Authority (NYSERDA) in an amount that is "the minimum necessary to stimulate development of" qualifying renewable energy facilities.[23] The New York RPS model has the advantage of providing more predictable incentive payments than other states' market-driven programs. NYSERDA's good credit standing as a state entity helps to bolster the credit profile of renewable energy projects that include New York RECs as a revenue source.

The RPS program in Texas is more typical of other states. The Texas renewable energy mandate requires increasing levels of renewable generating capacity up to 5,880 MW in 2015. The qualifying resources are largely the same as in New York, but Texas uses RECs to provide a market incentive for new renewable energy projects. Competitive electricity retailers in Texas are allocated a portion of the state renewable capacity objective based on their proportionate share of total Texas retail energy sales. The state's capacity target is converted into megawatt-hours by multiplying the capacity target by the number of hours in the year (8,760) and then by 35%, which is a representative capacity factor for the wind plants that make up the majority of the state's renewable energy facilities. RECs are credited to renewable generators based on metered output. One megawatt-hour of renewable energy generated in Texas equals one REC. Each REC is identified with a unique serial number, and a program administrator keeps an accounting of RECs held by all generators. RECs have a three-year shelf life, which allows them to be "banked" against future compliance obligations. A generator may sell its RECs to an electricity retailer, which then retires the credits to satisfy its compliance obligation, or the RECs may be held or traded for speculative purposes (within their three-year life). Penalties of up to $50/MWh can be assessed against an electricity retailer that fails to acquire RECs in an amount equal to its purchase requirement. Robust terrestrial wind power development in Texas adds to REC supply, which keeps the price of Texas RECs well below the penalty level cap.

RECs are most valuable in states where a dearth of renewable energy supply puts electricity retailers in jeopardy of having to pay a substantial penalty (the *compliance payment*) for failing to meet the renewable energy target. Low supply could be due to a restrictive definition of qualifying resources, difficulty in siting renewable energy facilities, or the high cost of renewable energy technologies suitable to that state's renewable resources. Figure 11–4 shows the source of RECs used to satisfy the Massachusetts RPS. The state relies substantially on landfill gas and biomass and currently very little on wind power. The role of renewable energy imports from neighboring states and provinces also has a large effect on the Massachusetts REC market.

Under the Massachusetts program, 4% of the electricity sold in the state in 2009 must come from renewable sources, with the percentage to increase by 1% annually in subsequent years. Retailers that fail to procure sufficient RECs to cover their renewables obligation will be liable for the compliance payment set under the RPS program rules. The compliance payment ($58.58/MWh in 2008, and adjusted annually for inflation) establishes an upward limit on the price of RECs in the market. The REC price floor is set by the supply of renewable generation

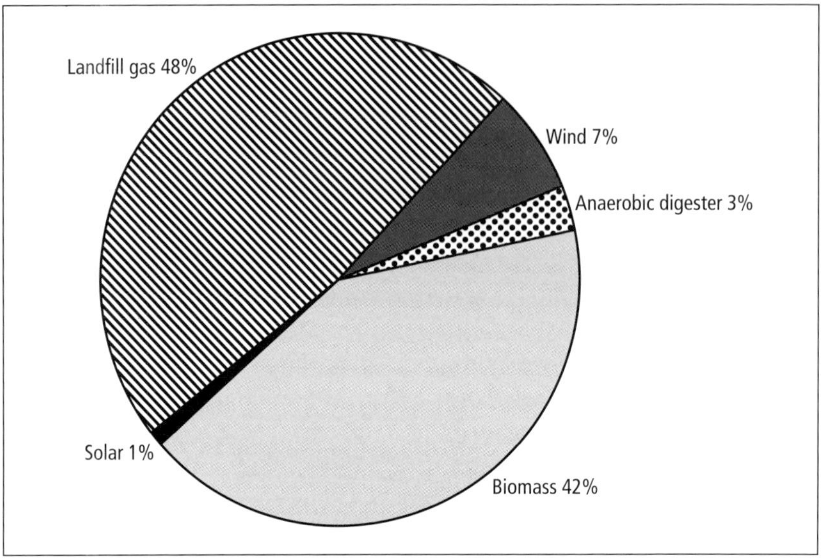

Fig. 11–4a. Massachusetts renewable generation by type (2006)

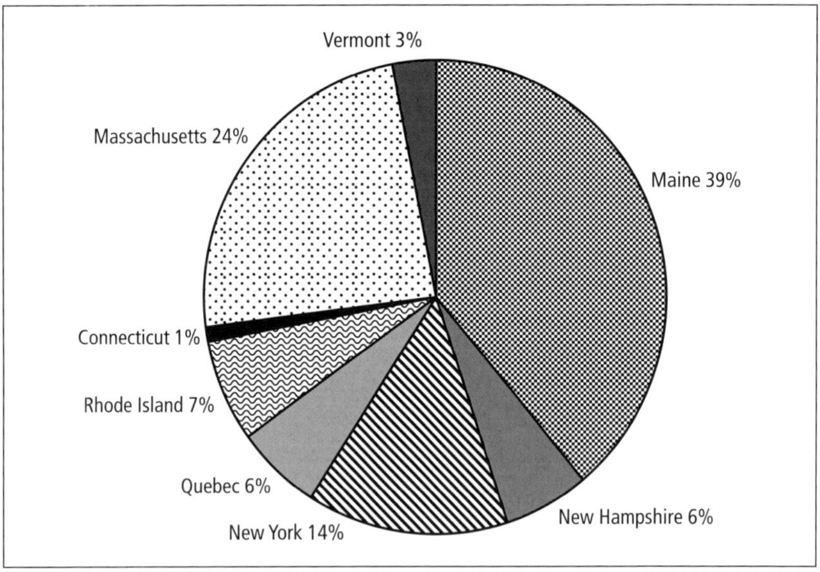

Fig. 11–4b. Massachusetts renewable generation by state/province (2006)
Source: Massachusetts Renewable Energy Portfolio Standard, Annual RPS Compliance Report for 2006 (Feb. 15, 2008)

and theoretically could approach zero if renewable energy becomes cheaper to produce than conventionally generated energy. The relative scarcity of renewable energy supply in Massachusetts has contributed to REC prices in the $50/MWh range.

Compliance payments go to the Massachusetts Technology Collaborative (MTC), which uses these and other funds to promote the commercial development of renewable energy plants. MTC may enter into REC price support contracts with renewable energy project owners. These contracts help place a floor on the value of RECs generated by a participating project and thus support third-party financing by providing a level of predictability to the project's sales revenue stream.

> *The absence of long term contracts for RECs with creditworthy entities makes it difficult for developers to secure equity and debt financing for their projects. The [Massachusetts Green Power Partnership] aims to address the lack of REC cash-flow certainty by providing long term REC contracts, assuming the risk that government mandated demand will persist under substantially similar terms and that a voluntary market for RECs will develop. By placing funds into an escrow account to support MTC's REC purchase commitments, we also provide the creditworthiness required by equity and debt investors.*[24]

REC price support contracts may take one of the following forms:

- A simple purchase agreement where MTC buys a fixed quantity of RECs at an agreed price over a given period
- A put option, which gives the project company the right to sell RECs to MTC at a given strike price
- A collar, where the project company holds the right to put (i.e., sell) RECs to MTC at a certain price, thereby protecting the seller from a fall in REC values, while MTC holds a call option to purchase RECs from the seller at a higher price, thereby allowing MTC to benefit from the upside of potential REC price increases

Although the efforts of MTC and NYSERDA have brought some predictability to the REC markets in Massachusetts and New York, the REC market in general remains speculative, immature, and fragmented. REC buyers are generally reluctant to take long-term positions. Buyers do not want REC purchase obligations that are longer than their underlying obligations to supply power to customers. Buyers also are unsure about long-term REC values. What would happen, for instance, to RECs earned under a state RPS after the implementation of a federal RPS? Will future greenhouse gas regulations permit RECs to be used interchangeably with carbon offset credits? Will states change the rules if utilities find it too difficult to acquire RECs at "reasonable" prices? Because REC values are so difficult to estimate over the life of a wind generating plant, the financial model in the Rhode Island wind study used a wide band of values ranging from \$20/MWh to \$90/MWh.

Estimating whether and when anticipated renewable energy projects would begin to sell RECs may provide useful insights into future REC market conditions. The Beacon Hill Institute found that Cape Wind's project in Nantucket Sound would supply a substantial part of the Massachusetts RPS requirement and that, in combination with other anticipated renewable energy projects, the additional supply would drive down REC prices. According to Beacon Hill, should the Nantucket Sound project be realized, long-term REC values would likely stabilize at approximately $34.50/MWh, providing roughly $48 million per year in revenue for the project. Developers should understand each state's unique REC market and evaluate the sensitivity of their project's revenues to the construction of other renewable energy projects in the state. How many projects would need to be built before that state's REC market would be flooded, and how likely is that level of development to occur?

A collapse in REC prices could lead to industry contraction, much as periods without the PTC have curtailed the construction of new terrestrial wind projects. REC oversupply may be avoided if renewables obligation percentages ratchet upwards at least as fast as new renewable energy projects are built. A federal RPS also could have a beneficial effect on REC values by lessening price volatility. A federal standard could create a deeper, national market for RECs that would not be overwhelmed by local concentrations of renewable generation.

Created as a market-based incentive, REC prices should respond to supply and demand. Yet it is hard to imagine a more dysfunctional method to foster a nascent renewable energy industry—one requiring the construction of long-lived, capital-intensive facilities—than the current unpredictable subsidy with its limited ability to support financing. The increased uncertainty that REC value fluctuation imposes on project financial models increases risk and capital costs. Projects could be financed at lower cost (benefiting consumers with a lower energy cost) if project revenues were less subject to uncertain REC values.

State governments could make RPS programs more effective by reconciling the differences among neighboring state programs, with the objective of creating regional RPS programs and multistate REC markets. States also should consider adding features of the U.K.'s renewable obligation, such as granting more expensive renewable energy technologies, like solar and offshore wind, a multiple of the standard grant of one REC per megawatt-hour generated. Each renewable energy technology is at a unique stage in its commercial development and requires a different level of support. Graduated incentives would promote the simultaneous development of multiple renewable energy technologies. Delaware has taken this path by adopting legislation that awards regulated electric utilities a 350% credit toward the RPS mandate for energy purchased from offshore wind energy installations sited off the Delaware coast. Efforts to place a floor on REC values also would provide some assurances to developers and financiers about future REC revenues, making RECs a more effective incentive.

The production tax credit

The PTC has greatly accelerated the expansion of wind power generation in the United States, and recently this incentive became available to hydrokinetic energy projects as well. Owners of facilities that produce electricity from *qualified energy resources* and sell the power to unaffiliated entities earn PTCs. Qualifying energy resources include wind, geothermal, and biomass, as well as certain improvements to existing hydropower impoundments that increase electricity production. In late 2008, Congress expanded the list of qualified energy resources to include hydrokinetic renewable energy. The facilities eligible for the credit include those that generate energy from waves, tides, ocean currents and ocean temperature differentials, and free flowing water in rivers, lakes, streams, canals, and man-made channels such as irrigation systems. Notably, facilities that use a dam or impoundment for electrical power production purposes are excluded.[26] This exclusion could extend to certain tidal barrages.

The PTC is available to wind facilities placed in service before January 1, 2013, and hydrokinetic renewable energy facilities placed in service before January 1, 2014. The credit is indexed to inflation and currently set at a rate of 2.1¢/kWh for wind electric generating facilities. Hydrokinetic renewable energy facilities earn credits at one-half the rate applicable to wind facilities. A project may earn PTCs for 10 years from the date the project is placed into service.[27]

The PTC is most valuable to a project owner with substantial current tax obligations that may be offset with the credit. PTCs are subject to the general business credit limitation under the Internal Revenue Code that is intended to ensure that a taxpayer's tax liability is not reduced below a minimum level.[28] Unused PTCs are allowed a one-year carryback and a 20-year carryforward period.[29] U.S. Internal Revenue Service (IRS) rules also affect how PTCs earned by partnerships flow through to the partners.

Efforts to maximize returns within the bounds of IRS rules have led to complex project ownership structures in which two classes of equity investors take interests in project cash flows that vary over time. Typically, a partnership agreement will provide that passive equity investors, often financial institutions (the *tax equity investors*), should receive 95% to 99% of the tax credits earned by the project (over the first 10 years) until the tax equity investors have realized an agreed rate of return. Other allocations, such as partnership income or loss, also may flow through to the tax equity investors. IRS regulations require tax credits to be allocated to partners in the same proportion as is allocated the ownership of the item giving rise to the credit. As energy sales give rise to the credit, so a tax equity investor must receive an allocation of project sales revenues proportionate to its tax credit allocation. Once the tax equity investor realizes the predetermined return, the economic allocation between the tax equity investor and the developer "flips," and the developer is then allocated as much as 95% of the economic returns, while the tax equity investor's share drops as low as 5%.

IRS guidelines for partnership structures that use the flip technique state generally that the tax equity investor should share in the economic risk of the project by maintaining (a) an investment in the project at least equal to 20% of the project's total capital contributions, and (b) a minimum interest in each material item of partnership income and gain, one that may vary but may not be less than 5% of the investor's largest percentage interest in the project at any time.[30] The developer may not loan funds to the tax equity investor to finance the acquisition of a project interest, nor may the developer guarantee the tax equity investor's investment in the project.

Notably, the PTC may be reduced by as much as 50% by grants, tax-exempt bonds, subsidized energy financing, and *any other credit* provided in connection with the project. IRS revenue rulings provide that the term *any other credit* is limited to federal credits; state and local tax credits for renewable energy projects or electricity production do not reduce the allowable amount of PTC.[31] It is also worth noting that the 10% business energy tax credit under I.R.C. section 48 is not available to property that is part of a facility the production from which is allowed a credit under section 45.[32]

Basic PTC planning should ensure that the tax equity investor acquires its interest no later than the in-service date so that the investor collects the maximum PTCs over the full 10-year period during which they may be earned. A project's in-service date starts the clock on the 10-year PTC accrual period. According to the IRS, a facility is placed into service when it is ready to generate and sell electricity to the grid, based on factors such as the following:

1. Receipt of necessary licenses and permits
2. The transfer of control over the facility from the developer to the taxpayer/owner
3. Completion of project testing

4. Commencement of regular operations
5. Synchronization of the facility with the electrical grid for export of power

To avoid tax benefit "leakage," project developers should plan the in-service date to coincide with the point at which the facility is ready to generate energy to its fullest extent. The issue of when a facility is deemed to be placed in service is important for projects consisting of tens and, perhaps, hundreds of turbines that could come online over a prolonged period. How should a situation be handled where 50 turbines out of a 100-turbine offshore farm are installed and ready to generate electricity, but installation of the others must wait until rough seas during fall and winter have passed? Should the sale to the tax equity investor occur in multiple steps as groups of turbines are connected to the grid and begin to export power? If the developer waits until the whole farm is complete, turbines that have been synchronized with the grid and have begun to export power could be considered in service. The subsequent transfer of the facility after its in-service date from the developer to the tax equity investors would not restart the clock. The IRS considers each wind turbine a separate facility because it can be operated, metered, and placed into service independent of other turbines in a wind farm.[33] If the sale is structured to occur in stages as individual turbines become operational, issues of hold-back and other contractual provisions that assure completion of the wind farm will arise.

The PTC period generally also continues to run even if the facility suffers significant damage (e.g., from a storm that halts production or due to a severed transmission cable). The facility must be almost completely destroyed and rebuilt before the PTC period would be reset. To reset the PTC period back to year one, the cost of repairing or rebuilding a storm-damaged facility must be such that the facility's value immediately after the loss is less than 20% of the facility's rebuilt value. In such a case, a rebuilt facility would be essentially new and treated as newly placed into service. "A facility would also qualify as originally placed in service even though it contains some used property, provided the fair market value of the used property is not more than 20 percent of the facility's total value (the cost of the new property plus the value of the used property)."[34] Consider a scenario where a storm damages the blades of a wind turbine but does not result in a loss of 80% of total value. In that case, the opportunity to earn PTCs would be lost for the duration of the turbine's downtime. Insurance coverage for offshore renewable energy projects should address the treatment of lost PTC value.

Temporary economic stimulus legislation, formally known as the American Recovery and Reinvestment Act of 2009, provides two other federal incentives that may be substituted for the PTC. Wind energy facilities are eligible for an investment tax credit (ITC) of 30% of the cost of new equipment for facilities placed in service before January 1, 2013. Hydrokinetic energy facilities placed in service before January 1, 2014 also are eligible for the ITC. A facility owner that chooses the ITC option must reduce the depreciation basis for the facility by one-half the amount of the credit (i.e., 85% of the cost of the facility may be depreciated). The Department of Treasury also may issue grants in lieu of the PTC or ITC for up to 30% of the basis of qualified facilities including wind and hydrokinetic energy facilities. The grants apply to facilities placed in service in 2009 or 2010, or to facilities that initiate construction in 2009 or 2010 and are completed by a *credit termination date* ranging from January 1, 2013 to January 1, 2017, depending on the type of facility. Unless an offshore renewable energy project is already well into the permitting process, it probably could not meet the construction and in-service date deadlines applicable to grants. The newly available ITC is likely to increase the financing options available to developers and improve project financial models by accelerating cash flows into the first years of a project, rather than extending the tax benefit over 10 years as with the PTC.

In addition to the federal PTC, state tax credits, sales and property tax exemptions, and grants for research and development may be available to public and private renewable energy developers. Florida has a renewable energy production tax credit of $0.01/kWh available to projects including wind, wave, tidal, and OTEC energy facilities. Maine provides matching funds up to $50,000 to encourage the development of renewable energy resources. New Jersey provides grants of up to $500,000 to developers of renewable energy technologies that are located in the state.

Accelerated depreciation

Wind project investment returns can be enhanced by the ability to deduct project capital costs on an accelerated basis over the project's first five years. This incentive is not currently applicable to investments in hydrokinetic renewable energy technologies. Under the Modified Accelerated Cost Recovery System of the Internal Revenue Code, a project's hard costs may be deducted over five years, instead of a period more closely related to the useful life of the facility (perhaps 20 years).[35] The deductions create tax losses in the early years of a project that may be used by project owners to offset income in other businesses. Of course, the tax burden will be commensurately higher once the project is fully depreciated, but the value of tax deferral to a tax-paying investor can be significant.

Naturally, there are other nuances to structuring renewable energy project investments that may help to maximize the value of a project to its investors. For example, a developer that contributes assets to a partnership may realize a greater return on its investment through tax deferral than if the same assets are sold to the partnership, triggering an immediate taxable gain.[36] For the most part, these tax strategies are likely to be the same whether the project is located onshore or offshore.

Financing Alternatives

Project financing

Terrestrial wind energy projects are usually built with project financing. Project finance underwriting focuses on the credit of the project, rather than the credit of the borrower. Lenders ask the question, "What could possibly go wrong?" and look only within the "four walls" of the project for an answer. Their opportunity for repayment will be confined solely to the project's performance and value as collateral. The task of the developer seeking to raise project financing is to identify, minimize, and allocate project risk, and structure the project agreements to provide reasonable assurance that the lender will be repaid.

Project financing is also commonly termed *limited recourse financing*, since the investment structure and project documents restrict the lender's ability to seek repayment from the developer/sponsor's other assets. The project itself is held in a *single-purpose entity (SPE)*, which limits the liability of the owners of the SPE unless they have undertaken to guarantee in whole or part the SPE's obligations. A guarantee could be required during the construction period, particularly if the technology lacks a long track record; such guarantees can be structured to fall away once commercial operations begin.

Lenders fashion the project documents both to ensure that the project is maintained and operated according to good practices and to preserve cash flow and asset values for debt repayment. Lenders want the ability to step into the owner's shoes to cure defaults or otherwise assume project operations if the owner fails to uphold its obligations. Therefore, they will require first-priority assignment of all key project agreements, including the PPA, O&M agreement, site lease, interconnection agreement, insurance policies, and construction contracts.

Waterfall and *lockbox* arrangements are a popular means of lender protection. The waterfall describes the order of priority in which project funds are used to pay obligations. In a typical waterfall, all cash payments to the project are deposited directly into a lockbox, which is an account controlled by the lender. Payment from lockbox funds is first made for expenses that are essential to maintain the project in good working order, then to service debt, and then to fund certain reserve accounts maintained as a cushion against adverse events (e.g., six-months' O&M expense reserve; major maintenance reserve; debt service reserve). Only after these needs have been adequately provided does money flow into a subaccount for distribution to the project owners.

Balance sheet and bond financing

Balance sheet financing is an alternative technique in which lenders place less emphasis on the performance of specific assets and instead rely more heavily on a borrower's general credit. The borrower's credit—its ability to repay debts—is determined by the cash flow of its various businesses and the value of the assorted assets held on the borrower's balance sheet. In contrast to project finance lenders, lenders to a balance sheet–financed project have the broad base of the borrower's assets as collateral. When a regulated public utility is the borrower, lenders can receive even greater security if regulators allow the utility to include the project in its rate base. In such cases, project assets will usually earn a steady regulated return through rates charged to the utility's customers and provide a certain source of cash to service debt.

When a project has substantial risks that are difficult to quantify, and the developer has substantial assets and cash flow, balance sheet financing may provide an alternative to project financing. Major oil and gas companies routinely finance billion-dollar offshore production platform investments on the strength of their balance sheets. Through asset diversification and sheer financial mass, these companies absorb risks that cannot be offloaded to other parties at a reasonable cost. At times of heightened anxiety in the insurance markets, for example, coverage for offshore production platforms has been so expensive that major oil and gas companies simply operated without it. The companies could assume the risk of catastrophic storm losses because their investments in vulnerable Gulf of Mexico platforms were part of a portfolio of assets located around the globe. It was unlikely that more than a fraction of their worldwide assets would be affected at any one time. In contrast, it would be inconceivable to obtain project financing for an offshore renewable energy project without property damage insurance coverage, since only the project's cash flow and asset value would secure the project's obligations.

Public utilities have traditionally financed new utility projects on their balance sheets. As noted above, prudent expenses for assets that are useful for utility service may be added to a utility's rate base with the consent of utility regulators. Amounts in the rate base earn a regulated return for utility investors secured by predictable revenues from utility customers under a *tariff* (i.e., a comprehensive contract with regulators and customers that sets forth the utility's approved rates and terms of service). The promise of a secure regulated return allows rate-based assets to be financed at a low cost compared to nonregulated power plant investments. Rate basing also may permit investment cost recovery over a longer period than traditional debt financing, possibly resulting in a lower cost of energy for utility customers. For utilities in deregulated markets, however, building generating plants for rate base would be a retreat from a commitment to competitive markets. Nevertheless, some utilities are encouraging regulators to consider rate-based investments for renewable energy and nuclear plants that are considered too difficult to finance on a project basis.[37]

Public entities like the Tennessee Valley Authority use bond issues to finance the construction of large capital-intensive projects, such as coal-fired plants, hydroelectric dams, and nuclear power plants. Public power authorities also may become a source of financing for offshore renewable energy projects. Given the trend towards very large offshore wind projects, it is possible that consortia of public and private utilities might form to pursue these projects. Each utility participating in a consortium would invest according to its appetite for risk. In return, participating utilities would receive a share of the project's generating output and other benefits such as RECs. PTCs could be allocated to taxable entity participants using the flip structure described above. A consortium approach would permit smaller utilities to capture the economies of scale associated with large projects while limiting each company's risk to a level that is manageable within the scope of its balance sheet. Wind projects developed for publicly owned utilities and cooperatives have occasionally featured power prepay contracts as a substitute for some or all of the project debt. The utilities prepay for a portion of the power to be produced by the project, and a taxable equity investor is allocated PTCs and accelerated depreciation tax benefits. This makes efficient use of the tax subsidy that otherwise could not be claimed by the public and cooperative utilities.

Structured portfolios

Structured portfolios are another option for financing renewable energy projects. As a first step, a large developer may build a portfolio of offshore renewable energy projects on a balance sheet basis. Next, the developer creates an SPE, such as a trust, for the purpose of holding the portfolio and attracting investors. Once the projects have been commissioned, the developer would then contribute the projects to the SPE in exchange for the proceeds of debt offered by the entity in a public or private transaction. In 2003, NextEra Energy Resources (d/b/a/ FPL Energy), the largest U.S. wind energy developer, became the first to demonstrate this approach. NextEra's subsidiary, FPL Energy American Wind, LLC, raised $380 million in a private bond offering. The proceeds of the 6.639%, 20-year senior secured bonds were used to repay a part of FPL Energy's investment in building or acquiring seven wind farms with an aggregate capacity of 680 MW.[38] The transaction followed the pattern of traditional asset securitization, where similar assets are grouped in a *bankruptcy remote vehicle*, and interests in that entity are sold to investors on the basis of the expected performance of the asset portfolio. Investors in structured portfolios avoid construction risks and receive a security that should be less risky than an investment in a single wind farm. The diversification of wind farm locations and technologies (e.g., turbines manufactured by different companies) helps to reduce risk. Hurricane damage to one farm or systemic gearbox failures affecting one turbine manufacturer should be offset by the performance of other assets in the portfolio. A wind project portfolio offering also may provide lower-cost funding as compared to project-by-project financing.

As the example above illustrates, financing may be easier to arrange if construction risk cans be avoided. A hybrid financing agreement that involves a separate construction loan, or construction financed by the developer's balance sheet, followed by a bank term loan, allows a long-term lender to bypass the risks associated with a project's construction phase. It also may provide the borrower with debt cost savings.

Loan guarantees

Wind energy consultant Garrad Hassan, reporting in 2004 on the development of offshore wind projects in Europe, found that "to date, the few existing commercial offshore wind projects have largely been developed on the balance sheet of the developers. As more and bigger offshore wind projects are developed, access to project finance will be essential in order to mobilise large-scale development. Without it, offshore wind will not emerge to fulfill its present potential."[39]

The recent collapse of so many financial institutions due to mortgage security–related risks that the institutions did not fully understand has made all financial institutions increasingly risk averse. This climate, should it persist over several years, could become a substantial handicap for offshore renewable energy projects. Lender hesitancy is worsened by inexperience with the risks particular to the industry. Banks may, for example, have teams of underwriting professionals with years of oil and gas industry experience who understand oil and gas technology and markets. These lenders can judge whether future oil and gas prices, political risk exposure, and other factors make project lending in any particular case a sound investment. A similar familiarity with offshore renewable energy is mostly lacking. The pace of change in technology and energy markets makes it extremely challenging to identify where offshore renewable energy fits within the spectrum of competing energy technologies. Financing an offshore renewable energy project today means taking a bet on an assortment of technical, market, and political risks. Finding bankers willing to take that bet has become harder.

This is where the loan guarantee program administered by the U.S. Department of Energy (DOE) can help. The program reduces bank exposure on loans to certain renewable energy projects with a pledge of the full faith and credit of the United States for the repayment of debt guaranteed under the program. The loan guarantee program was created under EPAct Section 1702, and the DOE has issued regulations describing program eligibility and the application process.[40] A guarantee of as much as 100% of project debt, not to exceed 80% of a project's cost, may be available for projects that employ new or significantly improved technologies, as compared to technologies now typically in service. To qualify for the program, the projects must avoid, reduce, or sequester greenhouse gases, a threshold offshore renewable energy technologies generally meet.

To obtain a loan guarantee, borrowers must first wait for a DOE solicitation. Applicants may respond to the solicitation with a preapplication that includes a business plan and certain information about the technology employed in the project. Information required includes how and why the technology is new or significantly improved compared to technology already in general use in the U.S. commercial marketplace.

Though offshore wind turbines have already been employed at numerous European installations, the technology is not currently in general use in the United States and may therefore qualify for the loan guarantee program. The DOE regulations provide that "a technology is in general use if it has been installed in and is being used in three or more commercial projects in the United States in the same general application as in the proposed project, and has been in operation in each such commercial project for a period of at least five years. The five year period shall be measured, for each project, starting on the in service date of the project or facility employing that particular technology."[41] By restricting loan guarantees to technologies not in general use, the program sets up a race among projects using similar technologies. The winners of the race would be the first three applicants to build commercial projects using the technology, plus any other applicants in the same field that are able to receive a DOE loan guarantee commitment before five years has passed from the time the last of the initial three projects enters service.

The preapplication also requires a description of off-take or other revenue generating agreements that will provide the source of revenues required to repay the project's debt. Other details of the project's financial model, including cash flow projections and an explanation of the "facts, assumptions, and methodologies in the financial model," also are required.[42] The DOE will review the preapplications to determine the following:

1. The commercial viability of the proposed project
2. The technology to be employed
3. The relevant experience of the principals
4. The financial capability of the project sponsor

If the DOE decides favorably on the viability of the project, it may invite the sponsor to submit a full application for a loan guarantee.

Applicants are required to provide significantly more detail in the full application than in the preapplication, including specifics regarding the project's location, principal contractors, construction schedules, and key development milestones. An independent engineering report addressing issues such as project design and environmental compliance also is required. The purpose of the loan guarantee program is to promote promising technologies to the level of general commercial use free of government support. Thus DOE regulations also require the application to explain how the applicant "intends to assure, to the extent possible, the further commercial availability of the technology(ies) in the United States."[43] Related to the issue of commercial viability is a requirement that applicants explain "the market for any product to be produced by the project, including relevant economics justifying the analysis."[44]

DOE will rank the applications according to factors such as the technology's commercial potential and efficacy in avoiding, reducing, or sequestering air pollutants and greenhouse gases. Preference will be given to projects that seek a smaller percentage of guaranteed funds as compared to total project capital. Projects that cost more than $25 million also must provide a credit rating from a nationally recognized rating agency reflecting the project without the proposed loan guarantee.

The stimulus legislation mentioned above also expanded the loan guarantee program and relaxed its requirements. Loan guarantees are now also available for transmission facility upgrades and renewable energy projects, even if they employ technologies available in the commercial marketplace, provided that they commence construction no later than September 30, 2011.

The guarantees are not free. Once DOE determines that the project is suitable for a guarantee, it will propose terms for the guarantee agreement, including the estimated credit subsidy cost to be paid by the project. This borrower-paid fee, which resembles an insurance premium, covers the risk of default on the guaranteed loan. The credit subsidy cost is determined under the principles of the Federal Credit Reform Act. The act generally requires a payment from the borrower at the time the loan guarantee agreement is executed equal to the net present value of two different cash flow streams, discounted to the point of disbursement. The first is payments by the government to cover "defaults and delinquencies, interest subsidies, or other payments." The second is payments to the government by the borrower "including origination and other fees, penalties and recoveries" (e.g., such as through the sale of collateral).[45] The terms and conditions of the guaranteed debt, the risks associated with the project, and any other factor that affects the amount and timing of cash flows will affect the credit subsidy cost calculation.

A lender's comfort with its decision regarding whether to finance a project will increase as project risks are assigned to creditworthy parties other than the borrower. Federal loan guarantees, equipment manufacturer warranties, the participation of experienced construction contractors, REC sales contracts, and many other arrangements each may remove certain risks from the project or the borrower, lessening lender exposure. Yet trading off risk requires giving up project value to those parties that assume the risk. The next section discusses how insurance may be used to allocate project risks.

Insuring Novel Projects and New Technologies

Insuring an offshore renewable energy project is in many ways similar to financing one. Insurers must understand and predict the impact of potential risks facing a novel technology. New technologies are difficult and expensive to insure. Gaps in the information will tend to be compensated for by a cushion in the premium. The significant underwriting effort required also may not be justified by the premiums that may be earned on small projects. As the initial, modest forays into offshore wind energy are now giving way to much larger projects, insurers have started to take note of this market. Large premium revenue potential has piqued interest among insurers, and insurance availability at competitive prices appears to be expanding for offshore wind projects. It will take time before a similar pattern plays out for other offshore renewable energy technologies.

Various insurance products may be useful to offshore renewable energy projects (table 11–2). *Construction/erection all risks insurance* protects against defects and problems arising during and out of the construction phase of a project. *Property damage insurance* covers damage from extreme weather events or collisions with offshore structures. *Machinery breakdown*

Table 11–2. Overview of insurance products available for renewable energy projects

Risk transfer product	Basic triggering mechanism	Scope of insurance/ risks addressed	Coverage issues/ underwriting concerns
Construction All Risks (CAR)/Erection All Risks	Physical loss and/or physical damage during the construction phase of a project.	All risks of physical loss or damage and third party liabilities including all contractors' work.[1]	Losses associated with cable laying such as snagging can be significant for offshore wind projects.[2] Quality control provisions for contractors.
Delay in Start Up (DSU)/Advance Loss of Profit (ALOP)	Physical loss of and/or physical damage during the construction phase of a project causing a delay to project handover.	Loss of revenue as a result of the delay triggered by perils insured under the CAR policy.	Cable laying risk. Loss of transformer. Lead times for replacement of major items. Offshore wind weather windows and availability of vessels.
Operating All Risks/Physical Damage	Sudden and unforeseen physical damage to the plant/assets during the operational phase of a project.	"All risks" package.	Explosion/fire concerns for biogas, geothermal. Increase in fire losses for wind. Lightning. Quality control and maintenance procedures.
Machinery Breakdown (MB)	Sudden and accidental mechanical and electrical breakdown necessitating repair or replacement.	Defects in material, design, construction, erection, or assembly.	Concern over errors in design, defective materials, or workmanship for all renewable energy technologies. Turbine technology risk. Scope and period of equipment warranties. Wear and tear (excluded from MB).
Business Interruption (BI)	Sudden and unforeseen physical loss or physical damage to the plant/ assets during the operational phase of a project causing interruption.	Loss of revenue as a result of an interruption in business caused by perils insured under the Operating All Risks policy.	Cable/transformer losses represent large potential BI scenarios. Lead times for replacement of major items. Offshore wind weather windows and availability of vessels. Supplier/customer exposure (e.g., biomass resource supply).
General/ Third-Party Liability	Liability imposed by law, and/or express contractual liability for bodily injury or property damage.	Includes coverage for hull and machinery, charters liability, cargo, etc.	Concern over third-party liabilities issues associated with toxic and fire/explosive perils.

[1]Scope of activities for insurance cover includes, but is not limited to: procurement; construction; fabrication; loading/unloading; transportation by land, sea or air (including call(s) at port(s) or place(s) as may be required); pile driving; installation; burying; hook-up, connection and/or tie-in operations; testing and commissioning; existence; initial operations and maintenance; project studies; engineering; design; project management; testing; trials; cable-laying; trenching; and commissioning.

[2]Cables for wind projects represent a high concentration of value for relatively horizontal risk exposure.

Source: United Nations Environment Programme. 2004. Financial Risk Management Instruments for Renewable Energy Projects, Summary Document: 23.

insurance protects against losses not covered by manufacturer warranties. *Marine insurance* would cover a cargo of turbines that sinks on its way to the offshore project site. *Business interruption insurance* protects the cash flow needed to service project debt if the project is damaged and not producing power sales revenues. Insurance also might cover the financial harm caused by construction delays. Coverage of this sort is widely available for fully commercialized and well-understood technologies, such as terrestrial wind, but may be unavailable or very expensive for marine renewable energy projects.[46]

The developers of the Pelamis wave energy converter were able to obtain machinery breakdown coverage only after a thorough vetting of the technology by independent engineers. The engineers verified that the design would withstand, and not pose a safety hazard in, a 100-year storm. Rick Gibbons, an executive with AIG Global Marine and Energy, shares a typical concern: "Any time you have a new technology, it's going to be protypical, it's going to be untested. It may have design problems, it may have performance [problems]."[47] When a construction vessel dropped anchor on an unprotected electrical cable at Denmark's Horns Rev project, insurers paid EUR 2 million to repair the damage. The project also experienced costly delays. Vestas, the turbine supplier to the Horns Rev project, had little experience with the logistics of offshore construction work at the time. Vestas significantly underestimated the harbor space required to marshal and assemble rotors, nacelles, and towers, creating a bottleneck in the construction process that delayed the project's in-service date. Insurers are appropriately cautious of newly developed technologies, as well as newly scaled-up versions of existing machines, and they are learning from early projects.

When a technology is so new that there is little or no experience upon which to base risk decisions, premium levels become prohibitively expensive, deductibles increase, and coverage may exclude high risk and little-understood perils. This dynamic is one reason offshore wind developers gravitate to turbines that have several years of offshore operating experience. To overcome the reluctance to adopt new technology, a manufacturer offering a new product must demonstrate the technology's clear economic advantage. In addition, the manufacturer must also back the product with a longer and more comprehensive warranty than the accepted and proven standard technologies.

Insurers, like lenders, also perceive significant risks in the construction, testing, and commissioning stages of a project. Risk factors include the following:

1. Inexperienced project contractors and other participants
2. Engineering and quality control problems

3. Poor design and workmanship
4. The lack of proper O&M practices
5. Uncreditworthy project participants
6. Exposure to major perils and natural catastrophes[48]

A survey by insurance broker Marsh Ltd. found that

> *44% of respondents identified a lack of sound operating and maintenance procedures and practices and lack of previous experience of project participants, contractors, subcontractors, etc., as a major concern for a small number of technologies. Again this is likely to be more applicable to those technologies which are based offshore (wind, wave, tidal) and require specialist marine equipment/vessels and contractors during construction, operating and maintenance periods. Specific mention was made of the concerns over cable installation and maintenance associated with marine renewables.*[49]

The additional risks present in offshore construction result in higher premiums and deductibles. A study by the U.N. Environment Programme found that "typically for an offshore wind energy construction project premium rates would be approximately 2 per cent of the estimated project cost compared with premium rates for onshore construction of 0.4–0.6 per cent."[50]

In an effort to minimize many of these risks (and premiums), developers of the initial round of U.S. offshore renewable energy projects will rely greatly on European expertise and technology. Bluewater Wind has assembled a team of contractors and suppliers, each of which has already participated in successful offshore renewable energy projects, to build its proposed 450-MW wind turbine farm off the coast of Delaware. The Bluewater team includes the following:

- Turbine supply: Vestas
- Engineering: Ramboll Engineering
- Foundations: Ballast Nedam
- Cabling/substations: ABB
- Electrical design: SEAS
- Wind resource consulting: Garrad Hassan
- Construction vessel services: A2Sea

Offshore construction standards and offshore equipment certification programs reduce underwriting uncertainties by creating an industry standard for the design, construction, and maintenance of offshore renewable energy facilities. Insurers will generally expect projects to be built to accepted standards, such as those published by the certification agencies DNV and Germanischer Lloyd. (See also chapter 5, noting MMS requirements for offshore project design and construction certification.) Insurers and lenders also may require that a third-party inspector supervise key aspects of construction, installation, testing, and commissioning.

In addition to following industry standards, developers may reduce losses and increase insurability by adopting industry best practices. "Increasingly, insurers require projects to demonstrate what loss control measures are in place to minimise losses from high wind, freak wave conditions, fire and lightning and vessel collision."[51] For example, an oil spill response plan that includes resources in place (e.g., oil spill containment and recovery equipment) to address a spill of oil stored on-site or spilled in an allision with an offshore structure will lessen losses from these accidents. Maintaining ready access to spare parts, such as replacement turbine blades, will lower lost revenue claims for business interruption by speeding repairs. Project design features also may reduce risks. Redundant transmission cabling lessens the risk of revenue losses in the event a cable is damaged. Assembling a renewable energy harvester at a dry dock and floating it to the project site may lessen the risk inherent in handling, erecting, testing, and commissioning facilities over the open water.

When sufficient accumulated operating data on an offshore renewable energy technology is unavailable, insurers draw analogies to other marine activities to measure the probability of loss events or correlations between events. The risk experience associated with cable and pipeline laid by the offshore oil and gas industry is relevant to evaluating the risk of failure or damage to transmission cables serving offshore renewable energy projects. The frequency of allisions with offshore oil and gas structures is relevant to predicting the likelihood of allisions with offshore wind turbines. Likewise, injury and accident rates in offshore construction should be similar whether the task is building offshore oil and gas platforms or wind farms.

On the other hand, analogies to other similar activities do not always work. Business interruption coverage for an offshore project should probably cover a longer period than that for a terrestrial project if there may be delays in bringing specialized machinery on-site for repairs. The likelihood and severity of allisions with offshore renewable energy facilities may vary significantly depending on traffic and weather conditions particular to the project site. The nature of local shipping traffic influences the severity of possible oil spills. A site that is adjacent to a major shipping lane frequently traversed by oil tankers will have a different risk rating than a project that would be exposed principally to allisions with pleasure boats and commercial fishing vessels. A developer should expect its insurance broker to make these distinctions. The insurance broker should tailor insurance coverage to the needs of the particular offshore renewable energy project, help educate underwriters, and improve the likelihood of a favorable reception in the insurance market.

Developers are likely to obtain more efficient insurance pricing by selecting an underwriting team with the technical expertise to evaluate the specific risks of individual offshore renewable energy technologies. When a project involves a substantial risk placement, a lead underwriter will direct the due diligence and negotiate the terms and conditions of the policy on behalf of a syndicate of insurers that collectively absorb the full placement.

Allocating specific risks to the consultant, contractor, manufacturer, or installer with the best ability to anticipate and manage each risk will usually control risks at the least cost. Manufacturers, for example, understand the durability and performance of their equipment better than others and are well placed to warrant performance, cover equipment failures, and limit the severity of claims. Manufacturers also benefit when their products attain a reputation for reliability. They have an interest in assuring the proper maintenance of their equipment through O&M contracts extending beyond the expiration of the warranty period. Manufacturers may guarantee the availability of machinery and cover maintenance and repair costs, including the replacement cost of expensive components. Pricing under an O&M contract may be based in part on project output to encourage the manufacturer to maintain high levels of production.

Agreement among project participants with regard to risk allocation also pays dividends during the contracting process. Once a subset of the total project risk is allocated to the several project participants, a developer could choose to insure any remaining risks under a single umbrella policy, known as a *wrap*, designed to cover every company and individual involved in the project. A wrap may reduce costs by avoiding disputes and litigation, since there would be less reason to resort to finger-pointing when something goes wrong. Overall contracting costs also may be lower, since contractor bids should not include the cost of any insurance that is duplicative of coverage provided under the wrap policy. A contractor that is unsure whether it will be obligated for certain losses will include an allowance for estimated losses or insurance costs in its bid, so it is important to communicate clearly about insurance coverage issues when soliciting bids.

The challenges of procuring efficiently priced insurance and financing are similar, and when insurance is unavailable, financing also is likely to be difficult to arrange. This is especially true for early commercial applications of offshore renewable energy technologies. For this reason early stage projects are usually small, built with significant government support, and/or financed by an entity with a strong balance sheet and an independent ability to absorb risk.

Protecting Revenues with Weather Derivatives

To this point, insurance has been discussed in terms of protecting the project from ordinary hazards such as property damage, construction misadventure, and business interruption. These risks are similar in that the events that require insurance result in a high potential loss but occur with a low probability.

There are other business risks that have a generally lower potential loss but a higher probability. Weather is one such risk. Weather risk management contracts (i.e., *weather derivatives*) in the energy industry evolved about 10 years ago as gas utilities sought to control their exposure to lower revenues from warmer-than-normal winters. The parties assuming the risk on the other side of the weather derivatives contracts were generally insurance companies. They sought to diversify their weather risk by not concentrating contracts in the same geographic area. Although the first weather derivatives involved insurance companies, weather derivatives are quite different from insurance. Insurance requires an insured to prove a loss before it can collect on a claim. Weather derivatives, in contrast, pay out upon the occurrence of certain measurable events, regardless of whether the holder of the contract incurs a loss.

The early derivatives were based on temperature, with payments determined by the number of degree days above or below a standard 65°F (18°C) average day. Additional products have since emerged to cover variability in precipitation and wind. The U.S. Futures Exchange lists wind futures contracts for locations in New York and Texas that are based on deviations from 20-year historical wind speed averages compared with the present. A standardized exchange-traded contract based on the weather history of a particular onshore location would probably be an imperfect hedge for an offshore project that could experience much different weather. Quantifying weather risk requires a significant amount of location-specific historical weather data, which can be particularly difficult to obtain for offshore sites.

Customized bilateral contracts might be an alternative for offshore projects. How might they work? The Sacramento Municipal Utility District (SMUD), which relies substantially on hydroelectric and gas-fired generation, entered into a derivative known as a *costless collar* to protect itself against the additional cost of gas-fired generation in the event that low precipitation reduced hydroelectric output. No payments would be made under the contract if precipitation remained within a band of normal variation. The contract would be triggered, however, if precipitation fell below a set number of inches during a 12-month period, and SMUD would receive a payment. The amount of the payment would be determined by reference to the price of natural gas, which would be used to generate power during a shortage of hydroelectric production. The contract also would be triggered if precipitation rose above a different set number of inches, requiring SMUD to make payment to the counterparty.[52]

The SMUD model would need revision to apply to an offshore project. The trigger could be a decrease of 10% or more below normal average wind speed measured over a month. If the winds are stronger than normal, the other party to the derivative could receive a payment commensurate with the extra production of the wind farm above normal levels. Payment could be based on the variation in wind speed above or below the "normal" band multiplied by a factor that reflects the relationship between wind speed and energy production. The energy production factor could be derived by combining site wind data with the power curve of the specific turbines used in the project. The wind speed alone cannot be used as a direct proxy for lost revenue because the relationship between power and wind speed is exponential, not linear. The International Swaps and Derivatives Association Master Agreement that underlies most commodity swaps transactions would typically form the contractual basis for a wind derivative agreement.

While the derivative described above would protect the project owner from lost energy sales revenue, it would not provide a perfect hedge. Without actual generation from the project, the owner will not accrue PTCs or RECs (and also might not earn carbon offset credits under any future greenhouse gas regulatory scheme). A weather derivative that also covered the lost credits might go a long way toward calming lender concerns about the variability of project revenues needed to service renewable energy project debt.

To finance and insure an offshore renewable energy project requires a robust financial model that provides sufficient leeway to manage around project risks. When developing a project with new technology, the financial hurdles are higher because lender and insurer uncertainty will be reflected in greater risk premiums. The last chapter of this book describes how industry and government might come together to lower the economic and regulatory obstacles that limit the broad adoption of offshore renewable energy in the United States.

12 Cultivating Disruptive Change: The Economic Challenge of Offshore Renewable Energy

For the world has changed, and we must change with it.

—*President Barack Obama*

Ask an engineer, "Can it be done?" and he or she may reply, "How much can you afford to pay?" Any conversation about offshore renewables quickly gets down to economics. "Can you build a wind turbine in water 50 meters deep that will withstand a category 5 hurricane?" "Yes," replies the engineer, "but it will take a lot of steel. What's the budget?"

People take various perspectives on project economics. Costs and benefits are weighed differently according to differing values. A project spreadsheet prepared for private investors will look different than an economic analysis prepared by a policy analyst from a societal perspective, because investors do not bear all the costs or collect all the benefits of a project. Timing also has its role. Project revenues will accrue to one entity, while some costs and benefits will be imposed on, or accrue to, another entity, group, or generation. State and federal governments must decide whether subsidies today for the first high-cost offshore projects are a wise investment to jump-start a new industry, provide employment, and earn export revenues. The value of a subsidy now must be weighed against the likelihood that the United States might need to spend billions on coastal defenses against the effects of global warming decades hence. Answering the question, "Can we afford to build it?" involves finding the economic sweet spot that harmonizes the interests of investors and society.

In chapters 5 through 8, the regulatory system as applied to offshore renewables was explained. Chapter 11 demonstrated that numerous risks and variable cash flows provide a challenging environment for privately financed offshore renewable energy projects. The chapter also considered the government incentives (RECs, PTCs, accelerated depreciation, and loan guarantees) used to promote renewables. Are the incentive programs, laws, and regulations working? According to some, the current lack of installed offshore renewable energy capacity means that the U.S. system is working. Offshore renewable energy technology is generally more expensive than many traditional energy alternatives, so it is appropriate that much of society's wealth has not been invested in offshore follies. Renewable energy advocates would disagree. The very fact that not one commercial offshore renewable energy facility has been built in the United States, despite the tremendous natural resource that awaits development, is proof that the system is broken. Cape Wind's Nantucket Sound project remains in permitting review after seven years, and Verdant Power could not even conduct limited testing of its tidal turbines in the East River without long and expensive permitting delays.

The failure of the regulatory system to process permit requests with an "appropriate" delay for thorough reflection signals a system that discourages innovation and favors the status quo. Similarly, the current patchwork of economic incentives biased towards production payments does little to stabilize the economics of offshore renewable energy projects and, in fact, exacerbates the variability of project cash flows. The way utilities purchase power is another obstacle. According to Navigant Consulting, "The current retail market structure in NJ, as in other deregulated states, is not conducive to long-term contracting. However, the ability to enter into long-term contracts (>10 years) for renewable energy and REC sales is critical to securing financing for projects."[1] The U.S. system is not efficiently delivering offshore renewable energy.

These failures are all the more alarming because action to address climate change is urgently needed, and building a robust and secure energy system requires diverse sources of supply. The United States advances its interest in solving global energy, climate, security, and economic development challenges when it promotes sustainable energy technologies at home that can be exported and implemented overseas. Efforts to find ways to control U.S. carbon emissions do little good if cost or other obstacles prevent the widespread use of these clean technologies abroad. The world's developing countries will simply continue to build polluting conventional generation at a frenetic pace.

The worldwide economic upheaval currently underway has led many to question the efficacy of unregulated and poorly regulated markets, and also to doubt whether the United States can afford to switch to renewable energy. Yet the United States also cannot afford to leave its environmental future on autopilot, subject to the invisible hand of the market. Oystein Dahle, the former vice president of Exxon in Norway, once observed, "Socialism collapsed because it did not allow the market to tell the economic truth. Capitalism may collapse because it does not allow the market to tell the ecological truth."[2] Markets have not allocated resources well because they have been blind to externalities.

This chapter describes European efforts to improve the economics of offshore wind and hydrokinetic energy and, in that connection, to put balance back into Europe's energy and environmental picture. Incentive payments and government technology accelerators, together with high prices for traditional energy sources, have been instrumental in quickly bringing about technological progress in alternative energy. The economic environment in Europe contrasts markedly with the challenges that U.S. developers face in building an economic case for their projects. Economic arguments also have been used by opponents of offshore wind energy, and this chapter discusses how developers might refute these attacks.

If the economic skeptics are believed and offshore renewable energy resources for utility-scale clean power are not developed, where will the United States get the clean energy needed for sustainable living? This chapter briefly considers the economics of "clean coal" and nuclear power, which are competing for the same limited pool of public subsidies. The chapter, and the book, close with some ideas concerning how to accelerate offshore renewables in the United States. Clear price signals and streamlined regulation are needed to create the environment for growth in offshore renewables. It is no longer sufficient, if it ever was, to muddle through with half-hearted, ineffective policies. Government, industry, environmental advocacy groups, and other stakeholders must engage to rapidly advance the United States' crucial interest in a clean energy future.

Offshore Wind Economics

Pioneering Europeans

Europe's lead in offshore renewable energy was not won simply by throwing subsidies blindly to assuage a liberal political fancy. Notably, the United Kingdom, Germany, and Denmark made a conscious choice to support renewables to attain greater energy security, lower national carbon emissions, develop green industry and jobs, and promote new export opportunities. Having made this choice with full knowledge of the existing cost disadvantage of renewables as compared to most traditional energy sources, these governments then set about designing the most efficient ways to drive renewable energy costs down.

A principal figure in this task was none other than a line; a curve, to be exact. The experience curve reflects the relationship between cumulative production and cost. Put simply, the more times that a manufacturer has made something, the better it gets at producing the product, and the faster and cheaper it becomes to make each successive item. How fast one moves from rough prototype to the latest optimized design built with just-in-time component delivery to a robot-assisted assembly line is a matter of the speed at which companies, industries, institutions, and governments learn (the *learning rate*). Collaboration among industry, academia, and government can accelerate learning rates, requiring fewer iterations of a design before it becomes practical and cost competitive. When the design-test-build process takes a short-cut, public support in the form of tax incentives, renewable obligation subsidies, and the like is needed for a shorter period than might otherwise be required, saving public treasury funds in the long run. Government support of collaborative efforts to accelerate the learning rate, therefore, is likely to be an efficient investment, since it lessens the need for prolonged subsidies. Europe has experience with the development of offshore wind turbine design, engineering, and maintenance standards, the development of best practices for public consultation, and collaboration in the analysis of the environmental impacts of offshore renewables. Europe's experience with joint planning of offshore transmission infrastructure for clustered wind parks and the establishment of wave energy technology incubators are further examples of cooperative efforts that lift all participants in the competitive marketplace.

The world's first offshore wind turbine was built in 1990 off the coast of Nogersund in Blekinge, Sweden. The small 220-kW WindWorld turbine had a rotor diameter of 89 ft (27 m) and was mounted on an expensive steel tripod foundation. Vindeby, the first offshore wind farm, soon followed. Built in 1991, in shallow (7 to 16 ft, or 2 to 5 m) waters off the coast of Denmark, the Vindeby project featured 11 450-kW turbines with an aggregate capacity of 5 MW. Vindeby's builders learned that lightning struck the offshore turbines more often than land-based turbines. Restricted maintenance access offshore during adverse weather contributed to turbine availability in the 95% range, somewhat lower than the 97% to 99% availability typically experienced by onshore turbine facilities.

By 2001, the Middelgrunden project, built with 20 2-MW turbines, each with a rotor diameter of 249 ft (76 m), was producing power. Located on a shoal near Copenhagen, Denmark, the shallow (10 to 16 ft, or 3 to 5 m) Middelgrunden site used concrete gravitational foundations, each weighing 2,000 tons, to support the turbine towers. The foundations were constructed in a nearby dry dock and floated out to the site.

The pioneering European offshore wind projects are each different in terms of turbine size and number, water depth, and distance from shore. Nevertheless, the projects demonstrate a few trends. Offshore wind turbines and wind farms are getting larger. Vindeby used turbines of less than one-fourth of the generating capacity used a decade later at

Middelgrunden. Today, 3.0-MW offshore turbines are in common use, and 4.5-MW and 5-MW units, with rotor diameters up to 126 m (413 ft), are being installed. The size of offshore wind farms has increased from tens of megawatts to hundreds. Installed offshore wind generating capacity has now surpassed 1,000 MW, and several proposed U.K. offshore projects are 500 MW or greater. These large projects represent substantial investments and a growing confidence among lenders and investors in the industry's ability to understand and manage project risks.

Europe's offshore projects illustrate a mixed picture with regard to project costs. Substantial increases in the price of raw materials, particularly steel used in foundations and towers, and copper used in transmission cables, have made offshore wind projects more expensive in recent years. Limited capacity for the manufacture of offshore turbines due to strong demand from onshore projects has kept prices high. Siemens and Vestas are the only two offshore turbine suppliers with at least 200 MW of turbines operating offshore. REpower, Multibrid, Clipper Windpower, GE Wind Energy, Gamesa, Enercon, and other turbine manufacturers have smaller fleets of installed offshore turbines or are developing them. They could potentially increase production if it appeared that the trend in offshore wind development could be sustained. Specialized offshore construction vessels also are in short supply. For these and other reasons, cost information should be considered highly suspect. It is, however, useful to understand the scale of investments and relative costs.

The effect of commodity cost volatility has been moderated by the beneficial effects of scale economies. Researchers from the University of Sunderland and Delft University built a cost model for offshore wind projects linked to a GIS database. The cost model used performance and cost information for a small (1.5-MW) and a large (4-MW) turbine to explore cost trends with various wind farm design options. The study considered variables such as the following:

1. Annual energy production
2. Cost of the support structure
3. Cost of the grid connection
4. Cost of the turbine
5. Availability of the wind farm
6. Annual operating and maintenance costs

The GIS database was populated with information about the waters of several European countries by assigning to cells information such as wind speed, water depth, and distance from shore, as well as factors that might preclude wind farm development such as shipping lanes, nature reserves, and regions with seabed slopes in excess of 5°.

The model confirmed the benefits of increased scale. As might be expected, foundation material costs increased with greater water depth. But the relationship was not direct. At a depth of 5 m, steel for the turbine foundation and tower would cost approximately EUR 850,000. At 15 m, or triple the depth, the support structure cost only increased to EUR 1,100,000; much less than a factor of three. Installation costs also were fairly independent of the size of the turbines or their distance from shore, depending most significantly on the overall number of turbines. Installation costs per turbine dropped dramatically as the size of the farm grew. Given a farm of just 25 turbines, installing each turbine would cost EUR 360,000. Increase the farm size to 100 turbines, and the per-turbine installation cost drops to EUR 280,000.[3] Much smaller cost decreases occurred after the 100-turbine threshold. Clearly, significant savings could be achieved by using multiple megawatt-sized turbines in very large farms rather than smaller machines and farms.

Looking at grid connection costs, the report identified the number of turbines in the farm, the turbine's individual power capacity, and the distance from shore as key cost factors. At a given distance from shore, connecting larger turbines was more capital efficient than connecting the same number of smaller turbines. The cost model estimated, for example, that at a distance of 12 miles (20 km) from shore, the cost of interconnecting a farm of 100 1.5-MW turbines was EUR 65 million, versus EUR 125 million for a farm of 100 4-MW turbines. A 167% increase in farm generating capacity (from 150 to 400 MW) required only a 92% greater investment in transmission infrastructure.[4]

Yet despite favorable economies of scale, offshore wind power continues to be more expensive than terrestrial wind energy. The harsh offshore environment and the costs of open water construction are partly responsible. Transit times to offshore sites and delays caused by hazards such as high winds, fog, and lightning stretch construction schedules and increase costs. Developers and engineers are attacking these challenges creatively. Onshore foundation construction at a dry dock is one innovation that permitted the builders of Copenhagen's Middelgrunden project to use conventional construction techniques and ordinary construction crews and machinery to reduce costs and avoid significant weather delays.

> *Avoiding offshore work is important for the economics of offshore wind farm projects, because the costs of work in factories compared to work at quayside and to offshore work is about 1:3:5 or more—up to 10. Offshore work can be avoided by extended testing before the serial production of wind farm elements starts: testing of training courses for personnel, testing of a full size model of a wind turbine, testing of service*

and maintenance of the main components (nacelle, blade, generator, gear box, transformer), and testing of access to the wind turbine.[5]

Learning effects are evidenced in numerous improvements in design and construction methods often adapted from other industries. Global positioning system technology was used to rapidly and accurately place Middelgrunden's precast foundations on the exact locations on the shoal that had been prepared in advance with crushed rock beds. Purpose-built ships capable of installing a steel monopile foundation and erecting a turbine in a day are another recent advance. Turbines have been designed to withstand the assaults of the marine environment and to minimize service visits. Advanced coatings, features such as sealed towers and nacelles, and dehumidifiers to maintain a dry environment around sensitive electrical equipment all help to minimize corrosion. Bearings may be equipped with automatic lubrication systems. To prolong gearbox life, oil preheating and cooling systems maintain lubricant temperature in the proper range. When repairs are necessary, built-in cranes allow certain heavy maintenance tasks to be completed without calling on crane vessels.[6] As more offshore projects are built, learning curve effects should further reduce costs. An 8% cost decrease for each doubling of production volume has been estimated for wind turbines.[7]

Although somewhat dated, a 1997 report by researchers from the Renewable Energy Centre, University of Sunderland, and the Institute for Wind Energy, Delft University of Technology, identifies the cost drivers of offshore wind energy projects.[8] The researchers found that the costs of grid connection, support structure, and the turbine each accounted for roughly 30% of an offshore project's total cost. In comparison, 58% of the cost of a typical onshore wind farm is attributable to the turbine, 21% is due to the foundation and tower, and grid connection accounts for only 11% of total cost.

More recently, New Energy Finance estimated the difference between onshore and offshore construction as shown in table 12–1.

Table 12–1. Typical costs per megawatt of wind energy capacity, onshore versus offshore

	Onshore ($ US)		Offshore ($ US)		Difference
Turbines	550,000	(55%)	950,000	(45%)	+72%
Foundations and towers	220,000	(22%)	460,000	(22%)	+110%
Grid & electrical connections	150,000	(15%)	530,000	(25%)	+250%
Other costs	8,000	(8%)	170,000	(8%)	+110%
Total	**1,000,000**	**(100%)**	**2,100,000**	**(100%)**	**+110%**

Source: Liebreich, Michael and William Young. 2005. Offshore Wind: Europe's EUR 90 Billion Funding Requirement. New Energy Finance: 3.

That offshore projects continue to be built despite the cost is a testament to both European vision and practical realities. Offshore winds are strong, less turbulent, and more consistent than terrestrial winds, leading to high capacity factors and greater energy production. Offshore wind in Europe is considered a viable contributor to meeting greenhouse gas reduction targets and a source of significant domestic employment and export earnings. Europe also is relatively crowded, and terrestrial wind development presents substantial land use conflicts. Offshore projects minimize visual impacts yet still can be built close to coastal loads. Finally, terrestrial wind turbines may be reaching a plateau in economies of scale as roadway capacity restricts further growth in turbine size and output, while the greater carrying capacity of ships permits very large scale offshore machines. Many of these reasons apply with equal force in the United States, particularly on the East Coast.

Recent developments in offshore wind energy

The U.K. government has set a target of 33 GW of offshore wind capacity by 2020, enough to generate about one-fifth of Britain's power requirements. Two rounds of offshore leases have been completed, representing almost 9 GW of offshore capacity if the proposed projects are fully developed (table 12–2). The U.K. Crown Estate has commenced its third round offshore leasing program for the delivery of up to 25 GW of new offshore wind farm capacity by 2020. More than 96 companies have registered to participate in this latest round. As compared to Europe's installed offshore wind capacity of approximately 1,100 MW in 2007, it is clear that the United Kingdom is aggressively developing its offshore wind resources and rapidly becoming the center of the global offshore wind industry.

Subsidies that are larger than those provided for onshore projects are partly responsible for this rapid progress. The United Kingdom's renewable obligation policy promotes diversity in renewable energy technologies by providing varying subsidies.[9] The U.K. program awards 1.5 renewable obligation certificates (ROCs) per megawatt-hour of offshore wind energy generation, compared to 1 ROC/MWh of onshore wind generation. The percentage of renewable energy that electricity sellers must acquire (the renewable obligation) increases in stages to 20%, which should support ROC prices by maintaining greater ROC demand than supply. Under this new program, given ROC values of about EUR 72/MWh and energy prices in the range of EUR 48/MWh, 1 MWh of offshore wind could be worth approximately EUR 156 (about $203 at 1 EUR = 1.298 USD).

Table 12–2. U.K. offshore wind energy projects built and proposed

Location	Generating Capacity (MW)	Developer
First Round		
Barrow	90	GB Gas Holdings
Burbo Bank	90	DONG A/S
Gunfleet Sands I	108	DONG Energy
Inner Dowsing	90	Centrica Energy / Amec
Kentish Flats	90	Elsam A/S
Lynn	90	Centrica Energy / Amec
North Hoyle	60	npower renewables
Ormonde	108	Eclipse Energy Company Limited
Rhyl Flats	100	npower renewables
Robin Rigg	216	E.ON UK Ltd
Scarweather Sands	108	E.ON UK Ltd
Scroby Sands	60	E.ON UK Ltd
Shell Flats	270	Shell WindEnergy Ltd CeltPower Ltd DONG A/S
Teesside	90	EDF Energy
Round One total	**1,570**	
Second Round		
Docking Shoal	500	Centrica Energy
Dudgeon	300	Warwick Energy Ltd
Greater Gabbard	500	Greater Gabbard Offshore Winds Ltd
Gunfleet Sands II	64	DONG Energy
Gwynt Y Mor	750	npower renewables
Humber Gateway	300	E.ON UK Ltd
Lincs	250	Centrica Energy
London Array	1,000	CORE Ltd E.ON UK Ltd Shell WindEnergy Ltd
Race Bank	500	Centrica Energy
Sheringham Shoal	315	Scira Offshore Energy Ltd
Thanet	300	Warwick Energy Limited
Triton Knoll	1,200	npower renewables
Walney	450	DONG A/S
West Duddon	500	Scottish Power Generation Ltd Elsam A/S Euros Energy Europe B.V.
Westermost Rough	240	DONG A/S
Round Two total	**7,169**	
Total, both rounds	**8,739**	

Source: The Crown Estate. 2009. Offshore Wind Farms Table. http://www.thecrownestate.co.uk/interactive_map_offshore_windfarms_table

In Germany, onshore wind development has proceeded at a robust rate due in part to a feed-in tariff. By setting a sufficiently high floor price for renewable energy, the tariff provides the revenue certainty that developers need to finance and build projects. Offshore wind has been slow to develop, however, due to a shortage of suitable nearshore sites. Much of Germany's nearshore coastal waters are protected, forcing offshore wind developers to install projects far from shore in deeper water than offshore projects being built elsewhere. To prevent the high cost of building long offshore transmission lines from blocking offshore wind development, the government requires the operators of Germany's transmission grid to assume the cost of the offshore interconnection. This policy distributes the cost among all grid users. Since the burden of interconnection has shifted to the grid operators, they are now thinking about a systematic, efficient approach to linking possible future offshore projects to the grid. A plan has emerged to use a transmission loop connecting several offshore wind farms to a single trunk line that will transmit wind energy to an interconnection point on land.

Incentives under Germany's feed-in tariff are adjusted for differences between renewable energy technologies. The tariff provides a more targeted level of support than the one-size-fits-all PTCs used in the United States. The tariff rate for offshore wind increases from EUR 0.0874/kWh (US$0.11/kWh) in 2009 to EUR 0.14/kWh (US$0.18/kWh) in 2011. Analysts expect that as large offshore turbines (5 MW or larger) become available, the per-megawatt expense of building deepwater foundations will fall. Coupled with the prospect of earning at least $180/MWh free of the high cost of building transmission ties, German offshore wind development should accelerate.

U.S. offshore wind projects

The headline read "Offshore Wind Farm Project Canceled: South Texas Venture That Would Have Been Largest in U.S. Wasn't Economically Viable, Developer Says." John Calaway, chief development officer for Babcock & Brown Ltd., said that the company backed down from the 170-turbine, 500-MW project because it was too expensive in the context of the local energy market. The AP reported, "'We just don't see the economics working offshore in Texas,' Calaway said, noting that project cost would have been 'in the billions.' He said East Coast offshore wind farms, such as a proposed project off the coast of Massachusetts, are more logical and potentially viable because of land constraints and higher energy prices in the region."[10]

Ironically, Babcock & Brown seemed to be on the fast track to success. The project was sited in Texas state waters and could partly avoid the complex federal permitting process. The Texas General Land

Office quickly issued an offshore site lease, and it seemed as if public acceptance was assured. Offshore renewable energy development builds on the region's long history of energy production in the Gulf of Mexico, and lease proceeds would provide millions for Texas schools. Yet Texas is an unlikely birthplace for U.S. offshore wind power because there are still many wind farm sites on land that could be developed, and expensive offshore power does not fit the moderately priced local power market. Contrast Texas with high-priced Rhode Island or New Jersey, two densely populated eastern states with limited terrestrial wind farm options (fig. 12–1).

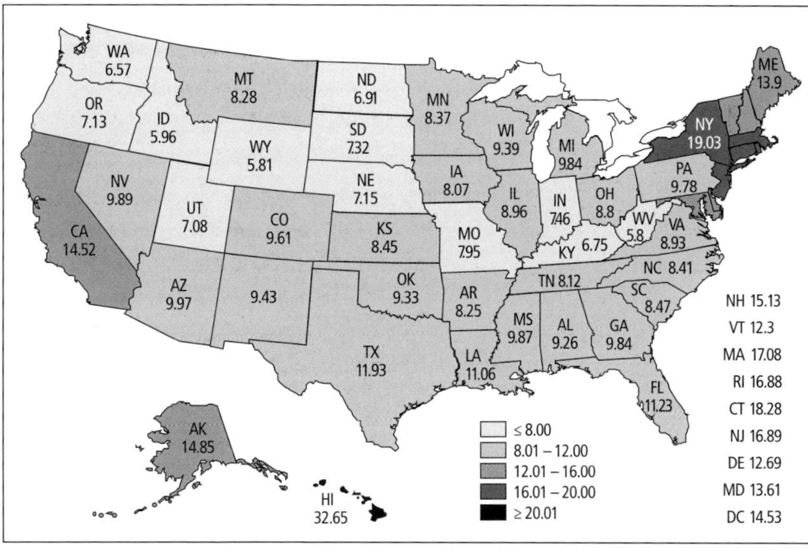

Fig. 12–1. Average retail price of electricity (cents/kWh) (August 2008)
Source: U.S. Department of Energy

Clearly, the market for offshore renewable energy is more attractive in New England and the Mid-Atlantic states. As Babcock & Brown's Calaway suggested, these markets have an ideal combination of strong offshore winds, high-cost electricity, and load situated close to the coast. The coastal states from Maryland to Massachusetts have renewable portfolio standards. The region shares the common geography of the gently sloping Atlantic continental shelf, limited land suitable for construction, weak onshore winds, and overall, less competitive pressure from terrestrial renewable energy resources.

Bluewater Wind's Delaware project illustrates one economic formulation of a U.S. offshore wind project. The project began as a response to a *request for proposals (RFP)* for new generation issued by Delmarva Power, which was required by law to develop additional supply for its *standard offer service (SOS)* customers. The law called for "a more diverse supply at the lowest reasonable cost." Bluewater's bid competed with a gas turbine project and a coal-fired integrated gasification combined-cycle facility proposed by two other energy developers. Previous sharp increases in the cost of fuel that had been passed on to ratepayers caused a consumer backlash. According to a staff report from Delaware's Public Service Commission, "The Bluewater-Delmarva PPA would reduce Delaware's reliance on wholesale market prices dictated by natural gas prices, and accordingly, would dampen SOS ratepayers' exposure to price volatility."[11] So it is not surprising that offshore wind energy, with its promise of price predictability, would be attractive. After a close review, the state regulatory commission directed Delmarva Power to negotiate a long-term PPA with Bluewater. Delmarva Power resisted, arguing that the cost of offshore wind energy would harm customers and could potentially leave the utility stranded with a high-cost contract as its customers migrated to alternative suppliers.

To understand Delmarva Power's opposition to offshore wind energy, a little background on the partial deregulation of the electric industry over the last decade or so is helpful. In Delaware and many other states, traditional vertically integrated utilities have been reorganized into *wires only* utilities that deliver energy at regulated rates. Separate power marketers own generation facilities (or have contracted to purchase energy at wholesale) and offer energy on competitive terms to consumers. Since all the customers of a wires company typically do not elect to be supplied by competing power marketers, the wires companies may retain a residual obligation to provide power. This is known as the *provider of last resort* obligation, and the default utility service, SOS, is regulated by state utility commissions.

A wires company shopping for its SOS customers is wary of entering into a 20- to 25-year fixed-price agreement for offshore wind energy at what may prove to be an above-market price. Purchases from generating plants fired by fossil fuels are often preferred if the state utility commission authorizes a tariff that includes a fuel-adjustment clause or permits pass-through of market-based purchased power costs. In such cases the utility's risk is lessened, because all or part of any increased fuel costs will automatically pass through to customers via higher electricity rates. A fixed-price wind contract, on the other hand, is comparatively risky. Should the cost of traditional electricity based on fossil fuels fall, the wires company's SOS customers could migrate to competitive marketers able to sell power at the lower cost then prevailing in the market due to the drop in fossil fuel costs. Migration would initiate a downward spiral as the high-cost wind energy purchased under the long-term contract becomes an increasing burden on fewer and fewer

SOS customers, who then have an increased incentive to flee to cheaper competitive suppliers. This is the classic *stranded asset* problem.

Wires companies like Delmarva Power object to being forced to pay for public benefits such as energy conservation or pollution reduction if it places them at a disadvantage in the competitive power supply market. In their view, clean renewable energy benefits the population as a whole, and the burden of promoting offshore wind power should not be placed on a subset of customers such as SOS customers. Renewable portfolio standards, which require all power sellers (including the wires companies fulfilling their SOS obligation) to sell a minimum percentage of renewable energy, are a better mechanism for pushing the above-market costs of renewable energy through to all consumers equally. Unfortunately, these programs promote only the least-expensive renewable energy, currently onshore wind. To promote a more diverse selection of renewable energy technologies, REC awards under RPS programs should be adjusted to reflect each technology's different costs.

Recognizing these obstacles, Delaware adopted additional legislation to sweeten the deal for Delmarva Power. The legislation authorized the public service commission to impose an adjustable nonbypassable surcharge allocating costs of the PPA to Delmarva Power's distribution customers.[12] The legislation also awarded Delmarva Power a 350% credit toward the Delaware RPS for RECs acquired from offshore wind farms like Bluewater Wind's.[13]

Granting offshore wind energy multiple RECs per megawatt-hour spreads the cost of supporting offshore wind technology across all participants in the Delaware REC market. Similarly, the surcharge removes the burden of supporting offshore wind power from SOS customers and spreads it evenly across all of the utility's distribution customers, even those purchasing power from competing suppliers. The surcharge gives Delmarva Power the assurance of full cost recovery under the PPA, removing the risk from the utility that electricity market prices might fall below the PPA price in the future. Finally, by making it possible for the utility to enter into the PPA, the Delaware legislation provided Delmarva Power's SOS customers security against future price increases.

The Delmarva/Bluewater PPA reflects an initial energy rate of \$98.93/MWh (i.e., 9.893¢/kWh) and \$15.32 per REC generated by the project. The PPA provides that Bluewater Wind earns payment for the RECs credited to Delmarva Power under Delaware's RPS, which means that the effective REC value of each megawatt-hour of energy generated by the project is \$53.62 (\$15.32 x 350%). All in all, the value of the energy, RECs, and the \$21/MWh PTC produced by the project is $173.55/MWh. The PPA also includes a capacity payment of $70.23/kW-year. This payment applies to 200 MW of wind farm capacity covered under the contract, as adjusted under PJM's capacity rules. Wind farms in PJM are assigned an initial capacity rating equal to 20% of their installed capacity, which rating may be subsequently adjusted according to actual experience. The PJM wind capacity "haircut" means that Bluewater Wind will earn the $70.23/kW-year capacity payment on 40 MW (200 MW x 20%), resulting in an annual capacity payment under the PPA of $2,809,200 ($70.23/kW-year x 1,000 kW/MW x 40 MW). Assuming a capacity factor of 37% and turbine availability of 95%, the value of the capacity payment distributed over the farm's estimated annual production (615,828 MW) is approximately $4.56/MWh. All three payments (energy, RECs, and capacity) are subject to a 2.5% annual inflation adjustment during the contract's 25-year term.

The Delmarva Power/Bluewater Wind PPA payments are low in comparison to payments for offshore wind energy in Germany ($180/MWh, with subsidized transmission) and the United Kingdom ($203/MWh). The European incentives for offshore wind energy may be very generous, or Bluewater may have underestimated its costs of developing the project. It remains to be seen whether the Delmarva/Bluewater PPA becomes the pattern for future offshore wind energy transactions in the United States. Delaware's legislature, responding to customer rate shock and desiring to lessen Delaware's future exposure to volatile fossil fuel prices, arranged a shotgun marriage. Assembling the right combination of motivated politicians, legislation, and incentives to back Delmarva into the PPA could not have been easy for Bluewater Wind. Negotiating offshore wind PPAs in other coastal states will require similar effort. Federal policies designed to promote offshore wind modeled after Germany's feed-in tariff could provide a faster path to building an offshore wind industry.

To date, no offshore wind energy project has been constructed under the unique conditions prevailing in the United States (e.g., hurricanes, U.S. law, U.S.- or foreign-sourced equipment, materials and labor, and U.S. incentives). Thus it is necessary to rely on rough translations of the cost of European projects to estimate U.S. project costs. Comparisons must correct for differences in subsidies, equipment cost and availability, mobilization costs, and other variables. Consider, for example, a wind farm's offshore collection and transmission system, which represents on the order of 20% to 30% of total project costs. In Germany and Denmark, offshore transmission infrastructure and grid reinforcement is paid for by the grid operators (who will pass the cost on to all grid users), not the project developers. Not so in the United States.

A comprehensive study of Rhode Island's wind power options concluded that a price in the range of $97/MWh to $120/MWh, including energy, capacity, and RECs (but excluding PTCs) was needed for offshore wind to provide a reasonable return to investors. Energy, capacity, and RECs sell for approximately $157/MWh under the Delmarva/Bluewater PPA. Terrestrial wind power in the United States has capacity and energy costs of approximately $2,000/kW and 7¢/kWh ($70/MWh), respectively. Studies by the National Renewable Energy Laboratory and others suggest that offshore wind is approximately 50% to 110% more expensive than terrestrial wind and that capacity and energy costs should fall into the range of $3,000 to $4,200 or more per kilowatt, and about 10¢/kWh to 15¢/kWh.[14] These crude estimates illustrate the wide variability in project costs due to siting, project size, shifting commodity costs, and other factors, and the general uncertainty of projections concerning a relatively novel technology like offshore wind energy.

Attack economics

The plain truth is that it will not be possible to really know with a high degree of certainty how much it costs to build offshore wind farms in the United States until a few have been built. However, it is likely that as more are built, the cost will drop over time. In the current environment of uncertainty, economic arguments that purport to prove that offshore wind energy is a beneficial or detrimental societal investment based on finely honed numerical analyses miss the larger point. Renewable energy should be promoted not as a short-term economic winner, but because of the importance of a long-term commitment to increased security in terms of energy supply, economic stability, and environmental/climate protection.

Nevertheless, dealing appropriately with the cost of offshore renewables is crucial, because consumer support for the technology may falter if ratepayers find themselves stuck with costs that appear much higher than market prices. Fear of high costs derailed the proposed 144-MW Long Island offshore wind park, which had otherwise enjoyed substantial public support. Memories of a previous Long Island fiasco, the Shoreham Nuclear Power Plant, lingered. Shoreham was built and tested but never entered commercial service due to concerns about evacuation in case of an accidental radiation release. Ratepayers burdened by the immense costs of Shoreham's construction and decommissioning were wary of perhaps another expensive mistake, and some Long Islanders questioned the economic case behind offshore wind power.

Regulators who will determine whether to issue project permits often apply a public interest standard. In determining whether a public resource (the seabed) should be used for a private project, these regulators will take an economic perspective, considering costs and benefits that would not appear in a purely financial analysis. The Beacon Hill Institute's economic critique of Cape Wind's Nantucket Sound project concluded that the costs of the Nantucket Sound project exceeded the benefits by $209 million (table 12–3). The critique illustrates how skillful economic analysis might be used to kill a project, and the importance of responding adequately to maintain public support.[15]

Table 12–3. Costs and benefits of the Nantucket Sound wind farm

Benefits		Costs	
Reduction in fossil fuel consumption:	$522 million	Project:	$888 million
Capital and operating cost savings:	$104 million	Grid integration:	$26 million
Emission reductions:	$108 million	Environmental effects:	$39 million
Greater energy independence:	$11 million		
Total benefits:	**$744 million**	**Total costs:**	**$952 million**

Source: Haughton, Jonathan, Douglas Giuffre, John Barrett, and David G. Tuerck. 2004. An Economic Analysis of a Wind Farm in Nantucket Sound. The Beacon Hill Institute at Suffolk University. http://www.beaconhill.org/BHIStudies/Windmills2004/WindFarmArmyCorps.pdf

Though a point-by-point response to Beacon Hill's analysis is beyond the scope of this chapter, several observations may help developers to anticipate economic arguments. One benefit of renewable energy is, of course, reduced consumption of fossil fuel. Projections of reduced fossil fuel consumption are based on a proposed wind farm's expected output and assumptions about the type of fossil-fueled generation that would be displaced by the farm's output. Beacon Hill used a conservative 20% capacity rating, assuming that about 91 MW of gas-fired generating capacity costing $500/kW would be avoided by the Nantucket Sound project. While Beacon Hill assumed that generation from a mix of natural gas, oil, and coal would be displaced by the wind farm's output, the cost of inexpensive gas-fired capacity was used to estimate capacity savings.

Beacon Hill minimizes the benefit of pollution avoidance. Under their *marginal emissions* approach, the Nantucket Sound project was estimated to avoid CO_2, SO_x, and NO_x emissions, but particulate emissions were ignored. Rather than taking a representative "slice" of emissions from the generation facilities currently used to supply the region, the economists used projections from the local transmission system operator, ISO-NE, to determine the marginal emissions that would be produced in the absence of the wind farm. The marginal generating resources in New England are, for the most part, relatively clean natural gas plants. Thus the avoided emissions estimates are lower than if the study's authors had assumed that the Nantucket Sound project might also displace emissions from baseload coal-fired plants.

Particulates from coal-fired generating plants have been demonstrated to be a significant cause of asthma and coronary disease. A connection between emissions avoidance and societal savings may be based on studies of human health effects and estimations of the mortality and morbidity attributable to power plants.[16] If wind energy achieves high penetration over the long-term, it could displace electricity made with coal, and society would have additional health cost savings from fewer particulate emissions.

Estimates of the value of avoided CO_2 emissions also appear low. Beacon Hill priced CO_2 emissions at only \$3.90 per metric ton. The cost of 1 ton of CO_2 emissions in European markets was in the EUR 20 to EUR 30 range in 2006. Economists at the Brattle Group expect that CO_2 prices between \$5 and \$40 per ton would be necessary for a future U.S. climate change regulatory program to drive the selection of new baseload capacity towards lower emitting technologies such as natural gas generation, nuclear, and renewables.[17] Future U.S. carbon regulations must cause utilities to adopt technologies with a lower carbon output if they are to have their intended effect. Under different reasonable assumptions, therefore, the value of avoided emissions attributable to the Nantucket Sound project could be many times Beacon Hill's estimates.

A project's contribution to energy independence is another measure of the social benefit of renewable energy. Beacon Hill looked solely at the quantity of oil that would not be burned to fuel electric generating plants because of displacement by wind energy and estimated the per-gallon cost of U.S. military expenditures in the Middle East. Based on that measure, the project's contribution to oil independence is worth merely \$11 million if all the oil saved by the Nantucket Sound project is imported. Yet gas dependence also is a growing issue, and imports of LNG may be used to supply the gas that heats New England homes and fuels its industry. Wind energy helps address both problems.

Moreover, a pure economic calculus misses the scope of the sacrifice that continues to be made to support the nation's oil and gas habit. Cumulative U.S. funding for the war against terror—principally the conflicts in Iraq and Afghanistan—is approximately \$864 billion as of fiscal year 2009, according to the Congressional Research Service.[18] Expenditures in Iraq alone run about \$10 billion per month, and a decades-long military involvement is possible.

In testimony before the Senate Committee on Foreign Relations, Dr. Gal Luft, executive director of the Institute for the Analysis of

Global Security, connects U.S. dependence on oil and gas with U.S. involvement in Middle East conflicts.

> *In November 2001, a barrel of oil was selling for $18. In less than four years the price jumped to $70. This means that Saudi Arabia, which exports about 10 mbd, receives an extra half billion dollars every day from consuming nations and Iran, which exports 2.5 mbd, an extra $125 million. This windfall benefits the non-democratic governments of the Middle East and other producers and finds its way to the jihadists committed to America's destruction as petrodollars trickle their way through charities and government handouts to madrassas and mosques, as well as outright support of terrorist groups... The U.S. is in an odd situation in which it is funding both sides in the war on terrorism. We finance the defense of the Free World against its sworn enemies through our tax dollars. And at the same time we support hostile regimes through the transfer of petrodollars.*[19]

The economic shock of oil surpassing $140 per barrel has without a doubt contributed to the country's present malaise. A crippled economy cannot support a strong military or a prolonged engagement overseas in defense of oil. Advances in renewable energy are part of breaking out of this military/security/economic quandary, and accordingly, projects like that proposed for Nantucket Sound have long-term strategic importance.

The detrimental environmental and economic effects of the Nantucket Sound project also enter into an economic analysis. According to Beacon Hill, homeowners believed the wind farm would reduce their property values by $1.35 billion in the aggregate, and realtors surveyed by were almost evenly split regarding whether values would fall. The mean response indicated a projected loss of approximately 4.6% in home values, leading perhaps to an annual property tax revenue loss of approximately $8.0 million.

It is also unclear whether tourism, which provides 41% of Cape Cod's employment and drives the region's economy, would suffer from the project. Overall, there is general indifference among tourists. An overwhelming 94% of tourists would not alter their future vacation plans on the Cape based on the presence of the wind energy project. After viewing simulated photos of Nantucket Sound with and without the wind farm, only about 5% of tourists either would not visit again or would shorten their subsequent visit by a few days, spending less and harming the local economy. One percent of tourists would stay longer.

The economic value of the visual intrusion of offshore wind farms was studied in Denmark. In a sample where the base case was a wind farm located 8 km (5 miles) from shore, people generally expressed a willingness to pay more in their annual electric bill to have the farm located farther from shore. The experience with Denmark's Nysted wind farm, located 9 to 10 km (6 miles) from shore, bears this out. Its closeness appears to have engendered strong negative perceptions among the population that actually sees the farm regularly, and a corresponding willingness to pay to have had the facility located farther away. The preferred distance for most people appears to be 18 km (11 miles), a distance at which wind turbines are not visually intrusive. Respondents were for the most part not willing to pay substantially more to have a wind farm located out of sight at a distance of 50 km (31 miles) (fig. 12–2). The general rule, therefore, seems rather sensible. People perceive a benefit from reducing the visual intrusiveness of a wind farm, but the benefit decreases with distance from shore. At approximately 11 km (7 miles) from certain points on the surrounding shore, it is understandable that some residents would consider the Nantucket Sound project an unwelcome visual intrusion.

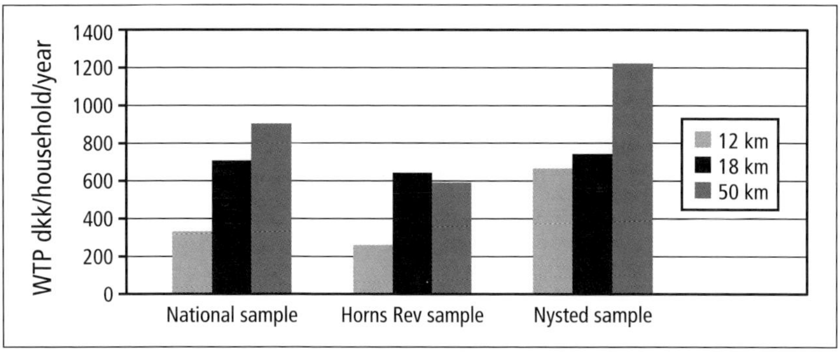

Fig. 12–2. Willingness in Denmark to pay for having future offshore wind farms located at the specified distances from shore—relative to an 8-km baseline

Source: Ladenburg, Jacob, et al. 2005. Economic Valuation of the Visual Externalities of Offshore Wind Farms. Food and Resource Economic Institute. Copenhagen: 15.

Yet it is far from certain that a wind farm would hurt the Cape's tourist economy. An educational center on the Cape that promotes the role of renewable energy and conservation in meeting the climate challenge might draw tourists. Afternoon cruises to the wind farm for a closer look could in time become another Cape Cod diversion. A wind farm might even serve as a catalyst to adopt other environmentally sensitive practices and technologies that would benefit the Cape communities. Beacon Hill's predictions of serious economic harm to the regional economy (that employment in the region would decrease [a loss of 1,173 to 2,533 jobs], earnings would fall [up to $61 million], and economic output would drop [$203 million per year]) seem unfounded.

Not surprisingly, Cape Wind's consultants, Global Insight, project benefits. They indicate that 135 jobs would be directly created in connection with the construction of the project and that operations and maintenance would provide an additional 45 permanent jobs for local residents. When indirect effects are considered, Global Insight believes that construction jobs would increase to approximately 600 to 1,000, and operations and maintenance related permanent employment would increase to 154.[20]

Beacon Hill's critique conveyed the message that subsidies totaling almost one-half of the project's revenues are promoting the environmental degradation of Nantucket Sound and damage to the region's economy.[21] Yet renewable energy subsidies are intended to promote socially desirable results such as expanded markets and reduced costs for cleaner forms of energy. Subsidies to clean energy level the playing field where subsidies also have been provided to other forms of generation. Developers must be prepared to explain and defend the subsidies that an offshore renewable energy project may receive in terms of the project's public benefits.

Hydrokinetic Energy Economics

At first glance, hydrokinetic energy should be more attractive than offshore wind energy. Hydrokinetic energy converters generally float or sit fully or partially submerged in the water, causing little to no aesthetic intrusion. They do not harm birds and seem to have little impact on sea life. Although generalizations are hazardous given the great variety of hydrokinetic machines, preliminary data suggests that they may have capacity factors similar to wind farms—between 20% and 40%, depending on the site.[22]

Ironically, the intensely energetic ocean environment is a key obstacle to the successful adoption of hydrokinetic energy technologies. Machines must withstand tremendous storm forces while achieving a cost of energy that is competitive with other energy options. The tale of the OSPREY, a 750-metric-ton wave energy converter that was destroyed by a storm before it could produce one watt of electricity, provides a lesson for all offshore energy project developers. OSPREY is an acronym for Ocean Swell Powered Renewable EnergY device, and despite its bird of prey namesake, the OSPREY looked more like a large steel crab sitting half-submerged in 14 to 15 m (46 to 49 ft) of water. Waves would flow into the arc of the crab's claws and be directed to a central chamber to drive an oscillating water column–powered 2 MW electric generator. Unfortunately, during installation in August 1995, a storm developed while the OSPREY was only partially ballasted, and large waves broke the OSPREY into pieces.[23]

Wave energy at a site may vary widely. Gentle summer swells with an energy density of 5 kW/m trade places with waves packing 1,000 kW/m of force in a winter storm. The broad potential variance in wave strength challenges designers to build machines that are efficient yet durable. Efficiency in terms of the percentage of energy in a wave that is captured and converted to electricity could be called *performance*. A point absorber designed to be tuned to oscillate at a frequency that coincides with the peak energy period of the prevailing waves embodies a performance improvement. In another example, Wavegen tested its onshore oscillating water column generator, the Limpet, and found that inaccurate seabed surveys and errors in constructing the gully that channeled waves to the Limpet resulted in significantly lower performance than expected. These errors reduced the plant's output to less than 40% of its capacity.[24] The company recognized that its next Limpet design iteration would have to change radically into an offshore machine to increase performance. "It is well accepted that, in the long term, offshore wave farms offer a much greater potential than shoreline stations. Waves lose energy through bottom friction and breaking as they migrate inshore and there is thus more energy available to an offshore device."[25]

Capital efficiency expressed in ratios such as *tons of steel per kilowatt of generating capacity* or *cost per kilowatt of capacity* is another way to evaluate a design. Capital efficiency is likely to be a more important measure than performance for renewable energy devices, because the fuel is free. Overall efficiency is a function of siting as well as design. A design can be made less robust and cheaper if the margin of extra strength needed to survive extreme waves can be reduced. A site with a high ratio of average wave strength to extreme wave strength is desirable, because the machine can be designed to withstand extremes that are closer to normal operating conditions. To illustrate, compare the design needs of two wave energy devices, each intended for a 40-kW/m average wave environment. Site A has storm waves of 150 kW/m and site B has storm waves of 200 kW/m (40/150 = 0.27; 40/200 = 0.20). Site A should permit a more efficient design and a smaller capital investment through, for example, less robust moorings. However, both devices will have similar electrical output, because both would be optimized for average strength waves, not storm conditions.

As is true with wind projects, the cost of the energy conversion components of a hydrokinetic generation device typically does not represent the majority of total project costs (table 12–4). Deployment and installation, moorings and foundations, and electrical works such as cabling and transformers are also significant costs. The cost of energy also is strongly influenced by ongoing project expenses such as maintenance and availability.[26] R&D efforts focused on making technologies more reliable also yield savings by lowering risks to lenders and investors.

Table 12–4. Cost centers and drivers for hydrokinetic energy facilities

Cost Center		Main Drivers	Example Measures and Variables
Capital cost	Structure cost	Material cost Extreme loads	Cost per ton, tons of material
	Mechanical and electrical costs	Rating of the machine (installed capacity)	Peak power output, mean power output
	Moorings	Water depth Tidal range Storm conditions Compliance	Cost per unit length
	Installation	Type and availability of vessels required Distance to port Time needed for installation Weather delays	Charter rates
	Grid connection	Power transmission level Distance to shore	Cost per mile
	Project management	Insurance Permitting	Proportion of the total capital cost
Operating cost	Planned maintenance	Cost of replacement parts Component design duty and known service intervals Time to complete service Distance to port Weather delays	Cost per kWh generated
	Unplanned maintenance	Cost of replacement parts Time to complete service Weather delays Personnel costs, including standby costs	Cost per kWh generated

Source: The Carbon Trust. 2006. Cost Estimation Methodology; The Marine Energy Challenge Approach to Estimating the Cost of Energy Produced by Marine Energy Systems: 17.

Lowering *balance of plant* costs (i.e., the costs other than the turbine/generator) was the impetus behind the Tidal Wing, a unique tidal range energy technology designed by Halcyon Marine Hydroelectric. The Tidal Wing uses modular, pile-supported construction to assemble precast minicaissons into the walls that enclose a small section of an estuary or bay. This construction technique, which uses less massive walls than traditional barrage construction, is claimed to reduce materials costs and construction time dramatically. Any advance that reduces the duration of a project's construction phase lowers risk by reducing exposure to weather delays and saves labor costs, equipment rental expenses, and interest expense during construction.

Accelerating technological progress

Improving the balance between performance and cost was the purpose of the Marine Energy Challenge (MEC), a program of the Carbon Trust, a U.K. government–sponsored company established with the goal of reducing carbon emissions. The MEC evaluated various wave and tidal energy technologies and found that the estimated cost of energy of an initial wave energy farm would be between 12p/kWh and 44p/kWh. Central estimates in the range of 22p/kWh to 25p/kWh represent the typical costs today. Tidal stream energy generally had a lower cost, estimated in the range of 9p/kWh to 18p/kWh, with central estimates in the range of 12p/kWh to 15p/kWh.[27] Converting MEC estimates into U.S. dollars (1 GBP = 1.97 USD), the central estimate for wave power is approximately \$0.43–\$0.49/kWh, and \$0.24–\$0.30/kWh for tidal energy.

According to the MEC, the installed cost for a wave farm ranges from £1,700/kW to £4,300/kW, while tidal stream farms range from £1,400/kW to £3,000/kW. Tidal range energy projects, which rely on the potential energy in a hydraulic head, much like traditional hydroelectric dams, may be less costly. Halcyon Marine Hydroelectric estimates that a Tidal Wing tidal range energy plant with a capacity of 3,740 MW can be built for approximately \$6 billion. At a location in the Bay of Fundy with a tidal range of 12 to 14 m, and running at a capacity factor of 35%, a Tidal Wing could be expected to generate approximately 10 million MWh annually. Such a plant could produce power for \$53/MWh to \$84/MWh, depending on capital costs, according to the company, which concludes that "hydroelectric Tidal Wings in the Bay of Fundy [would] provide large scale power at lower cost than new wind, coal, natural gas, or nuclear power."[28]

The MEC analysts predicted that wave and tidal stream energy will remain expensive until hundreds of megawatts of installed capacity have driven costs down through design optimization, economies of scale, and learning in production, construction, installation, and maintenance. In an effort to reduce the subsidies needed to attain this capacity level, whether in the form of direct government payments or higher energy prices paid by consumers, the British are looking to boost the learning rate. The learning rate for wave energy is presently about 10%–15%. This means that with every doubling in installed capacity, the industry should see a 10%–15% cost reduction. Tidal energy seems to have a lower learning rate, perhaps because the turbine design used in many in-stream tidal energy devices is somewhat derivative of wind power technology. T. W. Thorpe, a British wave energy researcher, described the progression of learning and cost reduction as an iterative process of design, testing, problem identification, costing, and redesign, with each iteration of design leading to a lower cost, more practical machine (fig. 12–3).[29]

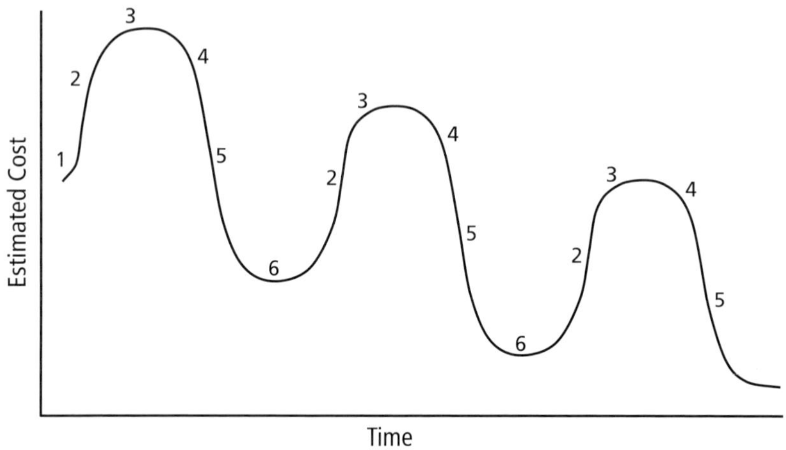

Key

1 Initial idea – looks promising
2 Idea researched, problems identified, predicted cost escalates
3 Design fully worked out, predicted cost too high
4 Radical design change or new approach
5 Changes lead to a reduction in predicted cost
6 New idea looks promising and so adopted – back to stage 2

Fig. 12–3. The iterative design process
Source: Thorpe, T.W. 1999. A Brief Review of Wave Energy*: 163.*

The European Marine Energy Centre is one place where Thorpe's design-redesign cycle happens at a fast pace. Located in Orkney, a group of islands in northern Scotland, the center hosts a technology accelerator containing marine laboratories and test platforms where devices can easily be connected for field trials. It is the hub of an intellectual community, populated with university researchers, private technology developers, and engineering and construction firms, that fosters collaboration and progress in hydrokinetic energy technology development. The U.K. Sustainable Development Commission has called the center "an impressive example of using public funds to create a generic resource to support and stimulate private sector investment."

Fortunately, the United States also is recognizing that the collaboration that occurs at such centers is an ideal way to drive development forward quickly. In 2008, DOE funded grants to the newly established National Renewable Marine Energy Center at the University of Hawaii in Honolulu and the Northwest National Marine Renewable

Energy Center at Oregon State University. The centers will advance the commercial viability, cost-competitiveness, and market acceptance of new marine renewable energy technologies by combining research in marine science, wave tank facilities, energy systems laboratories, and facilities for field experimentation. The latter would be available on a fee basis to private industry groups that want to test their technology. In the absence of these facilities, a developer seeking to test a hydrokinetic energy device in the United States would first need to obtain numerous federal, state, and local permits, spending funds on paperwork that could be better spent on engineering.

The Carbon Trust's MEC program was followed by its Marine Energy Accelerator project to further drive down the costs of hydrokinetic energy technologies. Deployment and handling costs were identified as an area in need of improvement. Savings here also lower maintenance and refitting costs (i.e., unit overhaul expenses). Deployment and handling costs are strongly influenced by the type of vessels required to move a device to the site and complete the installation, Modular designs easily installed and removed from the field without expensive specialized equipment should reduce installation and overhaul costs. Specialized vessels such as jack-up barges and heavy-lift vessels may have limited availability (extending downtime), and are typically costly to hire. These costs may then be magnified by weather delays. Electrical connectors that are easily mated and unmated underwater are one innovation of the oil and gas industry that has made cost-effective modular wave energy machines more practical. The Marine Energy Accelerator project sought designs suitable for towing behind a common tug and, once on site, able to be rapidly connected to mooring buoys and electrical cables. Improvements in materials and designs that resist corrosion and fouling provide other cost reduction opportunities (table 12–5).

The British hydrokinetic technology accelerator programs are driving down costs early in the development cycle, and the newly funded U.S. marine renewable energy centers are poised to do the same. These are smart investments from a societal perspective because they bring forward the point in time when a technology can compete without expensive public subsidies. The maturation of wind turbine technology provides a similar cost reduction example. Global installed wind turbine capacity increased from approximately 10 MW in 1980 to over 50 GW in 2005. During that time, wind power's cost per kilowatt-hour decreased from more than 80¢ to typically 6¢ to 9¢, and in some cases, to 3¢ to 4¢ at the best sites.[31] That trajectory is possible for hydrokinetic energy technology, but 25 years is much too long to wait. The urgent need to address climate change requires that the new technology accelerators receive a full measure of government and private sector support.

Table 12–5. Hydrokinetic energy device design considerations

Component	Design Principal
Absorber structure or other housing for the components that react to the wave or current and collect its energy	The machine's structural elements should be made of steel, concrete, or other relatively inexpensive materials and be simple enough to assemble locally near the project site. Testing should demonstrate the ability to survive in severe ocean conditions. Designs should minimize environmental issues such as entrainment of fish and larvae, potential hydraulic oil leakage, use of toxic anti-biofouling compounds, and seabed disruption.
Power take-off which converts the primary wave or current energy captured by the absorber structure into electricity	Power take-off should be simple, reliable, and efficient. Design for easy and infrequent maintenance to minimize down-time.
Mooring system or foundation which secures the floating or bottom-mounted energy harvester in place	Moorings and foundation systems should be robust and flexible to permit customization for different depths, wave environments, and seabed conditions. Deployment and recovery should be rapid and require minimal use of specialized equipment.
Electrical collection and transmission system which interconnects multiple energy harvesters to a transmission cable to shore, possibly through an offshore transformer station	Components should effectively feed power to the grid. The ability to smooth power output, perhaps with storage features, to maintain a high capacity factor is desirable.
Communication, command and control system which enables remote monitoring and control of each offshore energy machine	Employ remote monitoring and control to facilitate preventive maintenance.

Source: Previsic, Mirko, Omar Siddiqui and Roger Bedard. 2004. EPRI Global E2I Guideline: Economic Assessment Methodology for Offshore Wave Power Plants. http://oceanenergy.epri.com/attachments/wave/reports/002_Rev_4_Econ_Methodology_RB_12-18-04.pdf

Colocation and Combination Facilities

Colocating offshore renewable energy facilities with other offshore facilities provides another path to lower costs. As previously discussed, expensive offshore transmission interconnections have a lower per-unit cost when overall project size is increased. This thinking is behind a novel proposal that would combine a wind farm in the Irish Sea with an offshore natural gas production operation. The offshore gas field would be connected to an offshore gas turbine generating station as well as a 30-turbine wind farm. Energy from the wind farm and gas turbine would flow to shore along a common transmission cable. "As well as sharing common infrastructure and providing operational savings, the addition of gas fired generation capacity within the project will introduce predictability of power generation and revenue, countering the problem of intermittency typically associated with wind power."[32]

U.S. developers also have explored synergies between offshore oil and gas facilities and renewable energy. Some 3,900 platforms in the Gulf of Mexico service tens of thousands of oil and gas wells, and those located close to land may have electrical transmission cables in place. Other platforms are connected to old oil and gas pipelines that could be used as conduits for electric transmission cable. Wind Energy Systems and Technologies has proposed to mount wind turbines on platforms off the Louisiana coast that were no longer producing oil or gas. Platform owners would benefit by delaying the cost of decommissioning and removing their platforms. Despite its merits, the project has not advanced, possibly because Louisiana's below-average electricity prices and lack of an RPS make offshore wind power uneconomical in that market.

Some aspects of building wind and hydrokinetic energy farms are similar, and combination projects may succeed where individual projects may not. Opportunities for shared moorings and foundations, and communications and control infrastructure, should be explored. Wind turbine foundations or floating platforms could serve as electrical terminals for wave energy harvesters positioned nearby. Post-installation efficiencies in financing and maintaining offshore wind, wave, and tidal energy facilities also are possible. Although combining technologies would narrow siting options, this should not be an insurmountable obstacle given the breadth of the offshore environment.

A second generation design for the OSPREY oscillating water column plant previously mentioned incorporates a wind turbine mount into the OSPREY's reinforced concrete foundation. With a design life exceeding 60 years, the OSPREY's concrete structure has been considered for use in series as a harbor breakwater. A 1999 study of wave energy devices estimates that "by using the OSPREY2 as a base, approximately 31% of the [wind turbine] capital costs can be avoided. In addition, nearly 50% of the electrical costs can be saved by using the electrical transmission lines for the OSPREY2. Therefore, the predicted marginal cost of installing a wind turbine on an OSPREY2 is just over £ 1M."[33]

Wind and hydrokinetic energy facilities are likely to have different, and perhaps complementary, energy output characteristics. Ocean swells generated by distant storms can be reliably predicted several days in advance. Joint wind and wave predictions should allow the operators of combined facilities to better schedule output. With less intermittency, capacity credits earned by a mixed offshore renewable energy system may be higher and, with fewer spikes in output, the costs of balancing the regional transmission grid would be lower.

Offshore Renewables Compared to Alternate Energy Sources

Old notions of the costs of competing supply alternatives are beginning to yield to a new reality as governments adopt programs that to varying extents require industry to internalize the costs of carbon emissions. If offshore renewable energy technologies are ever to become a major energy source, they must outgrow subsidies and match other energy alternatives on a level playing field. Possible alternatives include nuclear power, fossil-fueled generation with carbon capture and storage/sequestration (CCS), fossil-fueled generation that pays for carbon emissions through a tax or the purchase of carbon credits, and alternate renewable energy technologies such as landfill gas, biomass, and solar energy.

Nuclear power is now experiencing a renaissance as its proponents see an advantage in an economy where carbon costs are internalized. John Rowe, CEO of Exelon, a utility with several nuclear plants, doubts that a pulverized coal plant could be built today at less than the cost of a nuclear plant, if the coal plant is built to incorporate CCS technology.[34] Since a large-scale plant using CCS technology has not yet been built, no one can say with certainty what it would cost. A proposal by Duke Energy Indiana and Vectren indicates that it would be costly. Their proposed 630-MW plant would have used gasified coal technology (the first step to separating the carbon dioxide for sequestration) and cost $2 billion ($3,175/kW). The proposed plant does not feature CCS, which would increase the capital cost even more and reduce operating efficiency.[35] Vectren subsequently abandoned the project after finding a variety of other avenues to satisfy its need for additional capacity, including conservation, wind energy, gas peaking plants, and purchased power. The FutureGen project, a proposed coal-fueled 275-MW power plant designed to produce near-zero emissions with CCS technology, is estimated to cost $1.8 billion ($6,545/kW) to construct.[36]

Given the high cost of cleaner coal-based generating technologies, Rowe claims that nuclear power provides a good value, with cost estimates in the range of $2,500 to $4,000/kW. There is good reason, however, to doubt Rowe's estimates. The pool of qualified nuclear equipment suppliers and skilled workers has shrunk due to the lack of new plant construction during the last two decades. France, which derives 76% of its electricity from nuclear power and has built new plants in recent years, should be expected to have well-developed nuclear capabilities. Yet French nuclear vendor Areva recently faced cost overruns and delays at a new plant it is building in Finland. Areva's loss on the Finnish contract is expected to fall between $700 and $900 million. The plant, which was supposed to take four years to build, is now 18 months behind schedule. According to one analyst, "The company had underestimated what it would take to reactivate the global supply chain."[37]

Interest expense during a construction period of six years or longer, and the trend in increased materials costs, has led others to predict much higher costs for new nuclear capacity. The *Wall Street Journal* reported that new U.S. nuclear plants could cost $5 billion to a stunning $12 billion each, "double to quadruple earlier rough estimates."[38] FPL Group is planning to add two new units at its south Florida Turkey Point plant that it expects will cost from $6 billion to $9 billion each. The Westinghouse reactor design that FPL proposes to use has a capacity of 1,154 MW, which suggests a cost of $5,199/kW to $7,799/kW. A GE reactor design estimated to cost $12 billion per unit was rejected by FPL. Progress Energy announced that its proposed two-unit, 2,200-MW nuclear plant in Florida would cost between $15 billion and $17 billion, about triple the company's earlier $5 billion estimate. This new estimate puts the cost between $6,818/kW and $7,727/kW.[39] The utilities are proposing that consumers, not shareholders, shoulder most of the risks associated with building the new generation of nuclear plants. Florida Governor Charlie Crist defends the nuclear projects, arguing that by shielding consumers from the rising costs of oil and gas, consumers will ultimately come out ahead.

At about $3,000/kW to $4,200/kW, offshore wind shares the virtue of shielding consumers from escalating fossil fuel costs and is only one-half as expensive as nuclear power on a capacity cost basis. Nuclear plants can be expected to run at a 75%–85% capacity factor, while offshore wind capacity factors are in the 40% range.[40] A study by the Georgia Institute of Technology of six years of offshore wind data off the coast of Georgia indicated that average annual wind speeds of 7.13 m/s could power a 2-MW wind turbine to a 40.5% capacity factor. This same study indicated that the capacity factors of comparable offshore wind farms in Europe varied from 22.8% to 45.6%.[41] Offshore wind energy appears competitive with nuclear power on a capacity basis alone, even before comparing nuclear operating and waste disposal costs with offshore wind power's O&M expenses. (Nuclear power's fuel cycle cost has been variously estimated at between 0.5¢/kWh and 2¢/kWh.) Offshore wind capacity also can be built faster and in smaller capacity chunks, exposing investors, consumers, and the government (that will be asked to provide both nuclear and offshore wind loan guarantees) to less risk.

DOE has estimated the costs of various generating technologies, many of which are new and therefore involve significant doubt regarding their ultimate cost (table 12–6). Technological uncertainties, siting difficulties, permitting and construction delays, and supply chain mobilization challenges are among the risks common to many of these alternatives. An accurate comparison of the relative cost of energy produced by each new technology will only be answered after several units (or farms) featuring each of the technologies are built in the United States.

Table 12–6. DOE estimates of the cost of competing generation technologies

Technology	Size (MW)	Total Overnight Cost in 2007 (2006 $/kWh)*	Variable O&M ($2006 mills/kWh)**	Fixed O&M ($2006/kW)	Heatrate in 2007 (Btu/kWh)
Scrubbed coal new	600	1,534	4.46	26.79	9,200
Integrated coal-gasification combined cycle (IGCC)	550	1,773	2.84	37.62	8,765
IGCC with carbon sequestration	380	2,537	4.32	44.27	10,781
Conventional gas/oil combined cycle	250	717	2.01	12.14	7,196
Advanced gas/oil combined cycle (CC)	400	706	1.95	11.38	6,752
Adv CC with carbon sequestration	400	1,409	2.86	19.36	8,613
Conventional combustion turbine	160	500	3.47	11.78	10,833
Advanced combustion turbine	230	473	3.08	10.24	9,289
Fuel cells	10	5,374	46.62	5.5	7,930
Advanced nuclear	1350	2,475	0.48	66.05	10,400
Distributed generation base	5	1,021	6.93	15.59	9,200
Distributed generation peak	2	1,227	6.93	15.59	10,257
Biomass	80	2,809	6.53	62.7	8,911
Municipal solid waste - landfill gas	30	1,897	0.01	111.15	13,648
Geothermal	50	1,110	0	160.18	35,376
Conventional hydropower	500	1,551	3.41	13.59	N/A
Wind - terrestrial	50	1,434	0	29.48	N/A
Wind - offshore	100	2,872	0	87.05	N/A
Solar thermal	100	3,744	0	55.24	N/A
Photovoltaic	5	5,649	0	11.37	N/A

* Total overnight cost excludes interest charges during plant construction.

**The standard industry measure is mills per kilowatt hour (mills/kWh; a mill is $0.001).

Source: US Department of Energy, Energy Information Administration. 2008. Electricity Market Module, Report # DOE/EIA-0554(2008). Table 38. Cost and performance characteristics of new central station electricity generating technologies.

Coal-fueled plants (with CCS) and nuclear plants also have issues other than high cost. The preferred method of CCS is to inject carbon dioxide into oil- and gas-bearing rock formations that have been largely pumped dry. Once injected, the carbon dioxide forces additional oil and gas from the porous rock. CCS technology is location dependent. Rock formations that would make for viable carbon storage sites in the eastern United States are not plentiful. The lack of suitable storage suggests either remote generation with large new investments in electrical transmission, or generation near load with long-distance transportation of carbon dioxide gas to appropriate geological storage formations.

In addition, even if success is achieved in cleaning coal-fueled power plants, the energy cycle of coal-based electricity remains environmentally destructive. CCS technology does not alleviate the ecological damage, pollution, and waste caused by coal mining and coal residue disposal. Coal producer Massey Energy Company, for example, recently entered a consent decree and agreed to a civil penalty of $20 million related to EPA claims that the company's West Virginia and Kentucky coal mining operations polluted the drinking water of Central Appalachia by illegally pouring pollutants into waterways numerous times within the last six years.[43] Constellation Energy Group paid $54 million to settle a class action suit alleging that coal ash it had disposed of in a quarry contaminated certain drinking water supplies to Maryland homes.[44]

New nuclear plants are likely to be located at a distance from population centers and will require additional transmission investment to move the power generated to the most congested markets. The proposed new nuclear plants for which license applications have been filed with the Nuclear Regulatory Commission, with the exception of one plant to be located in Maryland, would all be located in southern states (e.g., Virginia, North Carolina, South Carolina, Georgia, Mississippi, Alabama, and Texas). These plants do not seem to be a viable solution for densely populated East Coast and West Coast regions. Offshore renewable energy projects, on the other hand, could be built relatively close to populated coastal load centers. The issue of safe nuclear waste disposal likewise remains a challenge after decades of effort.

Nuclear proliferation is a further concern. A nuclear energy–driven solution to global warming would require the dissemination of nuclear fuel and related fuel handling and processing technology to countries around the world. Developing countries will assert their self-interest in energy independence and will not welcome nuclear generation if it leaves them vulnerable to western countries for fuel supplies. Many will aspire to own nuclear fuel processing, enrichment, reprocessing, and fuel fabrication capabilities and facilities. These facilities will make enriched nuclear materials, including plutonium that can be used for weapons, more widely available and more difficult to secure. This is an intractable social and political problem. Nuclear security is a considerable challenge in the United States, even with its well-developed public and private institutions. To expect developing country governments to adequately shoulder the burden of securing the nuclear fuel chain for peaceful purposes is probably more than one can reasonably expect.

A 2003 MIT study, *The Future of Nuclear Power*, opined that "the potential impact on the public from safety or waste management failure and the link to nuclear explosives technology are unique to nuclear energy among energy supply options. These characteristics and the fact that nuclear is more costly, make it impossible today to make a credible case for the immediate expanded use of nuclear power."[45] In contrast, renewable energy provides employment in the manufacturing and maintenance of numerous dispersed energy collectors, an energy/economic model much better suited to the developing world. If the United States joins with Europe to develop offshore renewable energy technology and succeeds in lowering its cost, the developed nations will have created a technology that can be fearlessly exported to the developing world to address global warming and promote economic growth.

Promoting Offshore Renewable Energy

The many advantages of offshore renewable energy, as compared with traditional energy technologies, have been described over the course of this book, yet the United States has no coherent national policy to guide the expansion of offshore renewables (table 12–7). This is partly out of a concern that the government should not "pick winners and losers" through its policies but rather let the market decide. In practice, the government's current haphazard policies often do not provide an effective stimulus to promising technologies. Further, as long as the externalities of the various traditional energy technologies remain uncorrected, a free-market approach will not work. The long-term interests of the United States require the implementation of a suite of renewable and low-emissions conventional technologies. Yet we have handicapped the development of these critical technologies by hooking them onto wildly swinging oil and natural gas markets and short-term, unreliable government incentive programs.

Table 12–7. Comparative strengths of offshore renewable energy

Level cost	Freedom from fuel price variability insulates customers from rate shocks over the long term.
Security	Dispersed renewable energy facilities are not subject to terrorist strikes, oil or gas embargoes or cartels, coal strikes, pipeline disruption, competition with other uses for fuel (e.g., gas heating), or water shortages. Renewables avoid the need for military intervention to protect access to fuel.
Job creating	Renewable energy provides more jobs per MWh than any other form of energy.
Modularity	Offshore renewable energy projects can be scaled to various sizes and built in increments, and the construction period is short (though environmental review may be prolonged).
Light environmental footprint	Offshore renewable energy projects do not generally alter the environment substantially and the environment should rapidly revert to its original condition once the facility is removed.
No carbon emissions	Offshore renewable energy helps to address global warming and is not subject to uncertain future carbon emissions regulations or the uncertain cost of carbon capture and sequestration.
No fuel; no fuel cycle	Environmental damage from fuel production, transport, combustion, and waste disposal is avoided.
No cooling water use required	Traditional thermal energy technologies require access to increasingly precious water resources for cooling.

The twin challenges of energy insecurity and climate change are too urgent, and the potential contribution of offshore renewables is too great, to muddle through with ineffective public policy. In the words of U.S. Energy Secretary Steven Chu, a Nobel-winning physicist, "We essentially need a second Industrial Revolution that can generate lots of energy cleanly, cheaply, sustainably."[46] Delay merely limits the options that will be available to the United States for addressing these threats, and most likely will increase the cost of any actions required in the future.

In 2005, *A Framework for Offshore Wind Energy Development in the United States* identified the need for a "focused, coordinated approach to planning, research and development, and policy development for this new industry."[47] The *Framework* set out objectives in the areas of: (1) technology development; (2) environmental compatibility; (3) economic and financial viability; (4) regulation and government policies; and (5) leadership coordination. Some progress toward these ends already has been achieved, but much more is needed. Following are a few specific actions to advance offshore renewable energy in the United States.

Expand energy research and development, especially collaborative efforts

Pre-commercial stage renewable energy technologies should receive research and development support through collaborations among industry, academia, and government. This is the intent behind the marine energy centers recently established in Hawaii and Oregon. These facilities save money for the private sector and government by providing sites for testing and access to expertise while they speed learning. Faster technological development lessens the aggregate subsidies required over the long-term.

A key recommendation of the *Framework*, to limit aesthetic objections and to gain access to areas of stronger winds, is to advance technologies and policies that could permit wind harvesting on deeper sites farther offshore. Finding practical designs for floating offshore wind platforms and addressing the high cost of offshore transmission are part of this challenge. The National Renewable Energy Laboratory has funded floating platform design studies. Prototypes now need to be built, tested, and improved. MMS should relax permitting requirements for field test installations.

Nuclear energy has long benefited from extraordinary levels of public investment. David Lilienthal, the first chairman of the Atomic Energy Commission, observed of nuclear power, "Never had there been, anywhere in the world, so broad and generously financed an attack on the technical and economic problems of bringing a new scientific discovery into the realm of application and of widespread use."[48] A similar, genuine commitment to advancing offshore renewable energy technology through cohesive policy and financial support is required.

Remove regulatory barriers

A broad array of federal and state regulations and a mix of incentives and economic conditions drive development decisions. The regulatory process is unpredictable and full of delay. More predictability in permitting is needed. All stakeholders would benefit from a deliberate process that proceeds at a reasonable pace to resolve project issues. Unanticipated delays raise costs without any real public benefit. Since environmental disputes are often at the center of these delays, the government could help by organizing coordinated long-term environmental studies of the effects of offshore renewable energy that reflect the participation of a broad group of stakeholders, including environmental groups, government agency scientists, and industry participants. As knowledge of the environmental effects of offshore renewable energy technology grows, the pace at which projects can be permitted should increase.

Energy market prices are set in many parts of the country by the marginal cost of gas-fired generators. The decision whether to invest in an offshore wind project with a 20- to 25-year life (or to enter into a long-term wind energy PPA) is based on a guess about the future price of gas. Yet in many places, utilities can automatically pass through fuel costs to consumers (either because they are included in the market price of purchased power or through a fuel cost adjustment clause in a utility tariff). Accordingly, utilities lack an incentive to control cost increases. In such a regulatory environment, the value of renewable energy as a hedge against fuel price volatility will be underemphasized in utility power purchasing decisions.

The economic fate of the nascent offshore renewable energy industry should be unlinked from natural gas prices. Just as it makes little sense to start and stop projects to develop electric vehicles as the price per gallon of gasoline cycles from $5 to $2 and back again, it also makes little sense to tie progress in the development of offshore renewable energy to volatile and cyclical natural gas markets. This is especially so since the natural gas market is becoming increasingly international and may in the future become vulnerable to manipulation by cartels.[49] Utility regulatory commissions should reconsider the perverse incentive created by fuel cost pass-through tariff clauses and take action, as Delaware has, to shield consumers from volatile electricity prices. A federal feed-in tariff, discussed in the following section, could uncouple offshore renewable energy facility investment from gas market price projections.

Rationalize the incentive system by implementing a federal feed-in tariff

The one-size-fits-all, intermittent PTC, and the uncertain value provided by state-by-state REC programs, should be replaced with a single federal feed-in tariff. A feed-in tariff would provide a long-term government-guaranteed floor price for renewable energy that would be much better suited to supporting long-term project debt. Like Germany's feed-in tariff and the United Kingdom's "banded" renewables obligation, the rate paid under a new federal feed-in tariff should be tailored to the cost structure of the several renewable energy technologies that are ready for commercial use. Offshore wind, for example, would earn higher tariff payments than onshore wind. The price offered under a feed-in tariff could be set through a federal auction where offshore wind project developers offer to provide a portion of the amount of energy desired under national renewable energy goals. The auction would introduce market discipline to the subsidy system to ensure that the public is not overpaying for renewable energy.

Currently, states without RPS programs free ride on the benefits provided by the RPS programs adopted in the other states. A federal feed-in tariff should provide a fair and consistent subsidy that spreads the burden of supporting clean energy across all the states. Appropriate price signals under a feed-in tariff would stimulate the large-scale deployment of renewables necessary to move down the experience curve, allowing government to progressively reduce the tariff subsidy. Figure 12–4 illustrates the reductions possible in the cost of offshore wind generating capacity with a learning rate of 10% and three doublings of global capacity between 2008 and 2018, according to MMS.

A feed-in tariff could be funded by tax revenues now lost to the PTC and with a national requirement on all electricity retailers to purchase a certain percentage of their energy supply under the feed-in tariff. The cost of the incentive provided through the tariff would in effect be shared among all electricity consumers in proportion to consumption, similar to a national RPS. Most importantly, a feed-in tariff would correct the boom-and-bust cycles that occur under the PTC and demonstrate long-term commitment to the offshore renewable energy sector. This would signal manufacturers to expand the supply chain and start a beneficial cost reduction cycle through economies of scale and learning effects.

Invest in renewable energy transmission

The current transmission grid was built to connect traditional generating resources. Renewable energy resources often are located in remote areas and require expensive interties to the existing grid.

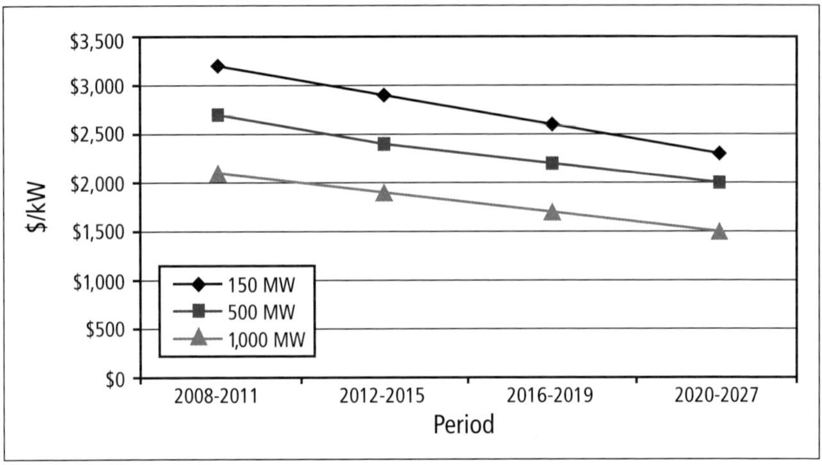

Fig. 12–4. Projected capital costs (per kilowatt) for U.S. offshore wind energy capacity for three wind farm sizes (US$[2007])
Source: U.S. Department of the Interior, Minerals Management Service. 2007. Fiscal cost-benefit analysis to support the rulemaking process for 30 CFR 285 governing alternative energy production and alternate uses of existing facilities on the outer continental shelf. OCS Study MMS 2007-050: 15, table 2–3

Government and industry should collaborate to identify the richest terrestrial and offshore resource areas, promote those areas for clustered renewable energy development, and build transmission trunk lines into such areas to efficiently connect renewable energy projects to the grid. Trunk lines should be considered network facilities, and their cost should be allocated across all users of the grid.

Industry, MMS, FERC, and the states should coordinate planning for offshore project siting on a regional basis to make it possible to cluster projects along one or more offshore transmission trunk lines. Shared costs and more efficient use of transmission capacity would result. FERC should promote the construction of offshore transmission trunk lines by classifying them as network facilities as it has done for the California wind energy trunk line project described in chapter 10.

Require traditional generation to internalize its pollution costs

The *Framework* authors saw offshore wind energy as vulnerable to, and perhaps benefited by, several factors.

> *Although today's costs of offshore wind energy are higher than onshore, expectations are that several factors working together will make offshore wind energy resources more cost effective. These factors include technology innovations, stronger wind regimes, economies of scale from*

> *large-scale development, close proximity to high-value load centers, and incentive programs responding to the public's growing demand for clean energy. Other influential factors are the uncertain price and supply of conventional energy sources, especially natural gas and oil, and the increasing regulatory pressures on emissions reductions by fossil fuel generation plants. The economic fate of offshore wind energy, therefore, rests on a combination of internal and external factors.*[50]

Government should restrict or tax greenhouse gas emissions and require traditional electrical generation to absorb an increasing fraction of its pollution and other fuel cycle costs. When traditional generation loses its pollution subsidy, the competitive playing field will become flatter, and the need for renewable energy subsidies will decrease.

Entrenched interests involved in traditional fossil-fueled generation will resist these new policies and claim that the nation cannot afford to abandon its existing generating infrastructure, or to lose the jobs connected to the old ways of generating power. These arguments for delay are likely to result in more wrenching changes later, as increasingly drastic measures will be required to adjust to climate change. If instead the United States begins now to implement coherent policies to promote a sustainable energy future, dislocation in the traditional generating sector will be gradual, since the transformation to clean renewable energy will take decades. Old polluting plants can be retired as they reach the end of their useful lives.

Many new jobs also will be created in the transformation to renewable energy, which will increase domestic employment in engineering, manufacturing, construction, maintenance, and many ancillary fields. In contrast to spending on imported fuels, money invested in offshore renewable energy projects will have a powerful local multiplier effect, since project evaluation and site work would, of course, be done in the United States, and much of the manufacturing also would be located in the United States. Efforts to promote offshore renewable energy could create a sizeable industry in the United States that could in time expand into substantial markets overseas.

Disruptive change

Offshore wind power and hydrokinetic energy are intermittent, variable, and at present, best suited for certain niche applications (e.g., wind energy in the high-cost Northeast and Mid-Atlantic states, and wave energy in the Pacific states). Not surprisingly, some take the view that it is just not worth the effort to develop offshore renewable energy projects that face the perils of winter storms and hurricanes, NIMBY opposition, uncertain public subsidies, and the general unwillingness of utilities and power marketers to purchase the output under long-term contracts.

In *The Innovator's Dilemma*, Clayton Christensen recounts instances of disruptive technological change that caused well-managed companies to fail. These companies seemingly did everything right. They listened to their customers' requests for more reliable, less-expensive, faster, more powerful devices, and to investors clamoring for growth in sales and profitability. They pursued large markets with existing technologies and employed their deep capabilities to develop newer, better, and cheaper iterations of their highly profitable existing products. But because they need large investments to significantly grow earnings, large companies often ignore small markets, underestimate the potential of transformative technologies, and get caught flat-footed during periods of rapid change.

Christensen's observations about corporate behavior also could apply to countries. Policy makers stick to what they know, follow the lobbying of entrenched interests, and change only around the margins. Hence the United States spends more on "clean" coal research, more on nuclear energy subsidies, and rallies behind chants of "Drill, baby, drill." Yet the dangerous energy dependence that President Nixon warned of decades ago has only worsened and now infects U.S. economic health and national security. When stalwart oilmen join with environmentalists to call for an end to business as usual, something transformative is in the air. Billionaire oilman T. Boone Pickens recommends an immediate sizeable expansion of wind energy to displace natural gas used in electricity generation. Pickens argues that the United States could use natural gas as a transportation fuel, divert funds spent on foreign oil into local investment, and create "more than 3.5 million jobs over a 10-year period."[52] There are no technical barriers to this path, merely a barrier of will.

Ukrainian and Russian brinksmanship over gas transiting Ukraine cut supplies to Europe, leaving many in the cold this past winter. It was not the first time supplies were interrupted. Scarce indigenous fossil fuel resources and a dangerous vulnerability to unreliable Russian and Middle Eastern supplies has driven many European countries to emphasize energy efficiency and supply diversity, including renewable energy, to reduce energy dependence. High prices for fuel and electricity, rather than hurting Europe's economy, have spurred demand for new energy solutions. Led by the United Kingdom and Denmark, Europe is ahead in offshore renewable energy technology and jobs. Farsighted government policies, and collaboration between industry and academia, are pushing these technologies down the experience curve and, as costs drop, it is likely that offshore renewable energy will emerge from its niche. The United States is wealthy with marine renewable energy resources. With appropriate policy support, the United States could build a substantial domestic offshore renewable energy industry and contain its energy vulnerability. It is time to commit the U.S. energy future to a new direction.

Notes

Chapter 1

1. Friedman, Thomas. 2008. *Hot, Flat, and Crowded.* New York: Farrar, Straus and Giroux. 31.

2. U.S. Government Accountability Office. 2005. Security assistance: Efforts to secure Colombia's Caño Limón-Coveñas oil pipeline have reduced attacks, but challenges remain. GAO-05-971. http://www.gao.gov/new.items/d05971.pdf

3. Amazon Watch. 2002. The real costs of pipeline protection in Columbia: Corporate welfare with deadly consequences. July.

4. U.S. Government Accountability Office. 2005. 4.

5. Fuller, Graham E., and Ian O. Lesser. 1997. Persian Gulf myths. *Foreign Affairs.* May-June: 42–53.

6. U.S. Department of Energy. Energy Information Administration. 2008. World proved reserves of oil and natural gas, most recent estimates. Table posted August 27, 2008. http://www.eia.doe.gov/emeu/international/reserves.html

7. Wescott, Robert F. 2006. What would $120 oil mean for the global economy? Securing America's Future Energy. http://www.secureenergy.org/reports/westcott_report.pdf

8. National Commission on Energy Policy and Securing America's Energy Future. 2005. *Oil ShockWave: Oil Crisis Executive Simulation.* June 23. http://www.energycommission.org/files/contentFiles/oil_shockwave_report_440cc39a643cd.pdf

9. Yergin, Daniel. 2005. Energy security and markets. In *Energy and Security: Toward a New Foreign Policy Strategy.* Jan H. Kalicki and David L. Goldwyn, eds. Washington, D.C.: Woodrow Wilson Center Press; Baltimore: Johns Hopkins University Press.

10. Luft, Gal. 2005. Oil puts Iran out of reach. *Baltimore Sun.* August 16. http://www.iags.org/baltsun081605.pdf

11. National Commission on Energy Policy and Securing America's Energy Future. 2005. 2.

12. Ibid.

13. Wescott, Robert F. 2006. 5.

14. Ibid.

15. Auerswald, Philip E. 2006. The myth of energy insecurity. *Issues in Science and Technology Online*. http://www.issues.org/22.4/auerswald.html

16. Auerswald, Philip E. 2007. Calling an end to oil alarmism. Op-ed. *Boston Globe*. January 23. http://belfercenter.ksg.harvard.edu/publication/1642/calling_an_end_to_oil_alarmism.html

17. Energy Security Leadership Council. 2006. *Recommendations to the Nation on Reducing U.S. Oil Dependence*. 17. http://www.secureenergy.org/reports/ESLC_Oil_Report.pdf

18. Morgan, Dan, Sarah Cohen, and Gilbert M. Gaul. 2006. Powerful interests ally to restructure agriculture subsidies. *Washington Post*. December 22.

19. Energy Security Leadership Council. 2006. 14.

20. Santini, Danilo J. 2007. Energy and petroleum consumption attributes of plug-in hybrids. Presentation at the CARB PHEV discussion meeting for the Office of Freedom Car and Vehicle Technologies. Sacramento, California. September 27. http://www.transportation.anl.gov/pdfs/HV/386.pdf

21. CNBC. 2008. Squawk on the Street. Transcript of interview with FERC Chairman Joseph Kelliher by Mark Haines and Erin Burnett. June 9. http://www.ferc.gov/about/com-mem/kelliher/cnbc0609.pdf

22. Yergin. 2005.

23. Ibid.

24. Musial, Walt. 2005. Offshore wind energy potential for the United States. Presentation at the Wind Powering America—Annual State Summit. Evergreen, Colorado. May 19. http://www.eere.energy.gov/windandhydro/windpoweringamerica/pdfs/workshops/2005_summit/musial.pdf

25. Energy for Sustainable Development Ltd. 2004. Offshore wind onshore jobs—a new industry for Britain. http://www.greenpeace.org.uk/MultimediaFiles/Live/FullReport/6702.pdf

26. Garrad Hassan and Partners Ltd. 2004. Sea wind Europe. Greenpeace. http://www.greenpeace.org.uk/MultimediaFiles/Live/FullReport/6204.pdf

27. U.S. Department of the Interior. Minerals Management Service. 2006. Technology white paper on ocean current energy potential on the U.S. outer continental shelf. http://ocsenergy.anl.gov/documents/docs/OCS_EIS_WhitePaper_Current.pdf

28. Verdant Power, LLC. 2003. Presentation to the NREL Industry Growth Forum. Austin, Texas. November 18.

29. U.S. Department of the Interior. Minerals Management Service. 2006. Technology white paper on wave energy potential on the U.S. outer continental shelf. http://www.ocsenergy.anl.gov/documents/docs/OCS_EIS_WhitePaper_Wave.pdf

30. Ibid. 3.

31. Harris, Kimberly. 2006. Puget wants 9.2% rate hike. *The Electricity Daily*. February 23.

32. Energy Security Leadership Council. 2006. 27.

33. Brown, Lester R. 2006. *Plan B 2.0: Rescuing a Planet Under Stress and a Civilization in Trouble*. New York: W.W. Norton & Company. 15.

34. Mills, Evan, and Eugene Lecomte. 2006. *From Risk to Opportunity: How Insurers Can Proactively and Profitably Manage Climate Change*. Ceres. 1. http://216.235.201.250//Document.Doc?id=75

35. Oreskes, Naomi. 2004. Beyond the ivory tower: The scientific consensus on climate change. *Science*. 306 (5702). December 3. http://www.sciencemag.org/cgi/content/full/306/5702/1686

36. Mills and Lecomte. 2006. 7.

37. Van Atten, Christopher, Thomas Curry, and Amlan Saha. 2008. *Benchmarking Air Emissions of the 100 Largest Electric Power Producers in the United States*. Ceres. 18, 64–65. http://www.nrdc.org/air/pollution/benchmarking/2006/benchmark2006.pdf

38. Gardiner, David, Miranda Anderson, and Lisa Jacobson. 2006. *Best Practices in Climate Change Risk Analysis for the Electric Power Sector*. Ceres. October. 1. http://www.dgardiner.com/doc/elec_power06.pdf

39. Mufson, Steven. 2007. Power plant rejected over carbon dioxide for first time. *Washington Post*. October 19.

40. Barrett, Devlin. 2007. More legal battles loom over pollution. *Washington Post*. October 10.

41. Van Atten, Christopher, Thomas Curry, and Amlan Saha. 2008. 3.

42. Ibid. 1.

43. Gardiner, David, Miranda Anderson, and Lisa Jacobson. 2006. 1.

44. Immelt, Jeff. 2005. Global environmental challenges. Speech at the George Washington School of Business. May 9. http://www.ge.com/files/usa/company/news/global_environmental_challenges.pdf

45. Ibid.

46. Mills and Lecomte. 2006. ii.

47. Merrill Lynch. 2007. Combating climate change—opportunities and risks. April 16. 30.

48. Gardiner, David, Miranda Anderson, and Lisa Jacobson. 2006. 18.

49. Eilperin, Juliet. 2006. California tightens rules on emissions. *Washington Post*. September 1.

50. Perera, Frederica, et al. 2006. Effect of prenatal exposure to airborne polycyclic aromatic hydrocarbons on neurodevelopment in the first 3 years of life among inner-city children. *Environmental Health Perspectives*. 114 (8). August. http://www.ehponline.org/members/2006/9084/9084.pdf

51. Seeyle, Katharine Q. 2002. Utility buys town it choked, lock, stock, and blue plume. *New York Times*. May 13.

52. U.S. Environmental Protection Agency. 2001. Mercury update: Impact on fish advisories. Fact Sheet EPA-823-F-01-111. http://www.epa.gov/waterscience/fishadvice/mercupd.pdf See also: U.S. Environmental Protection Agency. Office of Air Quality Planning and Standards. 1998. *Study of Hazardous Air Pollutant Emissions from Electric Utility Steam Generating Units—Final Report to Congress*. EPA-453/R-98-004a. Vol. 1. 7-1, 7-45. http://www.epa.gov/ttn/oarpg/t3/reports/eurtc1.pdf

53. U.S. Geological Survey. 2000. Mercury in the environment. Fact Sheet 146-00. October. http://www.usgs.gov/themes/factsheet/146-00/

54. U.S. Environmental Protection Agency. 2001.

55. U.S. Environmental Protection Agency. 2005. Mercury maps: Linking air deposition and fish contamination on a national scale. Fact Sheet. January. http://www.epa.gov/waterscience/models/maps/fs.htm

56. *New Jersey v. EPA*. 2008. 517 F.3d 574 (D.C. Cir.).

57. Reece, Erik. 2005. Death of a mountain: Radical strip mining and the leveling of Appalachia. *Harpers Magazine*. April. http://www.harpers.org/archive/2005/04/0080500

58. Moore, Patrick. 2006. Going nuclear, a Green makes the case. *Washington Post*. April 16.

59. Electric Utility Week. 2006. Xcel leads wind energy buyers, as FPL dwarfs other wind energy owners, AWEA report says. March 20. 11.

60. *Electric Utility Week*. 2006. Vermont wind project shot down over impact on landscape after five years of planning. March 20. 12.

61. Archer, Cristina L., and Mark Z. Jacobsnon. 2005. Evaluation of global wind power. *Journal of Geophysical Research*. Vol. 110. June 30. 8. http://www.stanford.edu/group/efmh/winds/2004jd005462.pdf

62. Sverrisdóttir, H.E.Valgerður. 2006. Iceland and the future hydrogen economy. Opening address at the International Seminar on the Hydrogen Economy for Sustainable Development. Reykjavik, Iceland. September 28–29. http://www.un.org/esa/sustdev/sdissues/energy/op/hydrogen_seminar/presentations/01_mof.pdf

63. U.S. Department of the Interior. Minerals Management Service. 2004. *History of the Offshore Oil and Gas Industry in Southern Louisiana, Interim Report*. OCS Study MMS 2004-049. Vol. 1. 75. (Quoting O.L. Brace.)

Chapter 2

1. Erickson, Jon. 1996. Marine geology: *Undersea landforms and life forms*. New York: Facts on File. 57.

2. National Oceanic and Atmospheric Administration's National Data Buoy Center. 2005. Reports from the National Data Buoy Center's stations in the Gulf of Mexico during the passage of Hurricane Katrina, 27–30 August. National Oceanic and Atmospheric Administration. http://www.ndbc.noaa.gov/hurricanes/2005/katrina/

3. Agence France-Presse. 2005. Hurricanes destroyed 109 oil platforms. TerraDaily. October 4. http://www.terradaily.com/news/energy-tech-05zzzzzp.html

4. U.S. Department of the Interior. Minerals Management Service. 2004. *History of the Offshore Oil and Gas Industry in Southern Louisiana: Interim Report.* OCS Study MMS 2004-049. Vol. 1. 13.

5. Ibid. 19–24.

6. U.N. Convention on the Law of the Sea. 1994. Entered into force, November 16, 1994. U.N. Doc.A/CONF. 62/122 (1982). Reprinted in United Nations, Official Text of the United Nations Convention on the Law of the Sea with Annexes and Index. U.N. Sales No. E.83. V. 5 (1983). Reprinted at 21 I.L.M. 1261 (1982).

7. Royal Swedish Academy of Sciences. 2008. *Statements on Energy from Moving Water by the Energy Committee at the Royal Swedish Academy of Sciences.* 6. http://www.kva.se/KVA_Root/publications/committees/energy_statements5.pdf

8. The Carbon Trust. 2006. *Future Marine Energy, Results of the Marine Energy Challenge: Cost Competitiveness and Growth of Wave and Tidal Stream Energy.* 7. (Citing to Building Options for UK Renewable Energy.) http://www.carbontrust.co.uk/Publications/publicationdetail.htm?productid=CTC601

9. Sverdrup, Keith A., Alyn C. Duxbury, and Alison B. Duxbury. 2006. *Fundamentals of Oceanography.* 5th ed. New York: The McGraw-Hill Companies. 184.

10. Hagerman, George. 2001. Southern New England wave energy resource potential. Paper presented at the Building Energy 2001 conference. Medford, Massachusetts. March 21–24. 3. http://www.ctinnovations.com/pdfs/S_New_Engl_Wave_Energy_Resource_Potential.pdf

11. Ibid.

12. Geist, Eric L., Vasily V. Titov, and Costas E. Synolakis. 2005. Tsunami: Wave of change. *Scientific American.* December 26. http://www.sciam.com/article.cfm?chanID=sa006&colID=1&articleID=000CDB86-32E0-13A8-B2E083414B7F0000

13. Prager, Ellen J., and Sylvia A. Earle. 2001. *The Oceans.* New York: The McGraw-Hill Companies. 97.

14. Weiss, John C., Brent B. Boehlert, and Jennifer R. Baxter. 2008. *Fiscal Cost-Benefit Analysis to Support the Rulemaking Process for 30 CFR 285 Governing Alternative Energy Production and Alternate Uses of Existing Facilities*

on the Outer Continental Shelf. OCS Study MMS. 2007-050. Herndon, VA: USDOI/MMS, Offshore Environmental Division. 10.

15. Mills, Evan, and Eugene Lecomte. 2006. *From risk to opportunity: How insurers can proactively and profitably manage climate change.* Ceres. 5. http://eetd.lbl.gov/ea/EMills/PUBS/PDF/risk-to-opportunity-082206.pdf

16. Ibid.

Chapter 3

1. Carlton, Jim. 2006. Kleiner's green investment machine: Why the fledgling clean-tech industry is gaining interest among venture-capital firms. *Wall Street Journal.* Technology Journal. December 14: B3.

2. Panja, Tariq. 2007. Scientists to vie for $25 M climate prize. *Washington Post.* February 9.

3. Intergovernmental Panel on Climate Change. 2007. *Climate Change 2007: Synthesis Report. Summary for Policymakers.* Geneva, Switzerland: IPCC.

4. Ibid. 6.

5. Sobel, David. 1995. Eco-phobia. *Earth Ethics.* Spring.

6. Offshore Wind Collaborative Organizing Group. 2005. *A Framework for Offshore Wind Energy Development in the United States.* http://www.mtpc.org/offshore/final_09_20.pdf

7. Norsk Hydro ASA. Floating windmills. Hydro oil and energy brochure. http://www.hydro.com/library/attachments/en/press_room/floating_windmills_en.pdf

8. Peace, Steven. 2004. Another approach to wind: Vertical-axis turbines may avoid the limitations of today's standard propeller-like machines. *Mechanical Engineering.* June.

9. Institute for Wind Energy, Faculty of Civil Engineering and Geoscience. 1998. Opti-OWECS final report. Vol. 0. Structural and economic optimisation of offshore wind energy converters—executive summary. *Delft University of Technology.* 14.

10. Offshore Wind Collaborative Organizing Group. 11.

11. Bittencourt Ferreira, Claudio. 2006. *DNV OSS-312 Certification of Tidal and Wave Energy Converters: Standards in the Renewable Marine Energy Sector.* IEA—Ocean Energy Systems. http://www.iea-oceans.org/_fich/4/DNV_Standards.pdf

12. The Carbon Trust. 2006. Future marine energy; results of the marine energy challenge: cost competitiveness and growth of wave and tidal stream energy. 5. http://www.carbontrust.co.uk/Publications/publicationdetail.htm?productid=CTC601

13. Previsic, Mirko, Roger Bedard, and George Hagerman. 2004. E2I EPRI *Assessment: Offshore Wave Energy Conversion Devices.* Electricity Innovation Institute. June 16. http://oceanenergy.epri.com/attachments/wave/

reports/004_WEC_Device_Assess_Report_Rev1_MP_6-16-04.pdf
See also: European Marine Energy Center Ltd. Wave energy developers. http://www.emec.org.uk/wave_energy_developers.asp

14. Previsic, Bedard, and Hagerman. 12.

15. International Water Power and Dam Construction. 2008. Wave power ready for take off. August 13. http://www.waterpowermagazine.com/story.asp?storyCode=2050584

16. Finavera Renewables Ocean Energy, Ltd. 2009. Application to surrender license. Federal Energy Regulatory Commission filing, Project No. 12751-000. February 6.

17. AWS Ocean Energy Ltd. Wave Swing. Technology. http://www.waveswing.com/

18. Pelamis Wave Power. Pelamis brochure. http://www.pelamiswave.com/content.php?id=161

19. Bedard, Roger. 2005. Final Summary Report, Project Definition Study, Offshore Wave Power Feasibility Demonstration Project. Palo Alto: Electric Power Research Institute. 7.

20. Ibid.

21. Previsic, Bedard, and Hagerman. 17.

22. Ibid. 38.

23. Marks, Paul. 2007. Eel feel helps wave power go with the flow. *New Scientist Magazine*. April 16. 28.

24. Wave Dragon. Principles. http://www.wavedragon.net/index.php?option=com_content&task=view&id=6&Itemid=5&limit=1&limitstart=3

25. Det Norske Veritas and the Carbon Trust. 2005. *Guidelines on Design and Operation of Wave Energy Converters: A Guide to Assessment and Application of Engineering Standards and Recommended Practices for Wave Energy Conversion Devices*. May. http://www.dnv.com/binaries/WECguideline_tcm4-270406.pdf

26. U.S. Department of the Interior. Minerals Management Service. 2006. *Comments of FPL Energy, LLC on MMS ANPR, Alternate Energy Related Uses of the Outer Continental Shelf*. RIN 1010-AD30. February 28. 15.

27. Environmental Change Institute and the Carbon Trust. 2005. *Variability of UK Marine Resources: An Assessment of the Variability Characteristics of the UK's Wave and Tidal Current Power Resources and Their Implications for Large Scale Development Scenarios*. Oxford, England: University of Oxford. http://www.carbontrust.co.uk/NR/rdonlyres/EC293061-611D-4BC8-A75C-9F84138184D3/0/variability_uk_marine_energy_resources.pdf

28. Llanos, Miguel. 2003. San Francisco to test tides for energy: Suction from underwater passageway to turn turbine. *MSNBC.com*. October 24. http://www.msnbc.msn.com/id/3339905/

29. Miller, Christine. 2004. A brief history of wave and tidal energy experiments in San Francisco and Santa Cruz. *Outsidelands.org*. August. http://www.outsidelands.org/wave-tidal.php

30. Ives, James. 2008. Orkney tides power National Grid. *Compute Scotland*. May 28. http://www.computescotland.com/1265.php

31. Clark, Robert H. 1995. *Encyclopedia of Energy Technology and the Environment*. New York: Wiley-Interscience. Vol. 4.

32. Bernshtein, Lev B. 1965. *Tidal Energy for Electric Power Plants*. Jerusalem: Israel Program for Scientific Translations. 284.

33. Sustainable Development Commission. 2007. *Turning the Tide: Tidal Power in the UK*. 101. http://www.sd-commission.org.uk/publications/downloads/Tidal_Power_in_the_UK_Oct07.pdf

34. Ibid. 89.

35. Atiya, Ramez. 2008. The Hydroelectric Tidal Wing: A new tidal range technology for the Bay of Fundy. Halcyon Marine Hydroelectric.

36. Freeman, Kris. 2004. Tidal turbines: Wave of the future? *Environmental Health Perspectives*. January. http://findarticles.com/p/articles/mi_m0CYP/is_1_112/ai_113455708

37. U.S. Department of the Interior. Minerals Management Service. 2006. Technology white paper on ocean current energy potential on the U.S. outer continental shelf. May: 3. http://ocsenergy.anl.gov/documents/docs/OCS_EIS_WhitePaper_Wave.pdf

38. Sverdrup, Keith, Alison Duxbury, and Alyn Duxbury. 2006. *Fundamentals of Oceanography*. 5th ed. New York: McGraw-Hill Publishers. 176.

39. Erling, Jonathan. 2005. A new model for district cooling: The Enwave deep lake water system. KPMG LLP, Canada. http://www.kpmg.ca/en/industries/enr/power/documents/WorldPower2005.pdf

40. Hoffman, Carl. 2005. The mad genius from the bottom of the sea: Unlimited energy. Fast-growing fruit. Free air-conditioning. John P. Craven says we can have it all by tapping the icy waters of the deep. *WIRED*. June. http://www.wired.com/wired/archive/13.06/craven.html

Chapter 4

1. Energy Information Administration. 2008. Electric power monthly. Table 5.6.A. Report released November 17. http://www.eia.doe.gov/cneaf/electricity/epm/table5_6_a.html

2. Global Energy Concepts, LLC. 2006. *A Catalog of Potential Sites for Renewable Energy in Hawaii*. Produced for the Hawaii Department of Land and Natural Resources and the Department of Business, Economic Development, and Tourism. Hawaii: Global Energy Concepts, LLC. http://www.hawaii.gov/dbedt/info/energy/publications/cpsre07.pdf

3. New York City Energy Policy Task Force. 2004. *New York City Energy Policy: An Electricity Resource Roadmap*. New York City Energy Policy Task Force. 13–14. http://www.nyc.gov/html/om/pdf/energy_task_force.pdf

4. New York City Energy Policy Task Force. 2004. 23.

5. New York State Energy Research and Development Authority. 2006. New York state renewable portfolio standard: Program performance report (program period ending December 2005). New York: New York State Energy Research and Development Authority. 3. http://www.dps.state.ny.us/2005_RPS_Status_Report_Final.pdf

6. State of New Jersey Blue Ribbon Panel on Development of Wind Turbine Facilities in Coastal Waters. 2006. *Final Report to Governor Jon S. Corzine*. 1.

7. Firestone, Jeremy, Willett Kempton, and Andrew Krueger. 2008. *Delaware Opinion on Offshore Wind Power*. Newark, DE: University of Delaware College of Marine and Earth Studies. 11, 21. http://www.ocean.udel.edu/windpower/docs/DE-survey-InterimReport-16Jan2007.pdf

8. AWS Scientific, Inc. 2003. *Long Island's Offshore Wind Energy Development Potential: Phase 2 Siting Assessment*. January 21. Prepared for the Long Island Power Authority. 1. http://www.fishermensenergy.com/dms/showfile.php?id=92

9. Ibid. 4.

10. Brower, Michael. 2002. *Wind Resource Maps of New Jersey*. Prepared for New Jersey Board of Public Utilities, Division of Energy, and U.S. Department of Energy, Philadelphia Regional Office. New Jersey: TrueWind Solutions, LLC. 1–2.

11. Dhanju, Amardeep, Phillip Whittaker, and Willett Kempton. 2007. Assessing offshore wind resources: An accessible methodology. *Renewable Energy*. Newark, DE: College of Marine and Earth Studies, University of Delaware. 5–7. http://www.ocean.udel.edu/windpower/docs/DhanjuWhitKemp-proof-RE07.pdf

12. AWS Truewind, LLC. *Wind Resource Explorer*. Maps available for New York, New Jersey, and Ohio. http://www.windexplorer.com/

13. Brower. 2002. *Wind Resource Maps of New Jersey*.

14. U.S. Department of Commerce. National Oceanic and Atmospheric Administration. National Weather Service. National Hurricane Center. http://www.nhc.noaa.gov/

15. AWS Scientific, Inc. 2003. 15.

16. *Code of Federal Regulations*. 2008. Navigation and navigable waters. 33 C.F.R. pt. 167.

17. Ibid. § 166.200.

18. Ibid. §§ 166.300–500.

19. U.S. Department of Commerce. National Oceanic and Atmospheric Administration. National Ocean Service. Data explorer. http://oceanservice.noaa.gov/dataexplorer/welcome.html

20. U.S. Department of the Interior. Minerals Management Service. 2007. *Programmatic Environmental Impact Statement for Alternative Energy Development and Production and Alternate Use of Facilities on the Outer Continental Shelf*. Vol. I, § 2.7.1. http://ocsenergy.anl.gov/eis/guide/index.cfm

21. *Code of Federal Regulations*. 33 C.F.R. § 165.20.

22. Ibid. § 147.1.

23. Ibid. § 147.15.

24. *New York Times*. 2006. Sneak attack on Cape Wind. February 28; Coleman, Jack. 2005. Oily bedfellows, analysis: Linking fossil fuel interests and the Alliance. *Cape Cod Today*. December 9. http://www.capecodtoday.com/modules.php?op=modload&name=News&file=article&sid=0288

25. Maritime Research Institute Netherlands. Platform allision risk. http://www.marin.nl/web/show/id=64579

26. Määttänen, Mauri. Numerical simulation of ice-induced vibrations in offshore structures. Helsinki: Department of Mechanical Engineering, Helsinki University of Technology. 13. http://konewww.hut.fi/courses/files/104/MMNSCM01.pdf

27. U.S. Department of the Interior. Minerals Management Service. 2007. Vol. I, § 2.7.2.15.

28. Ibid.

29. National Renewable Energy Laboratory. National Wind Technology Center. Wind-Wildlife Literature Database (WILD). http://nrelpubs.nrel.gov/Webtop/ws/avianlt/www/web_data/SearchForm ; Avian Reports Published by NREL. http://www.nrel.gov/wind/avian_reports.html

30. U.S. Department of the Interior. Fish and Wildlife Service. 2003. *Service Interim Guidance on Avoiding and Minimizing Wildlife Impacts from Wind Turbines*. 3. http://www.fws.gov/habitatconservation/wind.pdf

31. Lincoln, Frederick C., Steven R. Peterson, John L. Zimmerman. 1998. *Migration of Birds, Circular 16*. U.S. Fish and Wildlife Service. 54.

32. Ibid. 9.

33. U.S. Department of the Interior. Fish and Wildlife Service. 2003. Appendix 7. 50–51.

34. Fairly, Peter. 2007. Massive offshore wind turbines safe for birds: Infrared monitoring shows that savvy seabirds steer clear of wind turbines. *Technology Review*. February 12. http://www.technologyreview.com/energy/18167/

35. Guillemette, Magella, Jesper Kyed Larsen, and Ib Clausager. 1998. *Impact Assessment of an Offshore Wind Park on Sea Ducks*. NERI Technical Report No. 227. March. Denmark: Department of Coastal Zone Ecology, Ministry of Environment and Energy, National Environmental Research Institute. http://www2.dmu.dk/1_viden/2_Publikationer/3_fagrapporter/rapporter/FR227.pdf

36. AWS Scientific, Inc. 2003. iii.

37. U.S. Department of Commerce. National Oceanographic and Atmospheric Administration. National Ocean Service. Data Explorer. Environmental Sensitivity Index maps. http://oceanservice.noaa.gov/dataexplorer/welcome.html

38. U.S. Department of the Interior. Minerals Management Service. 2007. Vol. I, § 2.7.2.2.

39. Valberg, Peter A. 2005. *Memorandum Addressing Electric and Magnetic Field (EMF) Questions, Draft: Cape Wind Energy Project, Nantucket Sound.* Gradient Corporation. 6. http://www.mms.gov/offshore/PDFs/CWFiles/141.pdf

40. U.S. Department of the Interior. Minerals Management Service. 2007. Vol. I, § 2.7.2.3.

41. Fontana, Phil, and Chip Gill. 2002. Marine mammals and seismic vessels jockey for position. *World Oil.* December. http://www.worldoil.com/magazine/MAGAZINE_DETAIL.asp?ART_ID=1902&MONTH_YEAR=Dec-2002; *Code of Federal Regulations.* 2008. Mineral resources. 30 C.F.R. pt. 251.

42. Thomsen, Frank, et al. 2006. *Effects of Offshore Wind Farm Noise on Marine Mammals and Fish.* Hamburg, Germany: Biola (for COWRIE Ltd.). 5. http://www.offshorewindfarms.co.uk/Assets/BIOLAReport06072006FINAL.pdf

43. Ibid.

44. U.S. Department of Commerce. National Oceanic and Atmospheric Administration. National Ocean Service. Gerry E. Studds Stellwagen Bank National Marine Sanctuary. Ocean and dredged material disposal. http://stellwagen.nos.noaa.gov/about/sitereport/dredge.html

45. Marine Protection, Research, and Sanctuaries Act of 1972. 2006. 33 U.S.C. §§ 1401 et seq.; *U.S. Code.* Clean Water Act. 2006. 33 U.S.C. §§ 1251 et seq.; particularly §§ 1343 and 1344.

46. *Code of Federal Regulations.* 2008. Protection of environment. 40 C.F.R. pts. 220 through 230; specifically, 40 C.F.R. § 230.60.

47. Bearden, David M. 2006. *U.S. Disposal of Chemical Weapons in the Ocean: Background and Issues for Congress.* Congressional Research Service. 9 (footnote 10). http://www.fas.org/sgp/crs/natsec/RL33432.pdf

48. Bull, John M.R. 2005. Special report, part 2. The deadliness below: Overseas, fishermen have been hurt by chemical weapons the United States secretly sank, from the Riviera to Australia. *Daily Press.* October 31. http://www.dailypress.com/news/local/dp-02774sy0oct31,0,6036010.story

49. Bull, John M.R. 2005. Special report, part 1. The deadliness below: Weapons of mass destruction thrown into the sea years ago present danger now—and the Army doesn't know where they all are. *Daily Press.* October 30. http://www.dailypress.com/news/dp-02761sy0oct30,0,2199000.story.

50. Hagerman, George. 2004. *Offshore Wave Power in the U.S.: Environmental Issues.* E2I Global EPRI–07–US. Global Energy Partners, LLC. December 21. 14–15. http://oceanenergy.epri.com/attachments/wave/reports/007_Wave_Envr_Issues_Rpt.pdf

51. AWS Scientific, Inc. 2003. 18.

52. U.S. Department of the Interior. Minerals Management Service. 2007. Vol. I, § 2.7.2.10.

53. Oxford Archaeology with George Lambrick Archaeology and Heritage. 2008. *Guidance for Assessment of Cumulative Impacts on the Historic Environment from Offshore Renewable Energy.* COWRIE Ltd. http://www.offshorewindfarms.co.uk/Assets/ccowrie_ciarch%20web.pdf

54. U.S. Department of Commerce. National Oceanographic and Atmospheric Administration. Office of Coast Survey. Automated wreck and obstruction information system. http://www.nauticalcharts.noaa.gov/hsd/awois.html

55. New York State Historic Preservation Office. Underwater archaeology. New York: New York State Historic Preservation Office. http://www.nysparks.state.ny.us/shpo/archeo/underwater.htm

56. U.S. Department of the Interior. Minerals Management Service. 2007. Vol. I, § 2.7.2.14.

57. U.S. Geological Survey. Coastal and marine geology program. Internet map server and GIS data. http://coastalmap.marine.usgs.gov/; U.S. Geological Survey. usSEABED. http://walrus.wr.usgs.gov/usseabed/; U.S. Department of Commerce. National Oceanic and Atmospheric Administration. National Geophysical Data Center. NOAA Satellite and Information Service. Marine geology and geophysics. http://www.ngdc.noaa.gov/mgg/mggd.html

58. U.S. Department of the Interior. Minerals Management Service. 2007. Vol. 1, § 2.7.2.2.

59. Ibid. Vol. 1, § 2.7.2.4.

60. White, David. 2001. Fully protected marine reserves will promote sustainable fishing. Testimony given before the President's Commission on Ocean Policy, St. Petersburg, FL. (Reprinted in: Gerdes, Louise I. 2004. *Endangered oceans: Opposing viewpoints.* Farmington Hills, MI: Green Haven Press. 113–114.)

61. Warne, Kennedy. 2007. Blue haven: New Zealand marine reserves are a model for the world. *National Geographic.* April. http://ngm.nationalgeographic.com/2007/04/new-zealand-coast/warne-text

62. Safina, Carl. 2000. Renewing the world's fisheries. *People & the planet.* October 10. (Reprinted in: Gerdes, Louise I. 2004. Endangered oceans: Opposing viewpoints. Farmington Hills, MI: Green Haven Press. 41.)

63. U.S. Commission on Ocean Policy. 2004. *An Ocean Blueprint for the 21st Century: Final Report of the U.S. Commission on Ocean Policy.* Executive summary. Washington, DC: U.S. Commission on Ocean Policy. 20. http://www.oceancommission.gov/documents/full_color_rpt/welcome.html#final

64. *Globalsecurity.org.* State by state breakdown—U.S. military facilities. Global Security. http://www.globalsecurity.org/military/facility/state.htm

65. U.S. Department of Defense. Office of the Director of Defense Research and Engineering. 2006. *Report to the Congressional Defense Committees: The Effect of Windmill Farms on Military Readiness.* (As required by § 358 of the National Defense Authorization Act for fiscal year 2006 [public law 109–163].) http://www.defenselink.mil/pubs/pdfs/WindFarmReport.pdf

66. Ibid. 41–42.

67. Ibid. 45–47.

68. Vogt, Richard J., et al. Impacts of wind farms on WSR-88D operations and policy considerations. American Meteorological Society. 2. http://ams.confex.com/ams/pdfpapers/120352.pdf

69. University of Oklahoma. Atmospheric Radar Research Center. http://arrc.ou.edu/
70. U.S. Department of the Interior. Minerals Management Service. 2007. Vol. I, § 2.7.2.13.
71. Somashekhar, Sandhya. 2006. Landowners fear ruin from power line route. *Washington Post.* December 11: B01. http://www.washingtonpost.com/wp-dyn/content/article/2006/12/10/AR2006121000912.html

Chapter 5

1. Grays Harbor Ocean Energy Company. 2009. Regulation. http://www.graysharboroceanenergy.com/regulation.htm
2. U.S. Department of the Interior and Federal Energy Regulatory Commission. 2009. Memorandum of Understanding Between the U.S. Department of the Interior and Federal Energy Regulatory Commission. April 9.
3. Department of the Interior, Minerals Management Service. 2009. Renewable Energy and Alternate Uses of Existing Facilities on the Outer Continental Shelf. *Federal Register.* Vol. 74, no. 81. 19638-19871. April 29. To be codified at 30 C.F.R. Parts 250, 285, and 290. ("MMS Final Rules").
4. Department of Interior. 2009. Press Release: P*resident Obama, Secretary Salazar Announce Framework for Renewable Energy Development on the U.S. Outer Continental Shelf.* April 29. www.mms.gov/ooc/press/2009/press0422.htm
5. U.S. Code. Outer Continental Shelf Lands Act. 43 U.S.C. 1337, as amended by § 388(a) of the Energy Policy Act of 2005.
6. U.S. Department of the Interior. Minerals Management Service. 2007. Request for information and nominations of areas for leases authorizing alternative energy resource assessment and technology testing activities pursuant to subsection 8(p) of the Outer Continental Shelf Lands Act, as amended. *Federal Register.* Vol. 72, no. 214. 62673. November 6.
7. U.S. Code. 43 U.S.C. 1337 § 8(p)(3).
8. MMS Final Rules at 268.
9. Ibid. at 29. 30 C.F.R. § 285.300.
10. U.S. Department of the Interior. Minerals Management Service. 2008. Alternative energy and alternate uses of existing facilities on the outer continental shelf; Proposed rule. *Federal Register.* Vol. 73, no. 132. July 9. 39395–39396, proposing 30 C.F.R. § 285.300(b). http://www.mms.gov/federalregister/PDFs/PRAD30FRPUB7339376.pdf.
11. *Code of Federal Regulations.* 30 C.F.R. §§ 285.235 and 285.425.
12. MMS Final Rules at 29.
13. U.S. Department of the Interior. Minerals Management Service. 2008. 39392.
14. *Code of Federal Regulations.* 30 C.F.R. §§ 285.210 and 285.212.
15. Ibid. § 285.213.

16. Ibid. § 285.213(d).

17. *U.S. Code*. Freedom of Information Act. 5 U.S.C. § 552(b).

18. U.S. Department of the Interior. Minerals Management Service. 2008. § 285.213.

19. Ibid. § 285.211.

20. Ibid. § 285.216.

21. MMS Final Rules at 91.

22. Ibid.

23. Ibid. at 93.

24. 30 C.F.R. § 285.222(a)(2).

25. U.S. Department of the Interior. Minerals Management Service. 2008. 39399.

26. 30 C.F.R. §§ 285.506(c)(1) and 285.510.

27. Ibid. § 285.505.

28. Ibid. § 285.230.

29. Ibid. § 285.606.

30. Ibid. §§ 285.613(a)(1) and 285.614(b).

31. Ibid. 39422.

32. Ibid. 39417.

33. *U.S. Code*. Coastal Zone Management Act. § 307(c)(3)(B); U.S. Code. 16 U.S.C. § 1456(c)(3)(B).

34. U.S. Department of the Interior. Minerals Management Service. 2008.

35. Ibid. § 285.707(a).

36. Ibid. § 285.702.

37. Ibid. § 285.706.

38. Ibid. § 285.707.

39. Ibid. §§ 285.709 and 285.710.

40. Ibid. § 285.801.

41. Ibid.

42. Ibid. § 285.803.

43. Ibid.

44. Ibid. § 285.810.

45. Carter, J.M.F. 2007. North Hoyle offshore wind farm: Design and build. *Energy*. Vol. 160, no. 1. 21, 28. http://www.atypon-link.com/TELF/doi/abs/10.1680/ener.2007.160.1.21

46. British Wind Energy Association. 2008. BWEA *Guidelines for Health & Safety in the Wind Energy Industry*. October. http://www.bwea.com/pdf/HSGuidelines.pdf

47. IHS. 2000. *Recommended Practice for Planning, Designing and Constructing Fixed Offshore Platforms, Working Stress Design*. 21st ed. §§ 17.2.1–17.2.5 of API RP 2A-WSD. December; Errata and supplement 1. 2002. December; Errata and supplement 2. 2005. October. (Incorporated by reference in MMS regulations 30 C.F.R. § 285.825.)

48. U.S. Department of the Interior. Minerals Management Service. 2008. § 285.830.

49. Ibid. § 285.627(c).

50. Ibid. § 285.1018.

51. Ibid. §§ 285.913 and 285.400.

52. *U.S. Code*. Federal Power Act. § 4(e); U.S. Code. 16 U.S.C. § 797(e).

53. AquaEnergy Group, LTD. 2003. 102 FERC ¶61,242.

54. Energy Policy Act. 2005. § 388, amending § 8(p) of the OCS Lands Act at § 8(p)(9).

55. Katz, John. 2006. Comments from FERC Office of General Counsel at FERC technical conference held on December 6. Transcribed at *In the Matter of Hydroelectric Infrastructure Technical Conference*. Docket Number AD06-13-000. 157.

56. MMS Final Rules at 108.

57. Finavera Renewables Ocean Energy, Ltd. 2007. 121 FERC ¶61,288. n. 7

58. *Code of Federal Regulations*. 18 C.F.R. § 4.34(i).

59. *U.S. Code*. Federal Power Act § 4(f); 16 U.S.C. § 797(f).

60. California Wave Energy Partners I, LLC. 2008. 123 FERC ¶62,266. 2 and note 5.

61. Sperling, Gilbert. 2006. Comments at FERC technical conference. December 6. Transcribed at *In the Matter of Hydroelectric Infrastructure Technical Conference*. Docket Number AD06-13-000. 130.

62. Memorandum of Understanding between the Federal Energy Regulatory Commission and the State of Oregon. 2008. March 18.

63. Verdant Power LLC. 2005. 111 FERC ¶61,024. Order on rehearing. 112 FERC ¶61,143.

64. Walsh, M. 2008. Comment: A rising tide in renewable energy; the future of tidal in-stream energy conversion (TISEC). *Villanova Environmental Law Journal*. Vol. 19. 193–226.

65. U.S. Department of Energy. Federal Energy Regulatory Commission. 2008. *Licensing Hydrokinetic Pilot Projects*. White paper. April 14. http://www.ferc.gov/industries/hydropower/indus-act/hydrokinetics/pdf/white_paper.pdf

66. *Code of Federal Regulations*. 18 C.F.R. § 5.18

67. Ibid.

68. Finavera Renewables Ocean Energy, Ltd. 2007. 121 FERC ¶61,288.

69. U.S. Department of Energy. Federal Energy Regulatory Commission. 2008.

70. *Code of Federal Regulations*. Integrated license application process. 18 C.F.R. pt. 5. http://www.access.gpo.gov/nara/cfr/waisidx_04/18cfr5_04.html

71. Ibid. pt. 5.6.

72. U.S. Department of Energy. Federal Energy Regulatory Commission. Office of Energy Projects. 2006. *Ideas for Implementing and Participating in the Integrated Licensing Process (ILP)*. February. http://www.ferc.gov/industries/hydropower/gen-info/licensing/ilp/eff-eva/ideas.pdf

73. *Code of Federal Regulations*. 18 C.F.R. § 5.5.

74. Ibid. § 5.11.

75. Ibid. §§ 5.17 and 5.18.

76. Federal Power Act. § 15(b)(2).

77. *Code of Federal Regulations*. 18 C.F.R. § 11.1.

78. Ibid. 18 C.F.R. pts. 11 and 11.2, and appendix A.

79. MMS Final Rules at 157.

Chapter 6

1. *U.S. Code*. 2006. National Environmental Policy Act. Public law 91-190. Codified at 42 U.S.C. § 4331(a).

2. *Code of Federal Regulations*. 2008. Protection of environment. 40 C.F.R. § 1508.14.

3. *Robertson v. Methow Valley Citizens Council*. 1989. 490 U.S. 332, 351.

4. Ibid. 40 C.F.R. § 1508.23.

5. U.S. Department of the Interior and U.S. Department of Energy. Minerals Management Service and Federal Energy Regulatory Commission. 2009. *Memorandum of Understanding Between the U.S. Department of the Interior and Federal Regulatory Energy Commission*. See also *Pacific Gas & Electric Company*. 2008. 125 FERC ¶ 61,045.

6. *Code of Federal Regulations*. 40 C.F.R. §§ 1500–1508.

7. *Robertson v. Methow Valley Citizens Council*. 1989.

8. *Code of Federal Regulations*. 40 C.F.R. § 1508.4.

9. *Sierra Club v. Morton*. 1976. 427 U.S. 390; *Stewart Park & Preserve Coalition, Inc. ("SPARC") v. Slater*. 2003. 352 F.3d 545, 557 (2d Cir.); *Hudson River Sloop Clearwater, Inc. v. Dept. of the Navy*. 1988. 836 F.2d 760, 763 (2d Cir.).

10. *Code of Federal Regulations*. 40 C.F.R. § 1508.9.

11. U.S. Department of the Interior. Minerals Management Service. 2007. *Programmatic Environmental Impact Statement for Alternative Energy Development and Production and Alternate Use of Facilities on the Outer Continental Shelf*; U.S. Department of the Interior. Minerals Management Service. 2007. *Record of*

Decision: Establishment of an Alternative Energy and Alternate Use Program. Both documents are available at http://ocsenergy.anl.gov/eis/guide/index.cfm

12. *Marsh v. Oregon Natural Resources Council*. 1989. 490 U.S. 360, 375–78.

13. Ibid.

14. *U.S. Code*. 2006. 42 U.S.C. § 4321; *Code of Federal Regulations*. 40 C.F.R. § 1502.1; and *Marsh v. Oregon Natural Resources Council*. 1989. 490 U.S. 360, 371.

15. *Vermont Yankee Nuclear Power Corp. v. NRDC*. 1978. 435 U.S. 519, 558 (internal citations omitted).

16. *Robertson v. Methow Valley Citizens Council*. 1989. 490 U.S. 350; *Strycker's Bay Neighborhood Council v. Karlen*. 1980. 444 U.S. 223, 227–28 (citing *Vermont Yankee Nuclear Power Corp. v. NRDC*. 558).

17. *Code of Federal Regulations*. 40 C.F.R. § 1502.14.

18. *U.S. Code*. 2006. 42 U.S.C. § 4332 (C)(i)–(v).

19. *Code of Federal Regulations*. 40 C.F.R. § 1501.7.

20. Bisbee, Dorothy W. 2004. NEPA review of offshore wind farms: Ensuring emission reduction benefits outweigh visual impacts. *Boston College Environmental Affairs Law Review*. Vol. 31, no. 2: 349. http://www.bc.edu/schools/law/lawreviews/meta-elements/journals/bcealr/31_2/31_2_toc.htm

21. *Code of Federal Regulations*. 40 C.F.R. § 1508.25.

22. U.S. Department of the Interior. Minerals Management Service. 2009. *Cape Wind Energy Project: Final Environmental Impact Statement*. January. E-6. http://www.mms.gov/offshore/AlternativeEnergy/CapeWindfeis.htm

23. *Code of Federal Regulations*. 40 C.F.R. § 1508.8(a).

24. Ibid. § 1508.8(b).

25. Ibid. § 1508.7.

26. U.S. Department of the Interior. Minerals Management Service. 2007. Vol. II, ch. 5; Vol. III, ch. 7.

27. U.S. Department of the Interior. Minerals Management Service. 2009. E-7–E-9.

28. Executive Office of the President. Council on Environmental Quality. 1997. Draft guidance regarding consideration of global climate change in environmental documents prepared pursuant to the National Environmental Policy Act. 6. http://www.mms.gov/eppd/compliance/reports/ceqmemo.pdf

29. *Massachusetts v. EPA*. 2007. 549 U.S. 497, 520.

30. *Border Power Plant Working Group v. Department of Energy*. 2007. 260 F.Supp. 2d 997, 1028–1029 (S.D. California).

31. Ibid.

32. *Mid States Coalition for Progress v. Surface Transportation Board*. 2003. 345 F.3d 520, 532, and 550 (8th Cir.).

33. Ibid.

34. *Mayo Foundation v. Surface Transportation Board*. 2006. 472 F.3d 545, 555–56 (8th Cir.).

35. U.S. Department of the Interior. Minerals Management Service. 2007. Vol. III, §§ 7.5.1–7.5.5.

36. U.S. Department of the Interior. Minerals Management Service. 2008. *Cape Wind Energy Project: Draft Environmental Impact Statement*. January. 5–274, and § 5.4.6.

37. Erickson, Wallace P., et al. 2001. *Avian Collisions with Wind Turbines: A Summary of Existing Studies and Comparisons to Other Sources of Avian Collision Mortality in the United States*. Western EcoSystems Technology, Inc., on behalf of the National Wind Coordinating Committee. http://www.nationalwind.org/publications/wildlife/avian_collisions.pdf Peterson, Ib Krag, et al. 2006. *Final Results of Bird Studies at the Offshore Wind Farms at Nysted and Horns Rev, Denmark*. National Environmental Research Institute, Denmark, commissioned by DONG Energy and Vantenfall A/S. 141. http://www.ens.dk/graphics/Energiforsyning/Vedvarende_energi/Vind/havvindmoeller/vvm%20Horns%20Rev%202/Nysted/Birds%20final%202005.pdf

38. U.S. Department of the Interior. Minerals Management Service. 2008. Appendix 3-C. 2. (An edited revision of ESS, Inc. *Limitations of Long Transmission Cables for Offshore Wind Farms*. 2003. ESS, Inc., on behalf of Cape Wind Associates, LLC.) http://www.mms.gov/offshore/PDFs/CWFiles/41.pdf

39. U.S. Department of the Interior. Minerals Management Service. 2007. Vol. III. 7–11.

40. U.S. Department of the Interior. Minerals Management Service. 2007.

Chapter 7

1. *U.S. Code*. 2007. Fish and Wildlife Coordination Act. 16 U.S.C. § 662(a).

2. *U.S. Code*. 2007. Magnuson-Stevens Fishery Conservation and Management Act. 16 U.S.C. § 1855(b)(2). See also Code of Federal Regulations. 2008. Wildlife and Fisheries. 50 C.F.R. pt. 600.

3. *U.S. Code*. 2007. Marine Mammal Protection Act. Public Law 92-522, codified at 16 U.S.C. § 1361 et seq.

4. Ibid. 16 U.S.C. § 1362(18)(A).

5. See generally, *Code of Federal Regulations*. 2008. Wildlife and Fisheries. 50 C.F.R. § 216.101 et seq.

6. Ibid. 16 U.S.C. § 1371(A)(5).

7. (Examples of letters of authorization or incidental harassment authorization can be found at the NMFS Web site, along with an online application system for authorizations and permits for protected species under the Marine Mammal Protection Act and the Endangered Species Act.) U.S. Department of Commerce. National Oceanic and Atmospheric

Administration. NOAA Fisheries. National Marine Fisheries Service. Office of Protected Resources. Incidental take authorizations. http://www.nmfs.noaa.gov/pr/permits/incidental.htm; U.S. Department of Commerce. National Oceanic and Atmospheric Administration. National Marine Fisheries Service. Authorizations and permits for protected species (APPS). https://apps.nmfs.noaa.gov/index.cfm?

8. *U.S. Code.* 2007. Migratory Bird Treaty Act of 1918, as amended. 16 U.S.C. § 703 et seq.

9. Ibid. § 703(a).

10. *Code of Federal Regulations.* 2008. Wildlife and Fisheries. 50 C.F.R. § 10.13 (listing the birds covered under the MBTA).

11. Ibid. §10.12.

12. *Federal Register.* 2001. Executive Order No. 13,186. Vol. 66, no. 11. January. 3,853.

13. *U.S. Code.* 2007. Migratory Bird Treaty Act of 1918, as amended. 16 U.S.C. § 707

14. *United States v. FMC Corp.* 1978. 572 F.2d 902 (2d Cir.); *United States v. Corbin Farm Service.* 1978. 444 F. Supp. 510 (E.D. California), aff'd on other grounds. 1978. 578 F.2d 259 (9th Cir.).

15. *United States v. FMC Corp.* 1978. 572 F.2d. 902, 905.

16. (The question of liability under the MBTA for unintentional bird deaths was certainly contemplated by Congress historically. In 1918, Rep. Frank Wheeler Mondell [R-WY] asked, "What are you going to do in a case like this: A barefoot boy, as barefoot boys sometimes do, largely through inadvertence and without meaning anything wrong, happens to throw a stone at and strikes and injures a robin's nest and breaks one of the eggs, whereupon he is hauled before a court for violation of a solemn treaty entered into between the United States of America and the Provinces of Canada.)" *Congressional Record.* 1918. Statement of Representative Frank Wheeler Mondell (R-WY). 65th Cong. 2d sess. 56 Cong. Rec. 7454. June 6.

17. Corcoran, Larry Martin, and Elinor Colbourn. 1999. Shocked, crushed and poisoned: Criminal enforcement in non-hunting cases under the migratory bird treaties. *Denver University Law Review.* 77: 359.

18. *United States v. Moon Lake Electric Ass'n, Inc.* 1999. 45 F.Supp.2d 1070 (D. Colorado).

19. U.S. Department of the Interior. Office of the Solicitor General. 1992. Letter from C. Brester Chapman, Jr., associate solicitor general, to Spencer H. Smith of 7/20/1972. Subsequently reprinted in U.S. Congress. 1972. Senate. Committee on Commerce. Subcommittee on the Environment. *Bald Eagle Protection Act: Hearings on S. 2547, H.R. 12186, and H.R. 14731.* 92d Cong., 2d sess. June 20. 24.

20. U.S. Department of the Interior. Minerals Management Service. Offshore Environmental Studies Program. 2009. *Fiscal Years 2009–2011 Studies Development Plan, Alternative Energy.* http://www.mms.gov/offshore/AlternativeEnergy/Assets/PDFs/AE_SDP_2009_2011_FINAL.pdf

21. Sierra Club. Lone Star Chapter. 2006. Texas conservationists raise concerns about potential impacts of offshore wind energy project. Press release. March 2. http://lonestar.sierraclub.org/press/newsreleases/20060302.asp

22. Environmental Defense Fund. The facts on the TXU buyout. http://www.edf.org/article.cfm?contentID=6027

23. *U.S. Code*. 2007. Endangered Species Act of 1973, as amended. 16 U.S.C. § 1531 et seq.

24. Ibid. § 1532(3).

25. U.S. Department of the Interior. U.S. Fish & Wildlife Service; and U.S. Department of Commerce. National Oceanic and Atmospheric Administration. National Marine Fisheries Service. 1998. *Endangered Species Consultation Handbook: Procedures for Conducting Consultation and Conference Activities under Section 7 of the Endangered Species Act.* http://www.nmfs.noaa.gov/pr/pdfs/laws/esa_section7_handbook.pdf

26. *Code of Federal Regulations*. 2008. Wildlife and Fisheries. 50 C.F.R. § 412.12.

27. Ibid. § 412.02.

28. Perkins, Simon, Andrea Jones, and Taber Allison. 2003. *Survey of Tern Activity within Nantucket Sound, Massachusetts, During Pre-Migratory Fall Staging.* Division of Conservation Science and Ecological Management, Massachusetts Audubon Society. Final Report for Massachusetts Technology Collaborative.

29. U.S. Department of the Interior. U.S. Fish & Wildlife Service; and U.S. Department of Commerce. National Oceanic and Atmospheric Administration. National Marine Fisheries Service. 1998. 1-2–1-3.

30. *U.S. Code*. 2007. Endangered Species Act of 1973, as amended. 16 U.S.C. § 1536(o)(2).

31. Ibid. § 1536(h)(1).

32. Ibid. § 1532(19).

33. *Code of Federal Regulations*. 2008. Wildlife and Fisheries. 50 C.F.R. § 17.3.

34. *U.S. Code*. 2007. Endangered Species Act of 1973, as amended. 16 U.S.C. §§ 1533(b)(3)(a); 1540(g).

35. Ibid. § 1538(a)(1)(B)

36. Ibid. § 1539(a)(1)(B)

37. *U.S. Code*. 2006. Outer Continental Shelf Lands Act. 43 U.S.C. § 1337(p)(10).

38. Ibid.

39. *U.S. Code*. 2007. National Marine Sanctuaries Act. Public Law 92-532, codified at 16 U.S.C. § 1431 et seq. (Standards and guidelines for the preservation of shipwrecks and other historic and cultural artifacts published by the U.S. Department of the Interior are available at the http://www.nps.gov/archeology/submerged/intro.htm.)

40. Ibid. § 1434(d).

41. Ibid. § 1447(d).

42. *U.S. Code.* 2007. National Marine Sanctuaries Act. 16 U.S.C. § 1434(d)(4).

43. *U.S. Code.* 2006. Abandoned Shipwreck Act of 1987. Public Law 100-298, codified at 43 U.S.C. § 2101 et seq.

44. U.S. Department of the Interior. National Park Service. The National Register Information System. http://www.nps.gov/history/nr/research/nris.htm

45. U.S. Department of the Interior. National ParkService. Archeology program. http://www.nps.gov/archeology/submerged/NRShips.htm

46. *Code of Federal Regulations.* 2008. Navigation and Navigable Waters. 33 C.F.R. § 165.141.

47. *U.S. Code.* 2007. National Historic Preservation Act of 1966, as amended. Public Law 89-665, codified at 16 U.S.C. § 470 et seq.; Code of Federal Regulations. 2008. Parks, Forests, and Public Property, 36 C.F.R. pt. 800 et seq.

48. *U.S. Code.* 2007. National Historic Preservation Act of 1966, as amended. Public Law 89-665, codified at 16 U.S.C. § 470h-2(k).

49. *Code of Federal Regulations.* 2008. Parks, Forests, and Public Property, 36 C.F.R. § 800.5(a)(1).

50. Ibid. § 800.5(a)(2)(v).

51. *U.S. Code.* 2007. National Historic Preservation Act of 1966, as amended. Public Law 89-665, § 106, codified at 16 U.S.C. § 470f.

52. *U.S. Code.* 2007. Archaeological and Historic Preservation Act of 1974. Public Law 93-291, codified at 16 U.S.C. § 469a-1.

53. Ibid. § 469a-2(d); *U.S. Code.* 2007. National Historic Preservation Act Amendments of 1980. Public Law 96-515, codified at 16 U.S.C. § 470h-2(g).

54. U.S. Department of the Interior. Minerals Management Service. 2007. *Worldwide Synthesis and Analysis of Existing Information Regarding Environmental Effects of Alternative Energy Uses on the Outer Continental Shelf.* OCS Report. MMS 2007-038. July. http://www.mms.gov/offshore/AlternativeEnergy/PDFs/MMSAESynthesisReport.pdf

55. Dong Energy, Vattenfall, Danish Energy Authority, and Danish Forest and Nature Agency. 2006. *Danish Offshore Wind: Key Environmental Issues.* http://www.ens.dk/graphics/Publikationer/Havvindmoeller/havvindmoellebog_nov_2006_skrm.pdf

56. Ibid. 18.

57. Ibid.

58. Ibid. 106.

59. U.S. Department of the Interior. Minerals Management Service. 2007. *Worldwide Synthesis and Analysis of Existing Information Regarding Environmental Effects of Alternative Energy Uses on the Outer Continental Shelf.* 5-101.

60. Dong Energy, Vattenfall, Danish Energy Authority, and Danish Forest and Nature Agency. 2006. 18.

Chapter 8

1. *United States v. California*. 1947. 332 U.S. 19, 38–39.
2. U.S. Code. 2000. Submerged Land Act of 1953. 43 U.S.C. §§ 1301 et seq.
3. Ibid.§§ 1311(a), 1301(e).
4. Ibid. § 1311(d).
5. *United States v. Maine*. 1975. 420 U.S. 515, 525-26.
6. Submerged Land Act of1953 § 1302; *United States v. Maine*, 420 U.S. at 525–26.
7. McKinney. 2005. *N.Y. Environmental Conservation Law §15-0503(1)(b)*; West. 2001. *Massachusetts General Laws Annotated*. ch. 91. 14.
8. Walker, Kenneth, and Matt Arnn. 1998. *Preserving Waterfronts for Water Dependent Uses*. U.S. Department of Commerce. National Oceanic and Atmospheric Administration. State of the Coast Report. Silver Spring, MD: NOAA. http://state_of_coast.noaa.gov/bulletins/html/wdu_11/wdu.htm
9. *Illinois Central Railroad Co. v. Illinois*. 1892. 146 U.S. 387, 456.
10. U.S. Code. 2000. Coastal Zone Management Act of 1972. 16 U.S.C. §§ 1451-1465.
11. *Code of Federal Regulations*. 2008. Commerce and Foreign Trade. 15 C.F.R. § 930.11(g).
12. Ibid. §§ 930.62(a) and 930.63(a).
13. Ibid. §§ 930.63, 930.156, and 930.125(a).
14. Ibid. § 930.130(e).
15. *Islander East Pipeline Company v. Connecticut Department of Environmental Protection*. 2006. 482 F.3d 79 (2d Cir.).
16. Evans, Donald. 2004. *Decision & Findings of the U.S. Secretary of Commerce in the Consistency Appeal of Islander East Pipeline Co., LLC.; Code of Federal Regulations*. 2008. Commerce and Foreign Trade. 15 C.F.R. § 930.121.
17. U.S. Department of Commerce. National Oceanic and Space Administration. 2004. State coastal zone boundaries. April 22. http://coastalmanagement.noaa.gov/mystate/docs/StateCZBoundaries.pdf
18. *Federal Register*. 2000. 65 F.R. 77125 citing H.R. Conference Report No. 964. 101st Cong. 2d Sess. 970–71.
19. *Code of Federal Regulations. 2008*. Commerce and Foreign Trade. 15 C.F.R. § 930.154.
20. Ibid. §§ 930.151, 930.156; Federal Register. 2000. 65 F.R. 20270, 20280, 77124, 77152–77153.
21. Ibid.
22. *AES Sparrows Point LNG, LLC v. Smith*. 2008. 527 F.3d 120, 126 (4th Cir.).
23. Coastal Zone Management Act of 1972 §§ 1456(c)(1)(A) and 1456(c)(3)(A); *Decisions and Findings in the Consistency Appeal of the Virginal Electric and Power Company from an Objection by the North Carolina Department of Environment,*

Health and Natural Resources. 1994. May. 9–18, aff'd, *North Carolina v. Brown*. 1995. 1995 U.S. Dist. Lexis 22393 (D.C. Cir.).

24. Coastal Zone Management Act of 1972 § 1453(6)
25. Ibid. § 1455(d)(8).
26. Ibid. § 1452(2)
27. *Code of Federal Regulations*. 2008. Commerce and Foreign Trade. 15 C.F.R. pt. 930.
28. Ibid. § 930.56.
29. Ibid. § 930.58.
30. Ibid. § 930.57.
31. *U.S. Code*. 2006. Rivers and Harbors Act of 1899. 33 U.S.C.§ 403.
32. *U.S. Code*. 2006. Marine Protection, Research and Sanctuaries Act of 1972. 33 U.S.C. § 1413; *Code of Federal Regualtions*. 2008. Protection of the Environment. 40 C.F.R. pt. 228 promulgated pursuant to § 103 of the Marine Protection, Research and Sanctuaries Act of 1972; *Code of Federal Regulations*. 2008. Navigation and Navigable Waters. 33 C.F.R.§ 320.2(g).
33. *U.S. Code*. 2006. Clean Water Act. 33 U.S.C. §§ 1344, 1362; *Code of Federal Regulations*. 2008. Navigation and Navigable Waters. 33 C.F.R. 329.12.
34. U.S. Army. U.S. Army Corps of Engineers. Headquarters. http://www.usace.army.mil/cw/cecwo/reg/oceover.htm
35. *Code of Federal Regulations*. 2008. Navigation and Navigable Waters. 33 C.F.R. § 320.4.
36. *Code of Federal Regulations*. 2008. Protection of Environment. 40 C.F.R. pts. 220–230.
37. Ibid. § 230.21(b).
38. Rivers and Harbors Act of 1899 § 403.
39. *Alliance to Protect Nantucket Sound, Inc. v. United States Department of the Army*. 2005. 398 F.3d 105 (1st Cir.).
40. *U.S. Code*. 1953. Outer Continental Shelf Lands Act. of 1953. 43 U.S.C. § 1333(f).
41. *U.S. Code*. 1978. Outer Continental Shelf Lands Act Amendment of 1978. 43 U.S.C. § 1333(e).
42. Ibid. § 1333(a)(1).
43. *Alliance to Protect Nantucket Sound, Inc. v. United States Department of the Army*. 2005. 109.
44. *U.S. Code*. 1978. Outer Continental Shelf Lands Act Amendment of 1978. 43 U.S.C. §§ 1331(k), (l), (m), (q).
45. Ibid. See also *Alliance to Protect Nantucket Sound, Inc. v. United States Department of the Army*. 2005. 109–110.
46. U.S. Code. 2008. Clean Water Act. 33 U.S.C. §§ 1251–1387, as amended.

47. Clean Water Act. §§ 401(a), (d); 33 U.S.C. *National Wildlife Federation v. Gorsuch*. 1982. 693 F.2d 156 (D.C. Cir.).
48. *U.S. Code*. 2006. Clean Water Act. 33 U.S.C. §§ 1311(b)(1)(C), 1313.
49. Ibid. § 1313(d)(4)(B).
50. Ibid. § 1313(c)(2)(A).
51. Ibid. § 1313(c)(3).
52. Ibid. § 1370; *Code of Federal Regulations*. 2000. Protection of Environment. 40 C.F.R. § 131.4(a).
53. *National Wildlife Federation v. Gorsuch*. 1982. 693 F.2d 156 (D.C. Cir.); *South Florida Water Management District v. Miccosukee Tribe of Indians*. 2004. 541 U.S. 95.
54. *National Wildlife Federation v. Gorsuch*. 1982. 156, 165.
55. Clean Water Act. § 502(14); *U.S. Code*. 2006. 33 U.S.C. § 1362(14).
56. Clean Water Act. § 502(7); *U.S. Code*. 2006. 33 U.S.C. § 1362(7); and *United States v. Byrd*. 1979. 609 F.2d 1204 (7th Cir.).
57. Clean Water Act. § 502(6); *U.S. Code*. 2006. 33 U.S.C. § 1362(6).
58. Clean Water Act. § 502(19); *U.S. Code*. 2006. 33 U.S.C. § 1362(19).
59. *South Florida Water Management District v. Miccosukee Tribe of Indians*. 2004. 110. Quoting *Mountains Chapter of Trout Unlimited, Inc. v. New York*. 2001. 273 F.3d 481, 492 (2d Cir.).
60. *Code of Federal Regulations*. 2008. Protection of Environment. 40 C.F.R. § 122.3(i).
61. Ibid. § 122.3(a).
62. Ibid.
63. *U.S. Code*. 2006. Federal Insecticide, Fungicide, and Rodenticide Act. 7 U.S.C. §136a(c)(5)(D).
64. Clean Water Act. § 401(a)(1); *U.S. Code*. 2006. 33 U.S.C. § 1341(a)(1).
65. Clean Water Act. § 401(d). *U.S. Code*. 2006. 33 U.S.C. § 1341(d)
66. *PUD No. 1 of Jefferson County v. Washington Department of Ecology*. 1994. 511 U.S. 700.
67. Ibid. 700, 727–728, citing *Code of Federal Regulations*. 1990. Protection of Environment 40 C.F.R. § 121.2(a)(3).
68. *S.D. Warren Co. v. Maine Board of Environmental Protection*. 2006. 547 U.S. 370.
69. Ibid. 370, 376.
70. *Islander East Pipeline Company, LLC v. Connecticut Department of Environmental Protection*. 2006. 482 F.3d 79 (2d. Cir.).
71. *Islander East Pipeline Company, LLC v. McCarthy*. 2008. 525 F.3d 141, 151 (2d Cir.).
72. *Islander East Pipeline Company, LLC v. Connecticut Department of Environmental Protection*. 2006.

73. *Islander East Pipeline Co. v. McCarthy, Gina, et al.* 2008. Cert. denied. Case no. 08-367 (December 1).
74. Clean Water Act. *U.S. Code.* 2006. 33 U.S.C. § 1321(b)(1).
75. Clean Water Act. *U.S. Code.* 2006. 33 U.S.C. § 1321(j)(5); *Code of Federal Regulations.* Protection of Environment. 40 C.F.R. pt. 112.
76. Memorandum of Understanding (MOU) between EPA, MMS and the Department of Transportation. 1993. November 8. (Appendix B to the EPA regulations at 40 C.F.R. pt. 112.)
77. *Code of Federal Regulations.* 2006. Protection of Environment. 40 C.F.R. § 70.2.
78. Ibid. § 60.4200 et seq.
79. *U.S. Code.* 2006. Act to Prevent Pollution from Ships. 33 U.S.C. §§1905–1915.
80. *Code of Federal Regulations.* 2008. Protection of Environment. 40 C.F.R. pt. 55.
81. Ibid. § 55.4.
82. Ibid. § 55.6.
83. Clean Air Act. § 328(a)(4)(c)(iii); *U.S. Code.* 2006. 42 U.S.C. § 7627 (a)(4)(C)(iii)
84. *Code of Federal Regulations.* 2008. Protection of the Environment. 40 C.F.R. § 55.2.
85. Ibid.
86. Clean Air Act. § 176(c). U.S. Code. 2006. 42 U.S.C. § 7506(c)
87. *Code of Federal Regulations.* 2008. Protection of the Environment. 40 C.F.R. pt. 93.
88. Clean Air Act. § 176(c)(1). *U.S. Code.* 2006. 42 U.S.C. § 7506 (c)(1)
89. Clean Air Act. § 309; *U.S. Code.* 2006. 42 U.S.C. §7609. National Environmental Policy Act of 1969 § 102(2)(C) of Public Law 91-190, *U.S. Code.* 2006. 42 U.S.C. § 4332(2)(C)
90. *U.S. Code.* 2006. Jones Act of 1920. 46 U.S.C. § 55101 et seq.
91. *Code of Federal Regulations.* 2008. Customs Duties. 19 C.F.R. § 4.80.
92. *U.S. Code.* 2006. Jones Act of 1920 as amended. 46 U.S.C.. § 12112.
93. *U.S. Code.* 2006. Outer Continental Shelfs Lands Act Amendment of 1978. 43 U.S.C. § 1333.
94. *Guess v. Read.* 1962. 290 F.2d 622, 625 (5th Cir.), cert. denied, 368 U.S. 957, 7 L. Ed. 2d 388, 82 S. Ct. 394.
95. *Treasure Salvors, Inc. v. Unidentified Wrecked, Etc.* 1978. 569 F.2d 330, 339–40 (5th Cir.); *106 Mile Transport Associates v. Koch.* 1987. 656 F. Supp. 1474.
96. U.S. Department of the Treasury. U.S. Customs Service. 1983. *U.S. Customs Service Decisions.* 17 Cust. B. & Dec. 826; CUSBUL LEXIS 251. C.S.D. 83-52. File; VES-3-15-CO:R:CD:C 105909 PH (February 23).
97. Ibid.

98. U.S. Department of Homeland Security. U.S. Bureau of Customs & Border Protection. 2007. *U.S. Bureau of Customs & Border Protection Rulings*. HQ Series. U.S. CUSTOM HQ LEXIS 876; HQ H012082 (August 27).

99. MPI Offshore Ltd. *MV Resolution*. http://www.mpi-offshore.com/equipment/the_ship

100. *U.S. Code*. 2006. Jones Act of 1920 as amended. 46 U.S.C. § 55109.

101. U.S. Department of the Treasury. U.S. Customs Service. 1990. U.S. CUSTOM HQ LEXIS 5667; HQ 111275 (November 13).

102. U.S. Department of Homeland Security. Headquarters, U.S. Customs Service. 2007. HQ H012082 (August 27); 2003. HQ 115972 (April 22); U.S. Department of the Treasury. 2002. HQ 115823 (October 28); 1988. HQ 109412 (March 29); 1988. HQ 109882 (December 2); and 2002. HQ 115646 (August 12).

103. U.S. Department of the Treasury. U.S. Customs Service. 1998. U.S. CUSTOM HQ LEXIS 749; HQ 114487 (October 19); 1998 U.S. CUSTOM HQ LEXIS 265; 1998. HQ 114305 (March 31).

104. *Code of Federal Regulations*. 2008. Customs Duties. 19 C.F.R. § 177.9(b)(4).

105. Ibid. pt. 177.

106. *Code of Federal Regulations*. 2008. Navigation and Navigable Waters. 33 C.F.R. § 66.01-1(a).

107. U.S. Department of Homeland Security. U.S. Coast Guard. Private aids to navigation application. http://www.uscgboating.org/SAFETY/aton/CG2554.pdf

108. *Code of Federal Regulations*. 2008. Navigation and Navigable Waters. 33 C.F.R. pt. 62, subpt. B.

109. Ibid. § 66.01-30.

110. Ibid. § 67.01-5(a).

111. Ibid. § 67.05-1; 67.05-10.

112. Ibid. § 67.20-10.

113. *Code of Federal Regulations*. 2008. Aeronautics and Space. 14 C.F.R. § 77.17.

114. Ibid. § 77.13.

115. Ibid. § 77.19.

116. Federal Aviation Administration. 2007. *Obstruction Marking and Lighting*. Advisory circular. AC 70/7460-1K (effective February 1).

117. Ibid. 33–34.

118. *Code of Federal Regulations*. 2008. Aeronautics and Space. 14 C.F.R. § 77.23.

119. Ibid. § 77.35.

120. Ibid. § 77.13(c).

121. Outer Continental Shelf Lands Act. § 8(p)(4). U.S. Code. 2006. 43 U.S.C. §1331 et seq.

122. Righter, Robert W. 1996. *Wind Energy in America*. Norman: University of Oklahoma Press. 13.

Chapter 9

1. Trittin, Jurgen. 2005. Viewpoint: German nuclear phaseout. *BBC.com*. October 21. http://news.bbc.co.uk/1/hi/sci/tech/4357238.stm
2. Palfreman, Jon. Why the French like nuclear energy. *Frontline*. http://www.pbs.org/wgbh/pages/frontline/shows/reaction/readings/french.html
3. Public Opinion Strategies. 2006. Survey finds broad support for bold renewable energy goal. Executive summary. *Energy Future Coalition*. February. http://www.25x25.org/storage/25x25/documents/POS_executive_summary.pdf
4. Poremba, Sue Marquette. 2008. Experience drives advanced public outreach strategies. *North American Windpower*. February: 41.
5. British Wind Energy Association. 2002. *Best Practice Guidelines: Consultation for Offshore Wind Energy Developments*. http://www.bwea.com/pdf/bwea-bpg-offshore.pdf
6. Ibid. 8–9.
7. U.S. Commission on Ocean Policy. 2002. *Testimony of David White*. February 22. http://oceancommission.gov/meetings/feb_22_02/white_statement.pdf
8. British Wind Energy Association. 2002. 14.
9. Council on Environmental Quality. 1997. *The National Environmental Policy Act: A Study of Its Effectiveness after Twenty-five Years*. 19.
10. Williams, Wendy, and Robert Whitcomb. 2007. *Cape Wind: Money, Celebrity, Class, Politics, and the Battle for Our Energy Future on Nantucket Sound*. New York: Public Affairs. 90.
11. Firestone, Jeremy, and Willett Kempton. 2007. Public opinion about large offshore wind power: underlying factors. *Energy Policy*. 35 (March): 1584–1598.
12. Ibid. 1590.
13. Sørensen, Hans Christian, Lars Kjeld Hansen, and Jens H. Mølgaard Larsen. Middelgrunden 40 MW offshore wind farm Denmark—lessons learned. http://www.emu-consult.dk/includes/middelgrunden_munich.pdf
14. Nickerson, Susan. 2007. Public's role in Danish wind initiative. *Cape Cod Times*. January 1.
15. Firestone, Jeremy, et al. 2004. Regulating offshore wind power and aquaculture: Messages from land and sea. *Cornell Journal of Law and Public Policy*. 14: 98.
16. Harrington, Mark. 2006. High-profile campaign for planting turbines off Jones Beach wasn't just...Blowing in the wind. *Newsday*. November 2.
17. Farenthold, David A. 2007. Delaware energy debate could turn on the wind. *Washington Post*. May 7.
18. Ibid.
19. Forde, Tina. 2008. A step closer to reality for wind museum. *Tehachapi News*. June 10. http://www.tehachapinews.com/home/ViewPost/64430

20. Sterzinger, George, Fredric Beck, and Damian Kostiuk. 2003. *The Effect of Wind Development on Local Property Values.* Washington, D.C.: Renewable Energy Policy Project. May. http://www.repp.org/articles/static/1/binaries/wind_online_final.pdf

21. U.S. Department of the Interior. Minerals Management Service. *Rigs-to-Reefs Information.* http://www.gomr.mms.gov/homepg/regulate/environ/rigs-to-reefs/information.html

22. U.S. Commission on Ocean Policy. 2002.

23. D'Agnese, Joseph. 2007. Falling in love with the wind: How a small farm town traded its dairy cows for renewable energy. *OnEarth.* Summer. http://www.onearth.org/article/falling-in-love-with-wind

24. Fishermen's Energy of New Jersey. 2008. Fishermen respond to New Jersey BPU offshore wind solicitation. *Reuters.com.* March 4. http://www.reuters.com/article/pressRelease/idUS210775+04-Mar-2008+PRN20080304

25. Kammen, Daniel M., Kamal Kapadia, and Matthias Fripp. 2004. *Putting Renewables to Work: How Many Jobs Can the Clean Energy Industry Generate?* RAEL Report. University of California, Berkeley. April 13. http://rael.berkeley.edu/old-site/renewables.jobs.2006.pdf

26. U.S. Department of Energy. Office of Energy Efficiency and Renewable Energy. *Wind Energy for Rural Economic Development.* http://www.nrel.gov/docs/fy04osti/33590.pdf

27. Goldberg, Marshall R., Karin Sinclair, and Michael Milligan. 2004. *Job and Economic Development Impact (JEDI) Model: A User-Friendly Tool to Calculate Economic Impacts from Wind Projects.* National Renewable Energy Laboratory Report NREL/CP-500-35953. March. http://www.eere.energy.gov/windandhydro/windpoweringamerica/filter_detail.asp?itemid=707

28. Sterzinger, George. 2007. The Economic Promise of Renewable Energy. *New Labor Forum* 16:3: 80–91.

29. The Energy Policy Act of 2005. Public Law 109-58. 109th Cong. 1st Sess. August 8. Section 388, amending § 8(p) of the Outer Continental Shelf Lands Act (42 U.S.C. § 1337).

30. Firestone, Jeremy, Wilett Kempton, and Andrew Kreuger. 2007. Delaware opinion on offshore wind power. Interim report. University of Delaware College of Marine and Earth Studies. http://www.ocean.udel.edu/Windpower/docs/DE-survey-InterimReport-16Jan2007.pdf

31. U.S. Government Accountability Office. 2007. *Maritime Security: Public Safety Consequences of a Terrorist Attack on a Tanker Carrying Liquefied Natural Gas Need Clarification.* GAO-07-316.

32. Dewan, Shaila. 2008. At plant in coal ash spill, toxic deposits by the ton. *New York Times.* December 30.

33. U.S. Department of the Interior. Minerals Management Service. 2007. *Programmatic Environmental Impact Statement for Alternative Energy Development and Production and Alternate Use of Facilities on the Outer Continental Shelf: Final Environmental Impact Statement.* October. OCS EIS/EA MMS 2007-046: 7.6.1.4.2.

34. Clarke, Keith C., and Jeffrey Hemphill. 2002. The Santa Barbara oil spill, a retrospective. *Yearbook of the Association of Pacific Coast Geographers*. Hawaii: University of Hawaii Press. 64: 157–62.

35. Energy Information Administration. Office of Oil and Gas. 2005. *Overview of U.S. Legislation and Regulations Affecting Offshore Natural Gas and Oil Activity*. September. http://www.eia.doe.gov/pub/oil_gas/natural_gas/feature_articles/2005/offshore/offshore.pdf

36. Righter, Robert W. 1996. *Wind Energy in America*. Norman: University of Oklahoma Press. 250–251.

37. Erickson, Wallace, Gregory Johnson, and David Young, Jr. 2005. *A Summary and Comparison of Bird Mortality from Anthropogenic Causes with an Emphasis on Collisions*. USDA Forest Service. General Technical Report. PSW-GTR-191. 1035.

38. Nixon, Richard M. 1973. Address to the nation on the national energy policy. November 25. http://www.presidency.ucsb.edu/ws/index.php?pid=4051

39. Williams, Wendy, and Robert Whitcomb. 2007. xxiv.

Chapter 10

1. ICF Consulting. 2003. The economic cost of the blackout: An issue paper on the northeastern blackout. August. http://www.solarstorms.org/ICFBlackout2003.pdf

2. Electricity Consumers Resource Council (ELCON). 2004. The economic impacts of the August 2003 Blackout. February. http://www.elcon.org/Documents/EconomicImpactsOfAugust2003Blackout.pdf

3. Wright, Sally D., et al. 2002. Transmission options for offshore wind farms in the United States. American Wind Energy Association. http://www.ceere.org/rerl/publications/published/2002/Transmission_Options_For_Offshore_Wind_Farms_In_The_US,_AWEA_June_2002_.pdf

4. Deutsche Energie-Agentur GmbH (dena). 2005. Energy Management Planning of the Grid Integration of Wind Energy in Germany Onshore and Offshore up to the Year 2020. February. http://www.uwig.org/Dena-2005_English.pdf

5. Lundberg, Stefan. 2004. Evaluation of wind farm layouts. Paper presented at Nordic Workshop on Power and Industrial Electronics (NORpie 2004), Trondheim, Norway. http://www.elkraft.ntnu.no/norpie/10956873/Final%20Papers/009%20-%20Eval_WF_lay_009.pdf

6. U.S. Federal Energy Regulatory Commission. 2003. *Order No. 2003*. 104 FERC ¶ 61,103. July 24.

7. National Grid. 2006. *Transmission and Wind Energy: Capturing the Prevailing Winds for the Benefit of Customers*. September: 13. http://www.nationalgridus.com/non_html/c3-3_NG_wind_policy.pdf

8. U.S. Federal Energy Regulatory Commission. 2007. *California Independent System Operator Corp*. 119 FERC ¶ 61,061. April 19.

9. U.S. Federal Energy Regulatory Commission. 2007. Chairman Joseph T. Kelliher's statement on California Independent System Operator Corporation. April 19. http://www.ferc.gov/news/statements-speeches/kelliher/2007/04-19-07-kelliher-E-5.asp

10. Texas Senate. 2005. SB 20. An Act Relating to This State's Goal for Renewable Energy. August 20. http://www.capitol.state.tx.us/tlodocs/791/billtext/pdf/SB00020F.pdf

11. U.S. Federal Energy Regulatory Commission. 2006. *Proposed Order Directing Interconnection and Transmission Services and Ordering Further Procedures.* 115 FERC ¶ 61,128.

12. Federal Power Act. 1978. Public Law 617. 95th Cong. 1st sess. November 9. §§ 210(c) and 212(a).

13. *Code of Federal Regulations.* 2008. Conservation of power and water resources. 18 C.F.R. § 2.20.

14. U.S. Federal Energy Regulatory Commission. 2007. FERC Stats. & Regs. ¶ 31,241. Appendix C. Pro forma OATT. § 19.3.

15. Code of Federal Regulations. 18 C.F.R. § 2.20.

16. U.S. Federal Energy Regulatory Commission. 2007. FERC Stats. & Regs. ¶ 31,241. Appendix C. Pro forma OATT. §19.4.

17. U.S. Federal Energy Regulatory Commission. 2007. Order 890. *Preventing Undue Discrimination and Preference in Transmission Service.* February 16: 1370.

18. National Grid. 2006. *Transmission and Wind Energy: Capturing the Prevailing Winds for the Benefit of Customers.* September: 11, n. 43.

19. Smith, J.C., et al. 2004. *Wind Power Impacts on Electric Power System Operating Costs: Summary and Perspective on Work to Date.* National Renewable Energy Laboratory: 7; International Energy Agency. 2005. *Variability of Wind Power and Other Renewables: Management Options and Strategies.* June. http://www.awea.org/policy/regulatory_policy/transmission_documents/IEA_report_variability_wind.pdf

20. Smith, et al. 2004.

21. International Energy Agency. 2005. 33.

22. U.S. Department of Energy. 2008. 20% *Wind Energy by 2030, Increasing Wind Energy's Contribution to U.S. Electricity Supply.* DOE/GO-102008-2567. 11.

23. Hamaker, Christian. 2007. Money talks, thermal plants walk; why it pays for utilities to be more efficient. Public Utilities Fortnightly. June: 14.

24. U.S. Federal Energy Regulatory Commission. 2004. *In the Matter of Interconnection for Wind Energy and Other Alternative Technologies.* September 24. FERC Docket No. PL04-15-000. 10.

25. Jensen, Jytte Kaad, and Fritz Luxhøi. *New Challenges for the Transmission System Operator.* Abstract 227. Eltra, Denmark. http://www.wind-watch.org/documents/wp-content/uploads/eltra-challenges.pdf

26. GE Energy Consulting. 2005. *The Effects of Integrating Wind Power on Transmission System Planning, Reliability, and Operations.*

Report on Phase 2: System Performance Evaluation. March 4: 2.1. http://www.nyserda.org/publications/wind_integration_report.pdf

27. Ibid. Figure 5.7.

28. Gül, Timur and Till Stenzel. 2005 Variability of wind power and other renewables: management options and strategies. Working paper, International Energy Agency, Paris. 25.

29. GE Energy Consulting. 2005. Figure 5.22.

30. Redlinger, Robert Y., Per Dannemand Andersen, and Poul Erik Morthorst. 2002. *Wind Energy in the 21st Century: Economics, Politics, Technology, and the Changing Electricity Industry*. New York: Palgrave Macmillan. 65.

31. GE Energy Consulting. 2005. Figure 7.2.

32. International Energy Agency. 2005. 17.

33. GE Energy Consulting. 2005. Figure 7.14.

34. GE Energy Consulting. 2005. Figure 7.4.

35. Redlinger, Robert, Per Dannemand Andersen, and Poul Erik Morthorst. 2002. 67.

36. Zack, John. 2003. *Overview of Wind Energy Generation Forecasting*. Submitted to the New York Energy Research and Development Authority and the New York State Independent System Operator. http://www.uwig.org/forecst_overview_report_dec_2003.pdf

37. Cape Wind. 2007. Comparison of Cape Wind scientific data tower wind speed data with ISO New England list of top ten electric demand days. July 2. http://www.capewind.org/downloads/CWReport.pdf

38. U.S. Department of Energy. 2004. *Natural Gas in the New England Region: Implications for Offshore Wind Generation and Fuel Diversity*. June 6: 9.

39. U.S. Federal Energy Regulatory Commission. 2005. Order No. 661. 111 FERC ¶ 61,353, as amended by Order No. 661-A, 113 FERC ¶ 61,254.

40. U.S. Federal Energy Regulatory Commission. 2005. Order No. 2003-C. The large generator interconnection agreement. Appendix G. 111 FERC ¶ 61,401.

41. U.S. Federal Energy Regulatory Commission. 2005. Order No. 661. 41.

42. U.S. Federal Energy Regulatory Commission. 2007. FERC Stats. & Regs. ¶ 31,241. Appendix C. Pro forma OATT. § 30.2.

43. U.S. Federal Energy Regulatory Commission. 2007. Order 890. 808.

44. Ibid. 825.

45 Ibid. 958, 1043.

46. Ibid. 1024.

47. Ibid. 664–665.

48. Ibid. 666.

49. Ibid. 85.

50. Public Utilities Fortnightly. 2005. *Yet Another Subsidy for Wind; FERC Risks Going Overboard in Easing Penalties for Generation Imbalances.* August: 21.

51. U.S. Federal Energy Regulatory Commission. 2007. Order 890. 837.

52. Ibid. 848.

53. Ibid. 847.

54. *Electric Utility Week.* 2006. Allegheny follows AEP in proposing big line to transmit power eastward through PJM. March 6: 1.

55. *Electric Utility Week.* 2006. AEP expects CO_2 caps, though not quite yet; "firmly believes" IGC plants are a way to go. February 27: 8.

56. American Electric Power. Power plants and other assets. http://www.aep.com/about/powerplants/

57. PJM. 2008. 2010/2011 RPM Base Residual Auction Results. February 1. http://www.pjm.com/~/media/markets-ops/rpm/rpm-auction-info/20080201-2010-2011-bra-report.ashx

Chapter 11

1. Redlinger, Robert Y., Per Dannemand Andersen, and Poul Erik Morthorst. 2002. *Wind Energy in the 21st Century: Economics, Politics, Technology, and the Changing Electricity Industry.* New York: Palgrave Macmillan. 104–106.

2. U.N. Environment Programme. 2004. *Financial Risk Management Instruments for Renewable Energy Projects, Summary Document.* U.N. Environment Programme, Division of Technology, Industry, and Economics. 20.

3. Redlinger, Robert Y., Per Dannemand Andersen, and Poul Erik Morthorst. 2002. 108.

4. Applied Technology and Management, et al. 2007. *Final Report RIWinds Phase I: Wind Energy Siting Study.* Providence: Rhode Island Economic Development Corporation.

5. Ibid. 3-3, 4-2–4-3.

6. Haughton, Jonathan, et al. 2004. *An Economic Analysis of a Wind Farm in Nantucket Sound.* Boston: The Beacon Hill Institute at Suffolk University. 11. http://www.beaconhill.org/BHIStudies/Windmills2004/WindFarmArmyCorps.pdf

7. Applied Technology and Management, et al. 2007. 4-3.

8. KEMA, Inc. 2006. *Building a "Margin of Safety" into Renewable Energy Procurements: A Review of Experience with Contract Failure.* Prepared for the California Energy Commission.

9. Ibid. 15.

10. Enbar, Nadav, and Roberta Bigliani. 2008. *European Offshore Wind: Poised for Growth?* European Renewable Energy Strategies. Market Overview. February.

11. Pace Global Energy Services. 2007. *Assessment of Offshore Wind Power Resources*. Prepared for the Long Island Power Authority. 2-3.
12. Wright, Sally D., et al. 2002. *Transmission Options for Offshore Wind Farms in the United States*. Paper presented at the annual meeting of the America Wind Energy Association, Portland. http://www.ecs.umass.edu/mie/labs/rerl/pubs/2002/AWEA2002Transmission.pdf
13. Redlinger, Robert Y., Per Dannemand Andersen, and Poul Erik Morthorst. 2002. 85.
14. Applied Technology and Management, et al. 2007.
15. Redlinger, Robert Y., Per Dannemand Andersen, and Poul Erik Morthorst. 2002. 72.
16. Righter, Robert W. 1996. *Wind Energy in America: A History*. Norman: University of Oklahoma Press. 216–217.
17. Andersen, Erik. 2006. Risk rating wind generation assets. *EnergyPulse*. February 17. http://www.energypulse.net/centers/article/article_display.cfm?a_id=1210
18. Germanischer Lloyd WindEnergy. 2003. *Guideline for the Certification of Condition Monitoring Systems for Wind Turbines*.
19. Institut für Solare Energieversorgungstechnik e.V, et al. 2005. *Final Report: Advanced Maintenance and Repair for Offshore Wind Farms Using Fault Prediction and Condition Monitoring Techniques (Offshore M&R)*. Prepared for the European Commission, DG TREN. http://ec.europa.eu/energy/renewables/wind_energy/doc/offshore.pdf
20. U.S. Federal Energy Regulatory Commission. 2008. *Licensing Hydrokinetic Projects*. White paper. Appendix C, Standard Hydrokinetic Pilot Project License Articles. April 14: 28.
21. Climate Change Capital. 2006. *Offshore Renewable Energy Installation Decommissioning Study: Final Report*. Prepared for the U.K. Department for Trade and Industry. 1.3.
22. U.S. Department of the Interior, Minerals Management Service. 2009. Renewable energy and alternate uses of existing facilities on the outer continental shelf: Final rule. *Federal Register*. Vol. 74, no. 081. April 29. Adapting 30 C.F.R. § 285.516.
23. New York State Public Service Commission. 2003. *Order Regarding Retail Renewable Portfolio Standards*. Case 03-E-0188. September 24: 4.
24. Massachusetts Technology Collaborative. *Massachusetts Green Power Partnership*. http://www.masstech.org/renewableenergy/mgpp.htm
25. Haughton, Jonathan, and David G. Tuerck. 2006. *Subsidies to a Wind Farm in Nantucket Sound*. Boston: Beacon Hill Institute at Suffolk University. 5.
26. U.S. House of Representatives. 2008. HR 1424. Emergency Economic Stabilization Act of 2008, Division B—Energy Improvement and Extension Act of 2008, § 102, amending IRC § 45.
27. *U.S. Code*. Internal revenue code. 2007. 26 U.S.C. § 45..

28. Ibid. § 38(b)(8).

29. Ibid. § 39(a).

30. Department of the Treasury. Internal Revenue Service. 2007. Revenue Procedure 2007-65. *Internal Revenue Bulletin* 2007-45 (November 5): 967. http://www.irs.ustreas.gov/pub/ irs irbs/irb07-45.pdf; Ewing, William. 2007. IRS establishes safe harbor tax guidelines. North American Windpower. Vol. 4, no. 11. December: 1.

31. Department of the Treasury. Internal Revenue Service. 2006. Section 45 credit offset. *Internal Revenue Bulletin* 2006-9 (February 27): 519; IRS Private Letter Ruling No. 200311021. 2003. March 14.

32. *U.S. Code*. 26 U.S.C. § 45(b)(3); Wiser, Ryan, Mark Bolinger, and Troy Gagliano. 2002. *Analyzing the Interaction Between State Tax Incentives and the Federal Production Tax Credit for Wind Power.* Berkeley: Lawrence Berkeley National Lab. http://eetd.lbl.gov/EA/EMS/reports/51465.pdf

33. Department of the Treasury. *Internal Revenue Service.* 1994. Electricity produced from wind energy under section 45. Internal Revenue Bulletin 1994-31 (January): 4.

34. Ibid.

35. *U.S. Code*. 26 U.S.C. § 168(e)(3).

36. Moritz, Dennis. 2008. Modeling choices impact tax equity financing. *North American Windpower.* Vol. 5, no. 2. March: 1.

37. *Electric Utility Week*. 2007. New York utilities explore idea of putting renewable energy plants in rate base. October 8: 4.

38. FPL Energy LLC. 2003. FPL Energy announces completion of subsidiary bond offering. Press release. July 8.

39. Garrad Hassan and Partners, Ltd. 2004. *Sea Wind Europe*. Prepared for Greenpeace. 4.1.

40. *Federal Register.* 2007. Loan Guarantees for Projects that Employ Innovative Technologies, Final Rule, 72 Fed. Reg. 60116. October 23.; Code of Federal Regulations. 2009. Energy. 10 C.F.R. pt. 609.

41. *Code of Federal Regulations*. 10 C.F.R. § 609.2.

42. Ibid. § 609.4(d)(4).

43. Ibid. § 609.6(b)(5)(v).

44. Ibid. § 609.6(b)(12).

45. *U.S. Code*. 2007. The Congress. 2 U.S.C. § 661a(5).

46. Marsh Ltd. 2006. *Survey of Insurance Availability for Renewable Energy Projects.* Prepared for the U.N. Environment Programme.

47. As quoted in Childs, Dana. 2007. Insurers eye profits in Greentech. *Cleantech.com*. April 11.

48. Marsh Ltd. 2006. 10.

49. Ibid.

50. U.N. Environment Programme. 2004. *Financial Risk Management Instruments for Renewable Energy Projects, Summary Document.* 24.

51. Ibid.

52. Weather Risk Management Association. Great weather risk management transactions. http://www.wrma.org/risk_transactions.html

Chapter 12

1. Navigant Consulting Inc., Sustainable Energy Advantage LLC, and Boreal Renewable Energy Development. 2004. *New Jersey Renewable Energy Market Assessment.* Final report to Rutgers University Center for Energy, Economic and Environmental Policy. 14.

2. Friedman, Thomas. 2008. *Hot, Flat, and Crowded.* New York: Farrar, Straus and Giroux. 259.

3. Cockerill, Timothy T., et al. 1998. *Opti-OWECS Final Report.* Vol. 3: Comparison of cost of offshore wind energy at European sites. Institute for Wind Energy, Delft University of Technology. 3-8.

4. Ibid. 3-9.

5. Gerdes, Gerhard, Albrecht Tiedemann, and Sjoerd Zeelenberg. 2006. *Case Study: European Offshore Wind Farms—A Survey for the Analysis of the Experiences and Lessons Learnt by Developers of Offshore Wind Farms—Final Report.* Prepared for the Senator for Construction, Environment & Transport, Free State of Bremen. 145.

6. Stiesdal, Henrik, ed. 2001. *Middelgrunden offshore: The project.* BONUS Energy A/S. July 1: 36–37. http://www.middelgrunden.dk/projektinfo/01Bonus_Info_Middelgrunden_UK.pdf

7. Cockerill, T.T., et al. 1998. A-12.

8. Cockerill, T.T., et al. 1998.

9. Boyle, Godfrey. 2007. Offshore wind: The potential to contribute a quarter of UK electricity by 2024. *Wind Engineering.* 31(2): 69.

10. Poretto, John. 2007. Offshore wind project canceled. Associated Press. June 12.

11. Delaware Public Service Commission. 2007. *PSC Staff Report on the Power Purchase Agreement Between Delmarva Power and Bluewater Wind.* 9. http://depsc.delaware.gov/irp.shtml

12. *Delaware Code.* 2009. Public Utilities. 26 Del. C. § 364. According to the legislation, "All costs arising out of contracts entered into by a Commission regulated electric company pursuant to § 1007(d) of this Title shall be distributed among the entire Delaware customer base of such companies through an adjustably non-bypassable charge which shall be established by the Commission. Such costs shall be recovered if incurred as a result of such contracts unless, after Commission review, any such costs are determined by the Commission to have been incurred in bad faith, are the product of waste or out of an abuse of discretion, or in violation of law."

13. Ibid. § 356(c). "A Commission regulated electric company shall receive 350% credit toward meeting the Renewable Energy Portfolio Standards established pursuant to this subchapter for energy derived from offshore wind energy installations sited off the Delaware coast on or before May 31, 2017."

14. Enbar, Nadav, and Roberta Bigliani. 2008. *European Offshore Wind: Poised for Growth?* European Renewable Energy Strategies. Market Overview. February.

15 Haughton, Jonathan, et al. 2004. *An Economic Analysis of a Wind Farm in Nantucket Sound.* Boston: The Beacon Hill Institute at Suffolk University. http://www.beaconhill.org/BHIStudies/Windmills2004/WindFarmArmyCorps.pdf

16. Levy, Jonathan I., et al. 1999. Development of a new damage function model for power plants: Methodology and applications. *Environmental Science and Technology*. 33: 4,364–4,372.

17. The Brattle Group. 2007. *The Economics of U.S. Climate Policy: Impact on the Electric Industry.* March 2. http://www.brattle.com/_documents/UploadLibrary/ArticleReport2440.pdf

18. Belasco, Amy. 2008. CRS report for Congress: The cost of Iraq, Afghanistan, and other global war on terror operations since 9/11. *Congressional Research Service.* October 15. http://www.fas.org/sgp/crs/natsec/RL33110.pdf

19. U.S. Congress. Senate. Committee on Foreign Relations. Subcommittee on Near Eastern and South Asian Affairs. 2005. Testimony by Dr. Gal Luft. October 20. http://foreign.senate.gov/testimony/2005/LuftTestimony051020.pdf

20. Global Insight. 2003. *Impact Analysis of the Cape Wind Off-Shore Renewable Energy Project on Local, State, and Regional Economies.* Report prepared for Cape Wind Associates, Boston. Revised September 26. http://www.mms.gov/offshore/PDFs/CWFiles/68.pdf

21. Haughton, Jonathan, and David G. Tuerck. 2006. *Subsidies to a Wind Farm in Nantucket Sound.* Boston: Beacon Hill Institute at Suffolk University. May. http://www.beaconhill.org/BHIStudies/WindMills2006/BHIonPubSubsidiesforCapeWind.pdf

22. Callaghan, John. 2006. *Future Marine Energy—Results of the Marine Energy Challenge: Cost Competitiveness and Growth of Wave and Tidal Stream Energy.* London: The Carbon Trust. 11.

23. Weiss, Peter. 2001. Oceans of electricity: New technologies convert the motion of waves into watts. *Science News Online.* April 14. http://www.phschool.com/science/science_news/articles/oceans_of_electricity.html

24. Heath, Tom. *The Construction, Commissioning and Operation of the LIMPET Wave Energy Collector.* 13. http://www.wavegen.co.uk/pdf/Consruction,commission%20&%20operation%20of%20LIMPET.pdf

25. Ibid. 2.

26. Entek UK Ltd. 2006. *MEC Cost Estimation Methodology: The Marine Energy Challenge Approach to Estimating the Cost of Energy Produced by Marine Energy Systems.* Report prepared for The Carbon Trust, London. May. http://www.carbontrust.co.uk/NR/rdonlyres/B4410309-9EDB-4579-B0AB-D081B168D2BC/0/MECcostestimationmethodologyreport.pdf

27. Callaghan, John. 2006.

28. Atiya, Ramez. 2008. *The Hydroelectric Tidal Wing: A New Tidal Range Technology for the Bay of Fundy.* Halcyon Marine Hydroelectric.

29. Thorpe, T.W. 1999. *A Brief Review of Wave Energy.* ETSU Report Number R-120 for the DTI.

30. Sustainable Development Commission. 2007. *Turning the Tide: Tidal Power in the UK.* 37. http://www.sd-commission.org.uk/publications/downloads/Tidal_Power_in_the_UK_Oct07.pdf

31. Smith, Rebecca. 2007. The New Math of Alternative Energy. *Wall Street Journal.* February 12: R1.

32. Eclipse Energy Company, Ltd. 2007. Eclipse Energy granted consent to construct global first offshore gas-wind hybrid energy project. Press release. February 8. http://www.seapower-generation.co.uk/press%20releases/pr-11.htm

33. Thorpe, T.W. 1999. 47.

34. *Electric Utility Week.* 2007. Exelon's Rowe plugs importance of nuclear energy, competition to fight climate change. April 30: 4.

35. Electric Utility Week. 2007. Costs surge for Duke, Vectren coal plant. April 16: 8.

36. FutureGen Alliance. 2009. FutureGen project costs. http://www.futuregenalliance.org/costs.stm

37. Harding, Jim. 2007. Economics of new nuclear power and proliferation risks in a carbon-constrained world. *Nonproliferation Policy Education Center.* 4. http://www.npec-web.org/Essays/20070600-Harding-EconomicsNewNuclearPower.pdf

38. Smith, Rebecca. 2008. New wave of nuclear plants faces high costs. *Wall Street Journal.* May 12: B1.

39. Burr, Michael T. 2008. Debt renaissance: Eye-popping nuclear costs portent rising leverage. *Public Utilities Fortnightly.* July: 28.

40. Harding, Jim. 2007. 6–7.

41. Stewart, Susan. 2005. Offshore wind resources in the Southeast. Georgia Tech Strategic Energy Initiative Workshop. May 12.

42. Harding, Jim. 2007. 7, 11.

43. *United States of America v. Massey Energy Company, et al.* 2008. Consent Decree. Civil Action No. 2:07-0299 (U.S. Dist. Ct., Southern Dist. of W. VA). January.

44. Schultz, Sue. 2008. Constellation reaches $54M settlement on fly ash lawsuit. *Baltimore Business Journal.* December 31. http://baltimore.bizjournals.com/baltimore/stories/2008/12/29/daily26.html

45. Massachusetts Institute of Technology. 2003. *The Future of Nuclear Power: An Interdisciplinary MIT Study.* Massachusetts Institute of Technology. 22.

46. Tankersley, Jim. 2009. Obama energy plan reaches for the sky. *The Washington Post.* February 8.

47. Offshore Wind Collaborative Organizing Group. 2005. *A Framework for Offshore Wind Energy Development in the United States.* 2. http://www.mtpc.org/offshore/final_09_20.pdf

48. Lilienthal, David E. 1963. *Change, Hope, and the Bomb.* Princeton University Press. 96.

49. *Wall Street Journal.* 2008. Iran, Qatar, Russia form gas alliance. WSJ News Roundup. October 22. http://online.wsj.com/article/SB122460817038154673.html

50. Offshore Wind Collaborative Organizing Group. 2005.

51. Christensen, Clayton. 1997. *The Innovator's Dilemma: When New Technologies Cause Great Firms to Fail.* Boston: Harvard Business School Press.

52. Pickens, T. Boone. 2008. Job one. *Washington Post.* November 9: B2.

Index

B

C

F

G

H

J

K

L

M

O

S

T

U

V

W–X

Y–Z

About the Authors

Markian Melnyk has practiced in the field of energy law since 1991, first at the U.S. Securities and Exchange Commission and later in private practice at the international law firm Dewey & LeBoeuf. He has helped some of the largest multi-national utility holding companies with mergers, financings, and regulatory issues, and has worked with much smaller companies just breaking into the growing renewable energy arena.

As someone who has capsized a Sunfish more than a few times, Markian is not expert in all things marine. He has, however, dug deeply into the processes used to regulate and promote renewable energy activities in the U.S. offshore environment and combined that with learning from the development of renewable energy projects in European waters.

Markian lives in Maryland with his wife and three children.

Robert M. Andersen is a partner with Akerman Senterfitt LLP's Washington DC law offices, and practices in the Energy, Environment, and Natural Resources fields. He is co-chair of the Firm's Energy practice group and has assisted renewable energy companies with siting, environmental issues, and permitting for both offshore and terrestrial wind energy facilities. He also represents renewable and traditional energy companies in licensing and other proceedings before the Minerals Management Service, the United States Army Corps of Engineers, EPA, DOE, FERC, and NRC. Mr. Andersen previously served in a number of senior environmental positions with the federal government, including Chief Counsel with the United States Army Corps of Engineers, the first General Counsel of the Defense Nuclear Facilities Safety Board, Deputy General Counsel of the National Science Foundation with service as the Senior United States Representative in Antarctica, and various positions at EPA, Region V. He received his BS and JD degrees from the University of Iowa, and an MPA from Harvard University's John F. Kennedy School of Government where he studied as a Littauer fellow in 1985-86. He teaches environmental and energy management courses as an adjunct Professor at George Washington University.